WHEN THE LAND MEETS THE SEA: An ACUA and SHA Series

Series Editors: Annalies Corbin, *PAST Foundation, Columbus, OH, USA*
J.W. Joseph, *New South Associates, Inc., Stone Mountain, GA, USA*

More information about this series at http://www.springer.com/series/8370

Amanda M. Evans
Editor

The Archaeology of Vernacular Watercraft

Springer

Editor
Amanda M. Evans
Tesla Offshore, LLC
Prairieville, USA

ISSN 1869-6783 ISSN 1869-6791 (electronic)
When the Land Meets the Sea
ISBN 978-1-4939-3561-1 ISBN 978-1-4939-3563-5 (eBook)
DOI 10.1007/978-1-4939-3563-5

Library of Congress Control Number: 2016935497

Printed on acid-free paper

This Springer imprint is published by Springer Nature
The registered company is Springer Science+Business Media LLC New York

Foreword

The term "vernacular watercraft" has come into widespread use in North America in recent decades where a number of research students have written theses on various aspects of such vessels (Corbin 2010; Damien 2010; Evans 2005; Merwin 2000; Tolson 1992) and occasional consultancy cultural heritage management reports and book chapters have appeared (Goodwin et al. 1997; Langley 2011; O'Leary 1994). That the use of the term vernacular watercraft is by no means as common in places outside North America should come as no real surprise in a (sub)discipline that has regularly used slightly different, but partly (or largely) overlapping, definitions for its activities that are often based on their particular geographical or historical origins. The three most common examples of which are probably maritime archaeology, underwater archaeology, and nautical archaeology.

So is this simply a question of the meanings and usage of the term vernacular watercraft in the English language but as used in different countries such as English (US) or English (Aus)? To a certain extent it is; for example, for many years I have been involved in, and watched others, conduct research on what in the USA would probably be called vernacular watercraft. Here in Australia they are not usually referred to as such but, perhaps a little parochially, as "Australian-built" boats and ships (Bullers 2006, 2007; Coroneos 1991; Jeffery 1989, 1992; Nash 2003, 2004). In the UK, on the other hand, these types of vessels would commonly be called "traditional" or "working" watercraft (McCaughan 2008; McKee 1983). Throughout the world the term "indigenous" watercraft is often used for what in the USA may be termed vernacular. Finally, even in the USA, the term vernacular is not used by all researchers working on small, wooden-hulled, locally built watercraft (Moser 2011). Differing terminology is by no means unusual as you will not find the term vernacular watercraft used at all in Basil Greenhill's classic book *The Archaeology of Boats and Ships*, although the book is all about both traditional and indigenous watercraft (Greenhill 1995).

Vernacular watercraft may, in part, be defined as occurring or existing in a particular locality or, in other words, as "endemic" or local. Another aspect of a definition of vernacular watercraft might include an "indigenous" building style using local materials and traditional methods of construction, decoration, and style. Such a definition will have an overlap of meaning with traditional watercraft in terms of using traditional construction methods. It will also have overlapping meaning with working watercraft as opposed to recreational vessels although clearly some vernacular watercraft started life as working vessels and have ended up as recreational vessels. The balance to this is that some authors do include recreational vessels or "pleasure craft" in their definitions of vernacular watercraft (Damien 2010: 3). Varying terminology over time and the difficulties of developing classifications and rigid typologies for vernacular watercraft are recurrent themes in this book as they are, more broadly, in maritime history.

Evans and Smith (in Chap. 1) offer some definitions of vernacular watercraft and point out some of the overlaps between traditional and vernacular watercraft. They outline Doran's three criteria for recognition of vernacular watercraft: local community recognition, built by rule of thumb, and attributable by outsiders to a particular community. They also suggest that relationships between environment and social processes can be used to model the expected elements of vernacular watercraft.

In Chap. 2, Evans argues for vernacular watercraft as "common" boats and ships highlighting the distinctly local aspects of their design and construction and suggests that shipbuilding is a deliberate agent-based decisionmaking process. She also makes the important link between colonization and adaptation to local conditions and resources and identifies the importance of the incorporation of local timbers into ship construction and repair. She outlines Butzer's useful model of adaptive systems, which had three main components—technology, social behavior, and resource opportunities—for analyzing shipwreck sites.

Adaptation to the environment is clearly an important part of the development and evolution of vernacular watercraft in a colonial setting, and in Chap. 4 Turner considers colonization and the link between domestic and imported in the transition processes from Old World to New World. He suggests that vernacular watercraft in a colonial setting were often firmly based on preexisting traditions that were brought by colonists and immigrants as significant aspects of their parent culture and which can therefore be seen as examples of cultural continuity. Adaptation, on the other hand, is particularly important in terms of the use of endemic timbers in vernacular watercraft as ship and boat builders in northern Europe, specifically in Britain, had over many centuries developed detailed knowledge and understanding about shipbuilding including the most suitable timbers for particular tasks: oak for frames, beech for decks, ash for oars, and fir for masts and spars. When these familiar and traditional timbers were not available, the colonists had to use locally grown timbers, often of unknown suitability, instead.

So can we tease out some features of vernacular watercraft that somehow encapsulate what they are? In the purest sense vernacular watercraft are:

- Small boats and small vessels. While there is no accepted maximum size for a vernacular watercraft, the vast majority were probably under 100 ft long and/or 100 tons in size though these sizes grew as time went by and a vernacular watercraft built at the end of the nineteenth century may well have been larger than those built in previous centuries. In general, however, larger vessels tended to be built in shipyards by large teams of men that included one, or more, shipwright(s) who, until the nineteenth century, had learned his trade either from father to son or through the apprenticeship system (MacGregor 1997).

- Mostly wooden hulled and usually powered by sail—but by no means exclusively. The introduction of iron and steel hulls and steam-, and later coal- or oil-, powered vessels in many cases tended to result in these vessels being built in large shipyards by "professionals" or specialists. But there are clearly examples of steam-powered or iron-hulled or unpowered (barges) vernacular watercraft as Chap. 5 shows.
- Sometimes one-off or "unique" or at least only a small number of vessels of similar construction—again the construction of many vessels of similar size and shape tended to signify "mass" production, and vernacular watercraft were not generally built by commercial/industrial or specialized or large-scale shipyards. Certainly they often have unusual construction features, idiosyncratic design, and strange, or reused, fittings as Meide shows in Chap. 6.
- Not built from detailed plans and the involvement of a naval architect, or even a "trained" shipwright, signaled something other than a vernacular watercraft. In general vernacular watercraft were built by men who had little or no "formal" training (book learning) or trade qualifications as a naval architect or a shipwright. Some, but not all, were built by men who did not ordinarily make their full-time living building watercraft and only constructed watercraft on a part-time basis.
- Built for a local or regional trade or activity such as river, lake, inshore, and coastal trades or fishing rather than for the long-distance, offshore, or ocean-going transport of goods.
- Generally in daily use, which is not a terribly useful idea as most watercraft, at least until relatively recently, were designed and built to be in daily use.

Needless to say such definitions are difficult, overlapping, and often more a process of definition by exclusion—it is not something so it is (or just might be) a vernacular watercraft. For example, most definitions of vernacular watercraft would not include naval (or military) warships or "elite" vessels associated with royalty such as *Mary Rose* and *Vasa* which is a point made by Scott-Ireton and Horrell in Chap. 5.

Why then are we interested in vernacular watercraft? They reflect the values of the people who create and use them.

Vernacular watercraft are interesting to archaeologists because they are a class of vessel that has limited evidence available in the historical record and that has been largely ignored by mainstream historians. One reason for this is that most were small vessels that were often unregistered and therefore largely unrecorded, which is a point made by Damour in Chap. 7. To paraphrase others, vernacular watercraft can be called "ships without voice," yet these vessels were vital for the establishment and expansion of settlement throughout newly established colonies.

Clayton, VIC, Australia Mark Staniforth, Ph.D., F.S.A.

References

Bullers, R. (2006). Quality assured: Shipbuilding in colonial South Australia and Tasmania. *Flinders University Maritime Archaeology Monographs* (Series Number 8). Adelaide, South Australia, Australia: Shannon Research Press.

Bullers, R. (2007). *Zephyr*: A short-lived Australian-built schooner. *The Bulletin of the Australasian Institute for Maritime Archaeology, 31*, 11–17.

Corbin, S. R. (2010). *Square sails on the Chilkoot Trail: The vernacular watercraft of the Klondike Gold rush*. Master's thesis, Norwegian University of Science and Technology, Norway.

Coroneos, C. (1991). One interpretation for the short working lives of early Australian wooden sailing vessels in Victorian waters. *The Bulletin of the Australasian Institute for Maritime Archaeology, 15*(2), 7–14.

Damien, M. (2010). *Archaeology through Art: Japanese vernacular craft in late Edo-period woodblock prints*. Master's thesis, East Carolina University, Greenville, NC.

Evans, A. M. (2005). *Institutionalized piracy and the development of the Jamaica Sloop, 1630–1743*. Master's thesis, Department of Anthropology, Florida State University, Tallahassee, FL.

Goodwin, R. C., Robinson, D. S., & Seidel, J. L. (1997). *Documentation of several historic vernacular watercraft on Bayou Dularge, Terrebonne Paris, Louisiana*. Unpublished report for the U.S. Army Corps of Engineers, New Orleans District.

Greenhill, B. (1995). *The archaeology of boats and ships: An introduction*. London: Conway Maritime Press.

Jeffery, W. (1989). Research into Australian built coastal vessels wrecked in South Australia. *The Bulletin of the Australasian Institute for Maritime Archaeology, 15*(2), 37–56.

Jeffery, W. (1992). Maritime archaeological investigations into Australian-built vessels wrecked in South Australia. *The International Journal of Nautical Archaeology, 21*(3), 209–219.

Langley, S. B. M. (2011). Historic watercraft: Keeping them afloat. In T. F. King (Ed.), *A companion to cultural resource management*. Blackwell Companions to Anthropology.

MacGregor, D. R. (1997). *The Schooner: It's design and development from 1600 to the present*. London: Chatham.

McCaughan, M. (2008). Irish vernacular boats. In C. Mac Cárthaigh (Ed.), *Traditional boats of Ireland: History, folklore and construction* (pp. 3–11). Cork, Ireland: Collins Press.

McKee, E. (1983). *Working boats of Britain: Their shape and purpose*. London: Conway Maritime Press.

Merwin, D. E. (2000). *Gilbert M. Smith, Master boatbuilder of Long Island, New York*. Master's thesis, Department of Anthropology, Texas A&M University, College Station, TX.

Moser, J. D. (2011). *The art and mystery of shipbuilding: An archaeological study of shipyards, shipwrights and shipbuilding in Somerset County, Maryland, 1660–1900*. Ph.D. thesis, Florida State University, Electronic theses, treatises and dissertations, Paper 2232.

Nash, M. (2003). Convict shipbuilding in Tasmania. *Papers and Proceedings, Tasmanian Historical Research Association, 50*(2), 83–106.

Nash, M. (2004). The Australian-built schooner *Alert* (1846–1854). *The Bulletin of the Australasian Institute for Maritime Archaeology, 28*, 89–94.

O'Leary, W. M. (1994). *The Tancook schooners: An island and its boats*. Toronto, Ontario, Canada: McGill-Queen's University Press.

Tolson, H. (1992). *The vernacular watercraft of Isle Royale: A western Lake Superior boatbuilding tradition*. Master's thesis, Department of Anthropology, Texas A&M University, College Station, TX.

Contents

Contributors

Charles Beeker Center for Underwater Science, Indiana University, Bloomington, IN, USA

Brian Clayton Deep Sea Acoustic Surveyor, Fugro Chance Inc., Houston, TX

Charles Dagneau Underwater Archaeology Service, Parks Canada, Ottawa, ON, Canada

Melanie Damour Independent Researcher, Mandeville, LA, USA

Steve J. Dasovich, Ph.D. Anthropology & Sociology Department, Lindenwood University, St. Charles, MO, USA

Amanda M. Evans Tesla Offshore, LLC., Prairieville, LA, USA

John W. Foster California State Parks, Fain Oaks, CA, USA

Christopher E. Horrell, Ph.D. Bureau of Safety and Environmental Enforcement, New Orleans, LA, USA

Brad Loewen Département d'anthropologie, Université de Montréal, Montréal, QC, Canada

Deborah Marx Stellwagen Bank National Marine Sanctuary, Scituate, MA, USA

Chuck Meide Lighthouse Archaeological Maritime Program (LAMP), St. Augustine Lighthouse Maritime Museum, St. Augustine, FL, USA

Charles D. Moore Golder Associates Ltd., Vancouver, BC, Canada

Andrew Pietruszka Underwater Archaeologist, School of Marine Science and Policy College of Earth, Ocean, and Environment, University of Delaware, Lewes, DE, USA

Franklin H. Price Bureau of Archaeological Research, Florida Department of State Tallahassee, Florida, USA

Nathan Richards, Ph.D. Program in Maritime Studies, East Carolina University, Greenville, NC, USA

Bradley A. Rodgers Program in Maritime Studies, East Carolina University, Greenville, NC, USA

Della A. Scott-Ireton, Ph.D. Florida Public Archaeology Network, Pensacola, FL, USA

Sheli O. Smith PAST Innovation Lab, Columbus, OH, USA

Roger C. Smith, Ph.D. Division of Historical Resources, Bureau of Archaeological Research, Tallahassee, FL, USA

David C. Switzer Department of History, Philosophy, and Social Studies Education, Plymouth State University, Plymouth, NH, USA

Samuel P. Turner, Ph.D. Lighthouse Archaeological Maritime Program, St. Augustine Lighthouse & Maritime Museum, St. Augustine, FL, USA

Heather White Assistant Director for Assessment & Engagement, Joyner Library, East Carolina University, Greenville, North Carolina, USA

About the Editor

Amanda M. Evans, R.P.A., Ph.D. serves as Senior Marine Archaeologist with Tesla Offshore, LLC. She holds a Bachelor's degree from Indiana University, a Master's degree from Florida State University, and a Ph.D. from Louisiana State University. Amanda has published numerous peer-reviewed articles and book chapters on maritime archaeology, with a regional focus on the Gulf of Mexico and Caribbean, and recently co-edited the volume Prehistoric Archaeology on the Continental Shelf (Springer). Amanda currently serves on the board of directors for the Society for Historical Archaeology. She previously served on the board of directors for the Advisory Council on Underwater Archaeology and for the Register of Professional Archaeologists.

About the Editor

[illegible]

Chapter 1
Vernacular Watercraft: In Concept and in Practice

Amanda M. Evans and Sheli O. Smith

Introduction

By definition, *vernacular*, whether discussing language or architecture, refers to the common, ordinary, or domestic rather than the elite or monumental (OED 2013). Merriam-Webster (1990) defines *vernacular* as anything "of, relating to, or characteristic of a period, place, or group," it also refers to "the common building style of a period or place." Vernacular is also used in reference to maritime architectural styles when describing genres of ships and boats. Arguably, all vessels referred to as traditional boats or ships are in fact vernacular boats or ships that by definition have specific ties to the communities in which they are built and used.

Vernacular as Concept

Prior to the development of the railroad, ships were the largest and most technologically complex, means of transportation available, facilitating trade, exploration, interaction, and communication (Steffy 1994). Boats and ships transported people and cargos as well as intangibles such as ideas. According to Steffy (1994:23), ships were the mechanism that allowed ancient and historic populations the means to achieve "profit, convenience, security, or victory." Watercraft of all shapes and sizes were ubiquitous throughout history and still play an active role in modern culture. Today, supertankers and Panamax ships carry the lion's

A.M. Evans (✉)
Tesla Offshore, LLC., 36499 Perkins Rd., Prairieville, LA 70769, USA
e-mail: evansa@teslaoffshore.com

S.O. Smith
PAST Innovation Lab, 1003 Kinnear Rd., Columbus, OH 43212, USA
e-mail: sheli@pastfoundation.org

A.M. Evans (ed.), *The Archaeology of Vernacular Watercraft*,
When the Land Meets the Sea, DOI 10.1007/978-1-4939-3563-5_1

share of global commodities, while smaller freighters and ferries ply coastal waters moving cargo and passengers, and still smaller watercraft carry goods and people along rivers and across lakes.

In the archaeological record, the remains of watercraft are often found in isolation as wrecks on the bottoms of seas, lakes, and rivers. Through archaeological research, it is possible to link the vessel not only to its geographical source, but also its cultural origin. Vernacular boats, like all artifacts, are the products of the cultures and environments in which they were built (Maarleveld 1995). As such, they serve as physical embodiments of practical constraints and intended uses. They enlighten us regarding available resource materials, sea, lake or riverine conditions, technologies of construction, as well as cultural aesthetics. In short, each watercraft embodies all of the major aspects of culture; space, form, and time as well as technology, sociology, and ideology. According to McCaughan (2008:5), vernacular watercraft exhibit diversity because "regional and local boat types are responses in varying degrees to function, environment and tradition."

Prehistorically and historically as technology, social needs, and ideology vied for supremacy as cultural stimulus, vessel design, use, and purpose responded to these dynamic cultural factors. Since, these factors varied by location and people, so did the vessels they created, exhibiting the reciprocal relationships between human behavior and the surrounding environment (White 1949; Steward 1955; Rappaport 1968). Whether these adaptive strategies were "conscious or unconscious, explicit or implicit plans of action carried out by a population in response to either external or internal conditions" does not change the outcome (Moran 1982:325). These adaptive strategies created distinctive watercraft, specific to an area or culture, used for purposes that met the needs of the people.

Expanding on the themes of cultural ecology introduced by White, Steward, and Rappaport, Meltzer hypothesized that an essential element of successful colonization and growth was the ability to learn and master local resources (Meltzer 2003). Considering the adaptive strategies associated with vernacular watercraft, one would expect that traditional designs would evolve to meet the needs and respond to the available resources of a new location, creating over time a "new" or "modified" design. Thus, the concept of *vernacular*, which is based on a fundamental link between culture and environment, is dynamic. When the term is applied archaeologically to interpret shipwrecks and abandoned vessels, it must be considered in the light of the cultural ecology of the time, and should encompass as many aspects of technology, social need, and ideology as possible because vernacular watercraft are a cultural phenomenon.

Vernacular in Practice

Vernacular Watercraft in Culture History

Among the earliest written accounts of exploration are detailed drawings of vernacular watercraft (Churchill et al. 2011). With the advent of photography and the growing interest in culture at the close of the nineteenth century, anthropologists and cultural geographers began to collect visual evidence of watercraft around the world.

Photographs allowed for intensive comparison. Cultural geographers and Material Culturalists were further spurred as industrialization took hold around the world and indigenous populations began to vanish. With the advent of *Mariner's Mirror* in 1911, images of vernacular watercraft began to regularly appear. Needham, Worchester, Haddon, and Hornell were authors whose publications were synonymous with the study of vernacular watercraft of the Pacific and Indian Oceans (e.g., Worchester 1966; Hornell 1970; Haddon and Hornell 1975). Geographer Edwin Doran Jr. specifically examined local technologies regarding vernacular boat building and the use of local environmental knowledge and materials in their construction to more fully understand diffusion (e.g., Doran 1970, 1972, 1975, 1977, 1978, 1981). Initially drawn to vernacular ship research through his own experience with sailing, Doran (1970) was explicit that his approach to the study of boats was not the same as that taken by nautical historians or naval architects.

> Although the descriptive matter is much the same the study involves a full awareness and use of the concept of culture. Boats, their building, and their handling are part of man's vast cultural paraphernalia. Boat types are cultural complexes, which are learned and passed on to other members of one's cultural group. Ideas about boats, as about all other aspects of culture, may diffuse from one group to another (Doran 1970:4).

Despite the multifaceted development and diversity of vernacular boats, Doran (1970:6–8) identified three criteria by which watercraft could be associated with a distinctive vernacular type. The first criterion was that a vernacular type had to be recognized by local opinion; that is, the community in which the vernacular craft was built had to identify it as a part of the local community. Second, vernacular watercraft had to be built using generally accepted "rules of thumb" for determining dimensions and hull shape (Doran 1970:6–8). Finally, vernacular watercraft of a specific community had to be recognizable to outsiders as being of that given community.

In his cultural history of traditional boats of Indonesia, Horridge (1985) examined construction techniques, local language and names, functions, and the history and development of various types of vernacular craft using evidence from photographs, informant interviews, and direct measurements. Both Horridge and Doran documented traditional boats of living populations but also addressed historic craft, using any available evidence.

The concept of vernacular watercraft as practiced by Doran and Horridge is not restricted to modern vessels and communities. In his introduction to *Traditional Boats of Ireland*, McCaughan (2008:4) asserted that historic vernacular watercraft in Ireland were not highly technically advanced or standardized vessels but informally designed and "unconsciously reproduced by boat builders." He continued, stating that the design of the craft change over time within a community even though the "craftsmen often have little formal education, but they are knowledgeable and … experienced in building safe and serviceable boats, which serve their communities well" (McCaughan 2008:4).

By the very nature of being "common" most traditional boat types are not recorded in the historical record, and in these cases the archaeological record may retain evidence of vernacular watercraft in the form of shipwrecks and abandoned vessels. Just as with modern boats and historic vernacular craft, ships and boats in the archaeological record display technical and morphological attributes that reflect the culture responsible for their construction (Murphy 1983). Archaeological data recorded from vernacular watercraft can therefore provide information about the relationships between environment, technology, and the culture that made and used the vessel.

Vernacular Watercraft in Archaeology

In the archaeological record, the identity of a shipwreck may not be immediately apparent, nor is it guaranteed that a name can ever be attached to a site. It may be possible though, to recognize vessel types, including vernacular types, through the identification of specific features. According to Lenihan (1983:53), "the architectural elements of a sailing ship … represent a unique adaptive response to the demands for commerce, travel, exploration, or warfare, which are internalized by a particular society." As discussed in this volume, the dynamic relationships between the environment and social processes acting on a given community can be used to model the expected elements of a vernacular vessel type (e.g., Chap. 2). Subsequent vessels found in the archaeological record to have these characteristics may be tentatively identified as the vernacular type, and subsequent targeted testing can be conducted. Although the vessel may never have a name, its cultural, vernacular identity can be either refuted or confirmed.

Subsequent chapters in this book present a collection of case studies, and are organized generally by region, beginning in the Caribbean and Gulf of Mexico, moving northward along the Atlantic coast of North America, and then across the Great Lakes towards the Pacific coast of North America (Fig. 1.1). The case studies look at the topic from both the broader cultural implications and the more detailed interpretation of vessel form. Since watercraft are a reflection of culture and environmental pressures as well as adaptation, the studies reflect the diverse array of interpretation possibilities.

There are numerous ways to build a boat, but more interesting than how, is why a boat is built a certain way. As demonstrated in the following chapters, boats can be refined and modified to meet culturally specific functions, such as turtling (Chap. 3) or handlining (Chap. 15). In each case, different technical innovations were made to meet the same goal, and in both cases the vernacular watercraft serve as tangible evidence of the importance of a fishing industry to the culture and economy of each area.

Vernacular watercraft are cultural expressions, and as shown in cases from the Caribbean (Chap. 4) and the Gulf of St. Lawrence (Chap. 10) the vernacular watercraft reflect the transmission of ideas and manifestations of cultural identity through the form of the vessels. In each of these cases the vernacular watercraft were recognized as part of the cultural identity, both within the community and by outsiders who associated the vessel with an exotic culture.

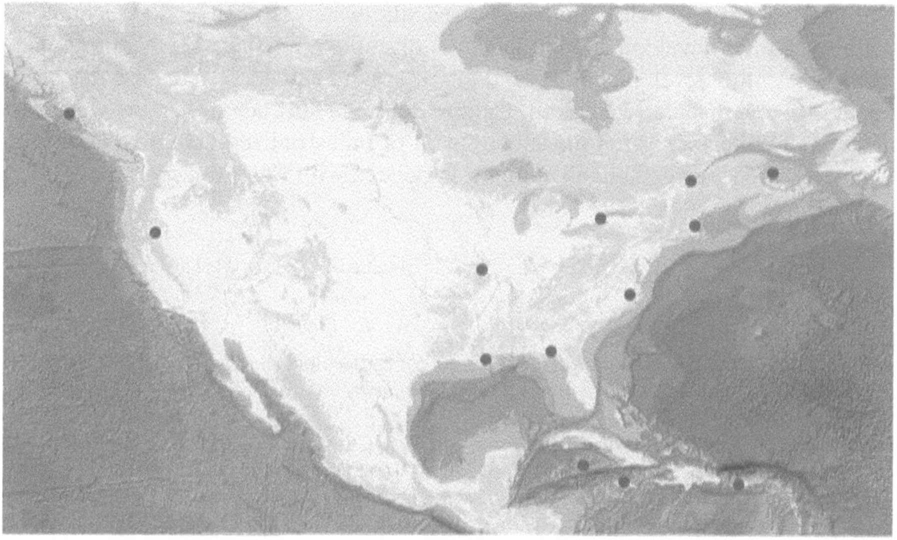

Fig. 1.1 Case studies from North America and the Caribbean are discussed in subsequent chapters

The concept of vernacular watercraft is not restricted to wooden boats. Nineteenth and twentieth-century vernacular watercraft include composite and metal-hulled ships. In cases from Florida (Chap. 5), North Carolina (Chap. 8), and the Great Lakes (Chap. 12), the vessels are part of a larger maritime landscape, and exhibit innovations to the environments in which the vessels operated. Abandoned boats, like the ones described in Chaps. 5 and 14, have a place within the archaeological record, and can provide a great amount of information concerning changing economies, or available materials (e.g., Richards 2008). Unlike previous publications on locally constructed vessels that performed regional analyses of vessel forms (e.g., Alford 2004; Fontenoy 1994; McGrail 2003), or analyses of specific ship types (e.g., Gardner 1987; Horridge 1985; Thomas 1989) these chapters discuss the vernacular in terms of their place. Previous studies have linked culture and local adaptation through the development of traditional boats (Alford 1998; Westerdahl 1994), but as demonstrated in Chaps. 6, 7, 8, 9, and 14, many studies remain unpublished or appear in the "gray" literature and do not enjoy widespread distribution (Cook 1997; Horrell 2005; Tolson 1992; Evans 2005; Jarvis 1990).

One of Horridge's (1985) more difficult concepts to illustrate concerning vernacular watercraft was the use of language in defining and associating vernacular watercraft. In his culture history of Indonesian vernacular watercraft, Horridge (1985:i) proclaimed that "the dictionaries are useless" in this regard because "names for the same boat are likely to be different in different places; sometimes the same name is used for several types." This phenomenon is discussed and illustrated through vernacular watercraft in Louisiana (Chap. 7) and Montreal (Chap. 11), where different types of vernacular watercraft often share the same name.

Finally, it is important to note that vernacular watercraft can lead to technological and cultural developments on land. As discussed in Chap. 13, there are many cases where the development of land-based structures and systems are the direct result of the vernacular watercraft. Culture does not start nor stop at the waterline, and ship architecture is not the sole domain of industrialized cultures. Vernacular watercraft are a universal to all cultures associated with water transport.

References

Alford, M. B. (1998). A small-craft typology: Tool for archaeological research. In L. E. Babbits & H. V. Tilburg (Eds.), *Maritime archaeology: A reader of substantive and theoretical contributions* (pp. 441–452). New York: Springer.

Alford, M. B. (2004). *Traditional work boats of North Carolina*. Beaufort, NC: North Carolina Maritime Museum.

Churchill, A., Churchill, J., Locke, J. (2011). A collection of voyages and travels: Some of the First Printed from Original Manuscripts, Others Now First Published in English. New York Public Library. Original Edition 1744, digitized in 2011.

Cook, G. (1997). *The readers point vessel: Hull analysis of an eighteenth-century merchant sloop excavated in St. Ann's Bay, Jamaica*. MA Thesis, Texas A&M University, Department of Anthropology, College Station, TX.

Doran, E., Jr. (1970). The Tortola boat: Characteristics, origin, demise. *Supplement to the Mariner's Mirror, 56*, 1–61.

Doran, E., Jr. (1972). Wa, vinta and trimaran. *Journal of the Polynesian Society, 81*, 144–159.

Doran, E., Jr. (1975). *Nao, Junk, and Vaka: Boats and culture history* [University Lecture Series]. College Station, TX: Texas A&M University Press.

Doran, E., Jr. (1977). The Laguna Madre Scow Sloop. In C. Reynolds (Ed.), *Proceedings of the North American Society for Oceanic History*. Salem, MA: NASOH.

Doran, E., Jr. (1978). The junk. *Oceans, 11*(3), 13–20.

Doran, E., Jr. (1981). *Wangka: Austronesian canoe origins*. College Station, TX: Texas A&M University Press.

Evans, A. M. (2005). *Institutionalized piracy and the development of the Jamaica sloop, 1630–1743*. MA Thesis, Florida State University, Department of Anthropology, Tallahassee, FL.

Fontenoy, P. E. (1994). *The sloops of the Hudson River: A historical and design survey*. Mystic, CT: Mystic Seaport Museum.

Gardner, J. (1987). *The dory book*. Mystic, CT: Mystic Seaport Museum.

Haddon, A. C., & Hornell, J. (1975). *Canoes of Oceania*. Honolulu, HI: Bishop Museum Press.

Hornell, J. (1970). *Water transport, origins & early evolution*. Newton Abbot, England: David & Charles.

Horrell, C. E. (2005). *Plying the waters of time. Maritime archaeology and history on the Florida gulf coast*. Doctoral Dissertation, Department of Anthropology, Florida State University, Tallahassee, FL.

Horridge, A. (1985). *The prahu: Traditional sailing boat of Indonesia*. Oxford, England: Oxford University Press.

Jarvis, M. J. (1990). 'The fastest vessels in the World': The origin and evolution of the Bermuda sloop, 1620-1800. *Bermuda Journal of Archaeology and Maritime History, 2*, 31–50.

Lenihan, D. J. (1983). Rethinking shipwreck archaeology: A history of ideas and considerations for new directions. In R. A. Gould (Ed.), *Shipwreck anthropology* (pp. 37–64). Albuquerque, NM: University of New Mexico Press.

Maarleveld, T. J. (1995). Type or technique. Some thoughts on boat and ship finds as indicative of cultural traditions. *International Journal of Nautical Archaeology, 24*(1), 3–7.

McCaughan, M. (2008). Irish vernacular boats. In C. Mac Cárthaigh (Ed.), *Traditional boats of Ireland: History, folklore and construction* (pp. 3–11). Cork, Ireland: Collins Press.

McGrail, S. (2003). *Boats of South Asia*. London: Routledge Curzon.
Meltzer, D. (2003). Lessons in landscape learning. In M. Rockman & J. Steele (Eds.), *Colonization of unfamiliar landscapes: The archaeology of adaptation* (pp. 222–241). London, England: Routledge.
Merriam-Webster. (1990). *Webster's ninth new collegiate dictionary*. Springfield, MA: Author.
Moran, E. F. (1982). *Human adaptability: An introduction to ecological anthropology*. Boulder, CO.: Westview Press.
Murphy, L. (1983). Shipwrecks as data base. In R. A. Gould (Ed.), *Shipwreck anthropology* (pp. 65–90). Albuquerque, NM: University of New Mexico Press.
Oxford English Dictionary (OED). (2013). *Oxford English dictionary*. Oxford, England: Oxford University Press.
Rappaport, R. A. (1968). *Pigs for the ancestors: Ritual in the ecology of a New Guinea people*. New Haven, CT: Yale University Press.
Richards, N. (2008). *Ship's graveyards: Abandoned watercraft and the archaeological site formation process*. Gainesville, FL: University Press of Florida.
Steffy, J. R. (1994). *Wooden ship building and the interpretation of shipwrecks*. College Station, TX: Texas A&M University Press.
Steward, J. H. (1955). The concept and method of cultural ecology. In J. H. Steward (Ed.), *Theory of culture change: The methodology of multilinear evolution* (pp. 30–42). Urbana, IL: University of Illinois Press.
Thomas, B. (1989). *Building the Crosby catboat*. Mystic, CT: Mystic Seaport Museum.
Tolson, H. (1992). *The Vernacular Watercraft of Isle Royale: A Western Lake Superior Boatbuilding Tradition*. Masters Thesis, Department of Anthropology, Texas A&M University, College Station.
Westerdahl, C. (1994). Maritime cultures and ship types: Brief comments on the significance of maritime archaeology. *International Journal of Nautical Archaeology, 23*(4), 265–270.
White, L. (1949). *The science of culture*. New York: Grove.
Worchester, G. R. C. (1966). *Sail and sweep in China*. London: Her Majesty's Stationary Office.

Chapter 2
Improvise, Adapt, Overcome: Vernacular Boats as Environmental Adaptations

Amanda M. Evans

Introduction

The year 2012 marked the centennial of the sinking of the RMS *Titanic*. Few vessels have achieved the same level of notoriety as that awarded to the ship once declared "unsinkable." Over the last hundred years *Titanic* has been the subject of both academic (e.g., Biel 1996; Annas and Elias 1999; Eustice et al. 2010) and popular publications (e.g., Archbold et al. 1997; Lord 2004). Famous vessels are sometimes well known because of their associations with tragic loss of life, or because they represented technological innovations or previously unknown levels of opulence and grandeur. For every infamous vessel like *Titanic* though, there are thousands of long-forgotten ships that were utilized in the development of human societies. Vessels have been built along every coastline using unique, locally available materials to meet different environmental conditions and serve specific functions. These common boats and ships, known collectively as vernacular watercraft, encompass a diversity of vessel types and were designed and constructed for use within local spheres of interaction and exploitation. Examples of vernacular vessels include riverboats, pirogues, catboats, and lumber schooners, as demonstrated by other chapters in this volume. The use and development of these vessels is not restricted to a single geographic place; they appear in a diverse range of communities over time. A vernacular watercraft is a unique combination of tangible and intangible aspects of place, and as such provides insight into the environment, available technologies, and economies in which it was built. Information about the construction, use, reuse, and discard of vernacular watercraft can contribute to an enhanced understanding of the communities in which they appear. In the colonial

A.M. Evans (✉)
Tesla Offshore, LLC., 36499 Perkins Rd., Prairieville, LA 70769, USA
e-mail: evansa@teslaoffshore.com

A.M. Evans (ed.), *The Archaeology of Vernacular Watercraft*,
When the Land Meets the Sea, DOI 10.1007/978-1-4939-3563-5_2

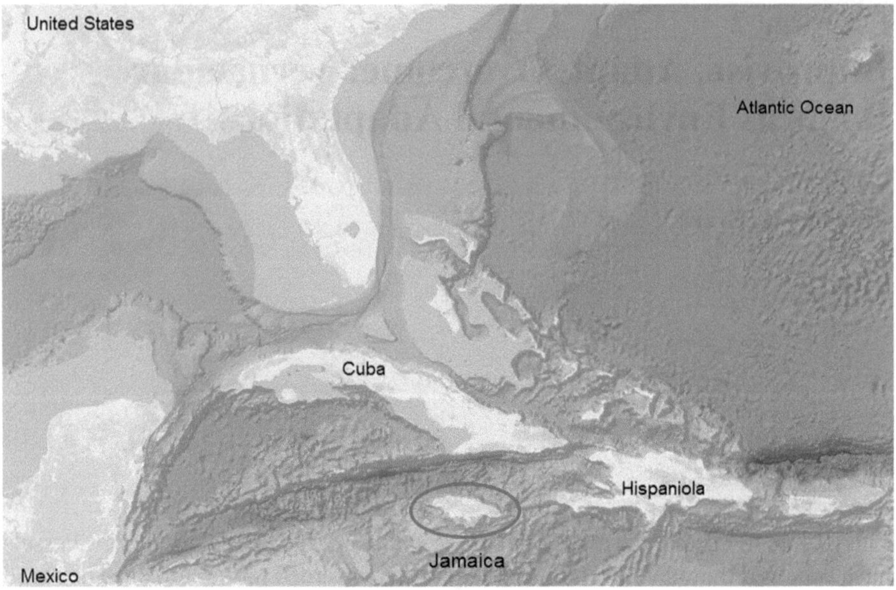

Fig. 2.1 Location map showing Jamaica and the Greater Antilles in the Caribbean; map is oriented north up

Caribbean, the Jamaica sloop is a vernacular watercraft that is defined by the cultural, economic, and technological systems in which it was developed (Fig. 2.1).

Ships as Cultural Artifacts

Humans have used ships for practical matters such as transportation, trade, and communication for over 5000 years (Ward 2004:19). Indirect evidence, however, suggests that humans have been making and using boats for over 60,000 years (Adams 2001:292). Despite this long history of use, only rare examples of shipwrecks sites, such as the Swedish warship *Vasa* (Cederlund and Hocker 2006), have been found relatively intact. More often, partial remnants of vessels are preserved. Archaeologists work with these remains to derive as much information as possible about the whole from the few preserved remains. Shipwreck sites can contain a diverse range of artifacts, but typically exhibit better states of preservation than contemporary terrestrial sites (Adams 2001:293). Despite the excellent preservation conditions, it can be difficult to justify the logistically complicated and expensive process of excavating shipwreck sites and conserving the artifacts. The value of shipwreck archaeology though, is in part, the ability to use idiographic data related to microprocesses apparent at the site and relate those to broader questions of social processes. As Gibbins and Adams (2001:280) described it, a ship's crew and the material culture reflected by the artifact assemblage represented "unique manifestations of society."

The metaphor of shipwrecks as time capsules containing well-preserved data has been used to describe the significance of shipwreck sites (Konstam 1999:10). The time capsule argument can be limiting though, since it reduces the role of the vessel to that of the setting for a static event. Ships were not merely platforms upon which activities happened. The use, design, construction, reuse, and discard of vessels all have implications related to human decision-making (e.g., Richards 2008). Shipbuilding is a deliberate, agent-based process, and decisions concerning the vessel's form and construction materials indicate the circumstances within which it was built. Wreck sites may contain a vast array of artifact types, but the ship itself should not be overlooked. As Gould (1983:6) states, "generalizations about various ways the human species has adapted to the conditions of voyaging and its use of the maritime habitat may be possible on the basis of evidence provided by shipwrecks." Ships are most often associated with transportation and trade, but their archaeological remains yield greater significance when viewed within the context of "the greater system at large" (Schmidt and Mrozowski 1983:143). The particular details of vernacular ship construction can provide evidence of agent-based adaptations to social and environmental conditions. Form and construction are essential features to consider when tracing the development of a vessel type, and hull data are likely to be preserved in inundated archaeological settings.

Archaeological investigations of shipwreck sites may result in data sets pertaining to the construction materials, form, and shape of at least a portion of the hull and its component pieces. Conceptual paradigms have been constructed to apply idiographic hull data to more general issues of economic systems, trade patterns, and local environments. According to Adams (2001:301) the seven "interrelated constraints on the form, structural characteristics, appearance and use of ships" include economics, environment, ideology, materials, purpose, technology, and tradition.

Vernacular watercraft were built for a variety of uses, and may or may not include specialized design features; however, constraints such as available technology, resources, and environmental conditions impact the final form of a ship. The interactive nature of the relationship between the constraints on ship design and appearance results in a dynamic system where changes in one factor necessitate changes in the others. This relationship is especially apparent in situations where people from one environment colonize a different geographic or climatic zone. Adapting to local conditions and resources is one aspect of colonizing new environments. Technological systems that offered maximum efficiency in one environment are not guaranteed to work as well in new areas. Changes from a traditional pattern of ship design, construction, use, and discard within these colonial situations may be interpreted as physical manifestations of human adaptation and variability. In the colonial Caribbean, it is hypothesized that English colonists developed a vernacular sloop type in response to new, available resources for ship construction and external environmental pressures.

Colonial Technology

Seventeenth-century European colonialism is characterized by the imposition of existing social and technological systems onto new landscapes (Blanton 2003:191). European colonists arriving in the Caribbean brought with them technologies and ideas developed over time in a European environment. New arrivals in the Caribbean faced several hurdles in order to succeed in the new settlements. Colonists first had to adjust to landfall after weeks or months at sea. According to Blanton (2003:193), overseas colonists cannot make gradual adjustments or process information in the same manner as overland travelers. English colonists to Jamaica were deposited into a radically different environment, but they had brought with them technology that was suited to European areas that featured established networks of improved overland roads and waterways connecting rural and urban areas (Blanton 2003:193). Caribbean islands at this point in time lacked improved waterways, and in some areas were not suited for overland travel. Caribbean colonies were established as peripheral, imperial outposts in the European competition for raw materials and wealth. The Caribbean colonies were part of an integrated economy, and ships were vital to both establishing the colonies and keeping them connected to global patterns of trade and conquest

In new environments, successful colonists "learn the local resources and adapt extant procurement strategies and technologies" (Meltzer 2003:238). Newly arrived colonists may also benefit from local knowledge that has been accumulated by preexisting populations (Meltzer 2003:225), but colonists must be willing to assimilate that local knowledge into their preexisting systems. In the Caribbean, for example, this meant successfully incorporating local flora and fauna as adaptations to the tropical climate. This phenomenon can be identified in the colonial Caribbean by the incorporation of local timber in shipbuilding or repairs. The identification of wood types used in hull construction allows archaeologists to make some initial observations about where the vessel was constructed, and where it might have been used, or even possibly repaired. Traditional European hardwoods used in ship construction during the seventeenth century were highly susceptible to shipworms present in warm Caribbean waters, and required frequent repair and timber replacement (Kaplan 1988:261; Smith et al. 1999:134–137). Timber stands of local Caribbean cedar were exploited as a readily available construction material and proved to be better suited for use on hulls operating in the region. In addition to being less expensive than imported English oak, local cedar had a higher tensile strength, meaning it was lighter and stronger than comparable oak timbers. Local cedar also had a high resin content that shipworms found bitter to the taste, resulting in less frequent careening (Jarvis 1990:35).

Colonists make decisions based on their experiences within an existing set of structures. Karl Butzer (1982:286) developed a model of these structures, defining them as interactive variables within an adaptive system (Fig. 2.2). In Butzer's model, an adaptive system has three main categories: technology, social behavior, and resource opportunities. If vernacular ships are viewed as an adaptive system, then based on Butzer's model, technology encompasses the design of a vessel, the assembly and form of the individual pieces, and the hydrodynamics of the hull-water interface. Again using Butzer's categories, resource opportunities are defined as those available at a particular place and time and subject to ecological processes such as biotic competition and disease. Caribbean colonists were restricted to materials that were either available locally, or items that could be imported or obtained through trade. In some cases, locally available timber stands were sacrificed to make room for plantation crops, a choice that stressed the

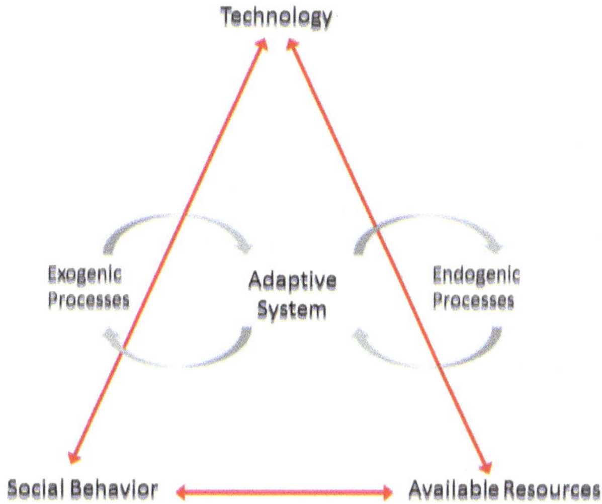

Fig. 2.2 Adaptive systems are feedback loops, created by the interaction of technology, available resources, and social behavior, and influenced by external pressures (after Butzer 1982:286)

economic viability of one crop over another. The shift from timber stands to plantation crops has also been used to infer a shift in colonial subsistence strategies (Adams 2001:303). Finally, Butzer's model includes aspects of social behavior, including existing political and economic systems, cultural attitudes and perceptions, and existing social organization. The variables defined by Butzer are stimulated by both exogenic and endogenic forces. Exogenic forces acting in the Caribbean included geological hazards such as offshore reef formations, and shallow water hazards. Other exogenic forces included climatic events such as hurricanes, ecosystemic imbalances (such as the shift from indigenous to exotic species), intergroup warfare or migration, the diffusion of new information through trade, and the subsequent colonization of new environments. Endogenic stimuli included processes such as innovation and acculturation.

Butzer wrote that the different stimuli and variables form feedback loops; changes in one part of the system necessitated adjustments in each of the others, creating a dynamic equilibrium. An effective examination of past human behavior requires a reflexive analysis of "structural events and patterns of practice" (Dornan 2002:325). When viewed within this framework, vernacular Jamaica sloops are an observable consequence of historically unique microprocesses of Caribbean geography and English colonization and broader macroprocesses of international economics and policy (Dornan 2002:235).

Agency and Cultural Ecology

Biophysical factors influencing Caribbean colonists included unpredictable hazards inherent to colonization of a new environment; sociocultural stimuli included intergroup competition from either warfare or migration, and diffusion of new information, generated outside of the ecosystem, through trade (Butzer 1982:289–283). Growing plantations within the Caribbean required trade goods and supplies, as islands at the periphery of European imperialism became more economically important within the global system of trade between 1600 and 1750 (Richardson 1992:4). Patterns of Caribbean trade demonstrated interdependence between nations, qualifying it as a world-system (Chase-Dunn and Grimes 1995:390). It is widely accepted that the triangular trade represented an international commodity flow linking the economies of Europe, Africa, and the Americas, but early interpretations of world-systems theory failed to recognize ships as an adaptive reaction within the system, ignoring them as static tools used to move trade goods between points. In 2004, Wallerstein (2004:17) wrote that the capitalist world-economy predicated "a need for constant technological change," and a constant expansion of geographical, intellectual, and scientific frontiers. Traditionally, nautical scholars such as Baker (1966:1) have argued that modifications in ship designs occurred gradually, based in large part on "changes in trade patterns that altered cargo carrying requirements, refinements of form, improvements in rig, availability of timber, and mere whim and fashion." This explanation of design modifications ignores economic or political stimuli, and more recent studies argue against such a simplistic explanation. In fact, it has been argued that technological change increases at a dramatic pace under capitalist systems, because the success of an adaptation is determined solely by its production efficiency, without regard to any potential disruption of existing social structures (Chase-Dunn and Grimes 1995:400). Capitalist systems allow the unchecked accumulation of wealth; therefore, social effects of technological change are generally ignored because the developers are primarily interested in raising their profit margins against aggressive competition (Chase-Dunn and Grimes 1995:400). Based on these principles, a ship design that can move trade items more quickly and more securely, while requiring less capital investment, or upkeep, will quickly be adopted regardless of social factors.

Maritime Technology and the Caribbean

Steele and Rockman (2003:132) state that it is a "truism that humans modify their habitats by technological means to make them more favorable to habitation" but humans also modify their technology to suit new habitats. Obviously the type and availability of construction materials will impact the finished product, but the actual form of a vessel has as much to do with its purpose as the area where it will be used. Travel by sea and across water was a calculated risk, and each venture was predicated on an evaluation of environmental, technological, and human factors (Adams 2001:293). Jamaica sloops were built by local colonists and used to transport goods across a volatile region. Pirates, privateers, vessels of enemy nations, and natural hazards were all factors that had to be considered when a ship went to sea. The decision to sail was the end result of an assessment of the risk of failure and the potential for success. Fast vessels that could outrun, and outmaneuver hostile vessels were more successful for their owners, and local shipbuilders experimented with designs that attempted to reach equilibrium between speed, maneuverability, and cargo capacity. Vernacular Jamaica sloops were relatively small vessels, characterized by design elements that included a raked mast slightly forward of the midships, a rounded bilge, raked sternpost, and sharply rising floor (Baker 1966:111; Chapelle 1967:65; Evans 2005:74). The design, combined with local cedar that had a higher tensile strength than heavier European oak, resulted in a vessel that earned an enviable reputation for speed and seaworthiness (Cordingly 1995:163). Jamaica sloops were adopted and adapted by other Caribbean colonists because the design was effective in providing safe and reliable transportation.

Social constructs do not exist in the archaeological record, but artifacts such as hull remains can inform hypotheses of these abstract behavioral patterns. The process of causation, or agency, as suggested by Ortner (1984:127) and reiterated by Dornan (2002:324), is a mechanism for identifying "the intersection of individual intention with resistance to or incorporation of particular social structures" at a given point in time. Dobres (2000:135) argues that agents and structures are relational. In a Caribbean example, the agents were English colonists in Jamaica. The structures can be defined as the material, social, and symbolic conditions within they existed and through which they reproduced and transformed themselves. As relational entities, any attempt to understand the agent must consider the structures in which it operates to avoid misinterpretations of the agent's, or more generally, human behavior.

> It is no sure bet that agents will act just because they can or should. In other words, not acting is still a form of agency [Kegan Gardiner 1995]—and context is the only means by which to understand these nuanced qualities of agency. What this generally means for technology studies in particular, but archaeology more generally, is a new and different interest in the question of material patterns of variability juxtaposed against norms (Dobres 2000:135).

Dobres' argument can be divided into two main points. The first is the idea that not making a decision, choosing not to decide, is still an active choice. The second part is that, from Dobres' perspective, the underlying question is not who made the object, or how they made it. Instead, an agent-based archaeological inquiry examines the degree of variability between the object and its normative equivalent. Agent-based studies are concerned with the process of causation. If seventeenth- and eighteenth-century English sloop data is used to create a normative equivalent for sloop construction, then this baseline may be compared with the stimuli necessitating the adaptation of existing ship designs by Jamaican colonists to identify areas of potential variability. As one interpretation of Bass (1983:97) implies, particularist data extracted from shipwreck sites are essential because the resulting datasets can be analyzed and used to reconstruct the behavior of the people associated with the ship. In several instances, limited excavation of portions of vernacular shipwrecks has produced particular data that resulted in a detailed analysis of the local community, economic systems, and trade patterns associated with the vessel (e.g., Horrell 2005; Stanbury 2003).

Shipwrecks, by definition, represent technological failure since the vessel did not survive to deliver its contents, but it should not be assumed that only unsuccessful designs result in shipwrecks (Adams 2001:293). English colonists in the colonial Caribbean were threatened by attack and their vulnerability stimulated the intentional development of an adaptive ship technology. Jamaica sloops were a unique product of spatial and temporal conditions, and they can be identified by their unique design and construction.

Conclusion

Vernacular watercraft is an academic term applied here to locally built ships. Colonists in new areas often found that technologies that were previously successful were not well suited for their new environment. In some cases, colonists benefited from knowledge obtained by preexisting populations, but in other cases, they learned through trial and error. In colonial Jamaica, shipbuilders drew from a common knowledge base to construct their vessels, but they made independent adaptations that are revealed by a particularistic analysis of the hull structure and materials used in the construction of Jamaica sloops.

Cultural ecological models, specifically Butzer's interactive variables of an adaptive system (Butzer 1982:286) are particularly useful for analyzing shipwreck sites. The particular details of construction materials or hull scantlings provide a picture of the vessel which can then be compared to normative equivalents, from which areas of adaptation can be identified. Although it can be easy to overlook the vessel involved in trade, it is not a static object moving trade goods. It is as much a part of the economic and cultural system that it was used in, and represents deliberate adaptations on the part of local colonists to their situation.

Early studies of ship design and innovation assumed that alterations were attributable to changes in trade patterns or stylistic whim (Baker 1966:1); however, more recent research demonstrates that ship design was influenced by a far greater number of variables. Despite historical references to Jamaica sloops as a specific type, no archaeological remains have been identified; however, it is possible to define the ship's form by modeling it as an adaptive system. Ship design was part of a larger tradition of human adaptations based on technological factors, available resources, and social behavior. These three variables, stimulated by exogenic and endogenic processes create an unstable environment, and adaptations to ship designs are therefore viewed as a coping mechanism. In addition to the colonial Caribbean, the concept of vernacular ships as adaptations to new environments can be seen in geographically and temporally distinct cultures who developed vernacular shipbuilding traditions to ensure their survival.

References

Adams, J. (2001). Ships and boats as archaeological source material. *World Archaeology, 32*(3), 292–310.

Annas, G. J., & Elias, S. (1999). Thalidomide and the *Titanic*: Reconstructing the technology tragedies of the twentieth century. *American Journal of Public Health, 89*(1), 98–101.

Archbold, R., McCauley, D., & Lord, W. (1997). *Last dinner on the titanic: Menus and recipes from the great liner*. New York: Hyperion Books.

Baker, W. A. (1966). *Sloops and shallops*. Barre, MA: Barre.

Bass, G. F. (1983). A plea for historical particularism in nautical archaeology. In R. Gould (Ed.), *Shipwreck anthropology* (pp. 91–104). Albequerque, NM: University of New Mexico Press.

Biel, S. (1996). *A cultural history of the titanic disaster*. New York: W.W. Norton.

Blanton, D. B. (2003). The weather is fine, wish you were here, because I'm the last one alive. In M. Rockman & J. Steele (Eds.), *Colonization of unfamiliar landscapes: The archaeology of adaptation* (pp. 190–200). London: Routledge.

Butzer, K. (1982). *Archaeology as human ecology*. Cambridge, England: Cambridge University Press.

Cederlund, C. O., & Hocker, F. M. (2006). *Vasa I: The archaeology of a Swedish royal ship of 1628*. Stockholm, Sweden: Statens Maritima Museer.

Chapelle, H. I. (1967). *The search for speed under sail 1700-1855*. New York: W.W. Norton.

Chase-Dunn, C., & Grimes, P. (1995). World-systems analysis. *Annual Review of Sociology, 21*, 387–417.

Cordingly, D. (1995). *Under the black flag: The romance and the reality of life among the pirates*. New York: Harcourt Brace.

Dobres, M.-A. (2000). *Technology and social agency: Outlining a practice framework for archaeology*. Oxford, England: Blackwell.

Dornan, J. L. (2002). Agency and archaeology: Past, present, and future directions. *Journal of Archaeological Method and Theory, 9*, 303–329.

Eustice, R. M., Singh, H., Leonard, J. J., & Walter, M. R. (2010). Visually mapping the RMS titanic: Conservative covariance estimates for SLAM information filters. *International Journal of Robotics Research, 29*, 941–957.

Evans, A. M. (2005). *Institutionalized piracy and the development of the Jamaica Sloop, 1630-1743*. Master's thesis, Department of Anthropology, Florida State University, Tallahassee, FL.

Gibbins, D., & Adams, J. (2001). Shipwrecks and maritime archaeology. *World Archaeology, 32*(3), 279–291.

Gould, R. (1983). Looking below the surface: Shipwreck archaeology as anthropology. In R. Gould (Ed.), *Shipwreck anthropology* (pp. 3–22). Albequerque, NM: University of New Mexico Press.

Horrell, C. E. (2005). *Plying the Waters of Time: Maritime Archaeology and History on the Florida Gulf Coast*. Doctoral dissertation, Department of Anthropology, Florida State University, Tallahassee, FL.

Jarvis, M. J. (1990). 'The fastest vessels in the World': The origin and evolution of the Bermuda sloop, 1620-1800. *Bermuda Journal of Archaeology and Maritime History, 2*, 31–50.

Kaplan, E. H. (1988). *A field guide to southeastern and Caribbean seashores: Cape Hatteras to the gulf coast, Florida, and the Caribbean*. Boston, MA: Houghton Mifflin.

Konstam, A. (1999). *The history of shipwrecks*. Guilford, CT: Lyons Press.

Lord, W. (2004). *A night to remember*. New York: Holt Paperbacks.

Meltzer, D. (2003). Lessons in landscape learning. In M. Rockman & J. Steele (Eds.), *Colonization of unfamiliar landscapes: The archaeology of adaptation* (pp. 222–241). London: Routledge.

Ortner, S. B. (1984). Theory in anthropology since the sixties. *Society for Comparative Study of Society and History, 26*(1), 126–166.

Richards, N. (2008). *Ship's graveyards: Abandoned watercraft and the archaeological site formation process*. Gainesville, FL: University Press of Florida.

Richardson, B. C. (1992). *The Caribbean in the wider world, 1492-1992: A regional geography*. Cambridge, England: Cambridge University Press.

Schmidt, P. R., & Mrozowski, S. A. (1983). History, smugglers, change and shipwrecks. In R. Gould (Ed.), *Shipwreck anthropology* (pp. 143–172). Albuquerque, FL: University of New Mexico Press.

Smith, R. C., Spirek, J., Bratten, J., & Scott-Ireton, D. (1999). *The Emanuel point ship: Archaeological investigations 1992-1995*. Tallahassee, FL: Florida Department of State, Bureau of Archaeological Research.

Stanbury, M. (Ed.). (2003). The Barque Eglinton: Wrecked Western Australia 1852. *Australian National Centre of Excellence for Maritime Archaeology, Special Publication* No. 6; *The Australasian Institute for Maritime Archaeology, Special Publication* No. 13. Fremantle: Western Australia Maritime Museum.

Steele, J., & Rockman, M. (2003). Where do we go from here: Modelling the decision-making process during exploratory dispersal. In M. Rockman & J. Steele (Eds.), *Colonization of unfamiliar landscapes: The archaeology of adaptation* (pp. 130–143). London: Routledge.

Wallerstein, I. (2004). *World-systems analysis: An introduction*. Durham, NC: Duke University Press.

Ward, C. A. (2004). Boatbuilding in ancient Egypt. In F. Hocker & C. Ward (Eds.), *Towards a philosophy of ancient shipbuilding* (pp. 13–24). College Station, TX: Texas A&M University Press.

Chapter 3
The Caymanian Catboat

Roger C. Smith

From the time of their discovery by European mariners in 1503, the Cayman Islands were associated with the sea turtle. Ferdinand Columbus reported that he saw so many of the swimming reptiles that their shells looked to him like a reef around the islands (Morison 1963:353). When his father gave the name Las Tortugas to the Cayman Islands he didn't realize that he was among the first Europeans to witness the annual gathering of the creatures to breed and to lay eggs. Soon the Islands became the New World's largest sea turtle fishery, as mariners discovered that captured sea turtles offered the prospect of fresh meat at sea because they could be kept alive aboard a ship for weeks without difficulty. Seasonal turtlers set up fishing stations and camps to exploit the resource, and not surprisingly for the early permanent settlers of the Islands in the eighteenth century, turtling became the paramount occupation (Smith 2000:67).

Caymanians not only fished their own waters but also regularly sailed to the southern coast of Cuba to take turtles. In the latter half of the eighteenth century, the waters around Cayman began to become depleted of turtles. By 1800, nine island vessels were turtling exclusively off Cuba, returning to Grand Cayman with their catch to stock the island crawls, or storage pens (Doran 1953:346). By the late 1830s, the turtling grounds on the south coast of Cuba began to deteriorate. Turtlers of Grand Cayman simply shifted fishing operations south to the Miskito Bank, a vast complex of cays, reefs, and shallow fishing grounds off the Central American coast, where green sea turtles were plentiful (Fig. 3.1). Fleets of 15–20 schooners and several small boats set sail from Grand Cayman for the Nicaraguan fishing grounds and were gone approximately 10 weeks at a time (Doran 1953:341).

When the fishing grounds were reached after a voyage of several days, the schooner carefully sailed among the shoals, selecting lone coral outcroppings on the

R.C. Smith, Ph.D. (✉)
Division of Historical Resources, Bureau of Archaeological Research,
R. A. Gray Building, 500 S. Bronough St., Tallahassee, FL 32399-02500, USA
e-mail: rsmith@dos.state.fl.us

A.M. Evans (ed.), *The Archaeology of Vernacular Watercraft*,
When the Land Meets the Sea, DOI 10.1007/978-1-4939-3563-5_3

Fig. 3.1 Map showing the Cayman Islands in relationship to Cuba and Nicaragua

seabed, narrow reef channels, or round sand holes ringed by coral—places that were likely to harbor sleeping turtles at night. Turtles, after feeding all day in submerged pastures of grass, generally sought these sheltered locations at night to avoid attacks by sharks while they slept. The fishermen marked their chosen spots with floats of light "bob wood" anchored to big chunks of fossil coral, or "kellecks." Before nightfall, the schooner anchored safely in the lee of the reefs, and the business of setting turtle nets began. The fishermen lowered their boats into the water and sailed to the marked fishing designations as the turtles prepared to roost under the rocks. Each wooden float quietly was approached upwind under oars, and a net carefully was paid out from the bow of the boat over the selected spot.

Between 10 and 30 nets were set horizontally over coral heads each evening (Doran 1953:344). As sleeping turtles slowly rose to the surface to breathe during the night, they would strike the overhead net, struggle, and become entangled. The wide mesh of the nets encouraged tangling, but permitted turtles to haul the net to the surface in order to breathe and not drown. As the sky grew light at the coming of dawn, the turtles grew restless and struggled to reach the open sea. But the turtlers soon were back over their nets, pulling them into the boats. The most difficult part of the entire voyage, the task of hauling an ensnared turtle aboard a small boat, was complicated by the size, weight, and clumsy thrashing of the entangled creature. Pulled over the gunwale by its front flippers and positioned on its back in the bilge, each turtle carefully was disengaged from the net. When first taken aboard, the turtles slapped their flippers on their bellies, then subsided and lay still. After the crews collected their catches, they made their way back to the schooner (Fig. 3.2).

As the boats came alongside the schooner, a cable was lowered and the turtles were swung inboard over the rails. To avoid excess commotion on deck, the turtles

Fig. 3.2 A Caymanian catboat with a load of sea turtles (photo by Wright Langley 1964)

sometimes were "spancelled" by tying their flippers together with thatch string, fore limb to rear limb. Often, the men cut the initials of their schooner's name into the bottom plate of each turtle as a brand that would distinguish their catches later at market (Duncan 1943:183; Langley 1964:62). Left right side-up out of the water, a green turtle's lungs would collapse from the pressure of its top shell and body weight against its soft bottom shell, so the turtles were placed upside-down on deck. Wedges were kicked under the shells of the upturned reptiles to keep them from sliding in rough seas, and a wooden pillow was placed beneath their heads for support (Matthieson 1975:206–207).

The captured turtles were taken to a central cay, where they were confined in a crawl fashioned of long mangrove saplings stuck into the soft seabed (Doran 1953:346; Matthieson 1975:300). The penned reptiles were fed turtle grass or other greens for the duration of their confinement. When a sufficient number of turtles had been gathered in the crawls, they were rounded up, lassoed, wrestled from the water, and loaded in the hold of the schooner for the passage home. In Grand Cayman, the catch was crawled again until the turtles were bartered or sold, either live or butchered.

Turtles also were taken to Kingston, Jamaica; in later years, they were marketed at Key West or Tampa, Florida, at the end of each season (Craton 2003:219–220).

Caymanian turtlers developed a unique maritime folklore based on the swimming reptiles. Over seasons of hunting, catching, and crawling, they became intimate with the ways and wiles of turtles. To make good sets, the fishermen studied the animals' routine behavior in combination with the topography of the sea bottom, viewing both through their "water glasses." They learned the subtle underwater signs to locate the best turtle roosting rocks and reefs; they spent years watching and waiting for the right combination of water and weather in order to outwit their quarry (Smith 2000:74).

In the Sister Islands (Little Cayman and Cayman Brac), inhabitants also gradually began turtling outside the islands. First permanently settled as late as 1833, isolated Cayman Brac required its people to build ships to survive. Vessels were constructed to fish the remote southern banks of Serrana, Quita Sueño, Seranilla, and Pedro. Brackers, as local inhabitants were known, fished primarily for the hawksbill turtle, preferring to trade in turtle shell rather than meat. Brac schooners and sloops of 50–60 tons served as transport vessels for small fishing boats. As many as ten boats could be carried on the deck of a 50-ton schooner, arranged on edge between the hatches and the wheelhouse (L. Jervis, personal communication, 1980).

The boats figured prominently in the hunt for turtles, each carrying two men and a single net. The "puller" manned the oars, propelling the boat toward a turtle spotted on the surface of the water. He was directed by the "trapper," who sat in the bow and followed the turtle's progress if it dived and swam away. A "water glass" made from a box with a clear glass bottom helped the trapper to see under the water and enabled him to guide the puller, who maneuvered the boat directly over the turtle. A conical-shaped "trap net" was lowered quietly into the water. Weighted at the base by an iron ring, the net was open at the bottom and attached to a line at the top, which was controlled by the trapper so that the net could be dropped over the turtle at the appropriate moment (Hirst 1910:273–274). Once encompassed by the trap net, the turtle attempted to rise, became entangled, and then quickly was pulled to the surface of the water. What could weigh 300 pounds of fighting reptile with a sharp hooked beak was heaved into the boat by hand.

As many as 60 boats were employed by Brackers during the turtling season (Hirst 1910:276). Most of the hawksbill turtles were slaughtered soon after being netted, rather than being kept alive in a crawl. The meat was scored with a knife, rubbed with salt, and hung in the sun to dry. Sometimes it was "corned," or pickled in brine. The shells were marketed primarily in Jamaica, where they were fashioned into decorative items or were shipped directly to North America or Europe.

Between the years 1929 and 1939, the value of turtles per head dropped from an average of 40 shillings to 20 shillings (Rebel 1974:122–123). During the same years, the export of turtle shell declined by almost half, reflecting the gradual replacement of "tortoise shell" by synthetic products such as plastics. As green turtles approached extinction in the Caribbean, Caymanian turtling began a sharp decline; in 1950, the annual fleet sailing to Nicaragua numbered only ten vessels, whereas not long before there had been 25. By 1970, the government of Nicaragua had severely limited the number of fishing permits issued, and not a single registered turtler sailed from Cayman (Considine and Winberry 1978:53; Williams 1992/1970:88).

The boat that characterized and sustained this unique maritime pursuit was the Caymanian catboat, a single-masted, double-ended sailing vessel with a fore-and-aft sail that was propelled efficiently by sail, oars, paddle, or pole, and became the specialized utility vehicle of the Islands. In 1904, Daniel Jervis, a turtling captain from Cayman Brac, decided to build a shorter and wider boat that would be easier to maneuver than a canoe (L. Jervis, personal communication, 1980). The resulting double-ended vessel, which he named *Terror*, was to be the prototype of the traditional Caymanian catboat, an ultimate adaptation of small craft designed especially for turtling. Only 14 ft. in length and 3 ft. 8 in. in breadth, the new vessel was constructed by attaching four temporary frames to the keel, planking the entire boat, and then inserting permanent framing and thwarts.

Equipped with sails and oars, the design was so successful that it soon was adopted by the islanders of Grand Cayman (Smith 1985:332). Jervis' stern differed from most other double-enders in that it was sharp where the others are usually full bodied. Jervis found that the sharpness allowed the vessel to reverse course instantly, and that when leaving reef channels less drag was created aft and it was easier to get clear on the rise of a wave. Unfortunately, *Terror* was lost at sea on a schooner that went down with all hands in the hurricane of 1932 (McLaughlin 1994:101).

The double-ended catboat's features soon became standardized. Typical dimensions were 16 ft. in length and 4 ft. in breadth, although some boats were slightly larger. The first step in catboat construction was to carve a half-model of the intended shape of the boat. The model usually was scaled 1:10 or 1:12. No plans or drawings were considered necessary, since the builder invariably knew exactly the desired shape and dimensions that his boat should have from long experience building or fishing in catboats. Framing shapes were taken directly from the half-model, which occasionally was cut into station sections to facilitate obtaining the desired curves (Smith 2000:123). Frames and compass timbers for the keel, stem, and sternpost carefully were selected from local hardwood forests. Mahogany, cedar, pompero, jasmine, whitewood, plopnut, fiddlewood, sea grape, wild sapodilla, and Jamaica walnut all were used (Burton 1997). Curved timbers were selected by "curve-stalkers" searching the buttonwood swamps; the partially buried roots that buttress the mahogany tree were ideal. Occasionally, a hardwood sapling was bent over to the lee of the prevailing wind and tied to the ground to be "trained" to the desired curvature.

Floor frames (called "dunnage") and half frames, numbering from 11 to 15, were either sawn or steamed to fit the internal shape of the hull and were fastened to the keel, stem, and sternpost with treenails. Seven or sometimes eight strakes of Cayman cedar or white American or Honduran pine were carvel-fastened with trunnels and screws to each side of the boat. The planking was caulked with a mixture of paint and ashes collected from burning termite hives for more than a week. The uppermost, or sheer strake, was slightly thicker than the others. A caprail and sometimes a rubrail were added to the gunwale. The sheer strake and rails reinforced the gunwale of the boat, which took the brunt of the action when a thrashing turtle was hauled aboard (Smith 1985:333).

A longitudinal stringer was inserted on each side of the internal hull, serving to support the four thwarts. The forwardmost thwart—a heavy member often made up of three or more planks fitted edge-to-edge—supported the mast, which was passed through the thwart and into a keyhole-shaped socket of the mast step atop the stem scarf. This forward thwart also served as a breasthook, reinforcing the bow. As in the typical catboat rig, the mast was stepped far forward in the vessel, but it had no stays or shrouds; the boom and mast assembly commonly was unshipped quickly and laid in the boat when the turtling grounds were reached. The mast was rigged with a sprit or Marconi mainsail and had a distinctive rake aft to distribute the weight of the sail toward the center of the shallow craft. Masts and booms usually were made of Douglas fir or Spanish elm.

Cayman Brackers used to put their sail high in the foot off the rail, because they looked for turtles when they fished, and could see better overboard. Catboats on Grand Cayman had sloop rigs, sometimes with a small jib. The mainmast was stepped in the sailing thwart, because it was thought to sail better that way. They were rigged with a lighter, taller mast, sometimes with light shrouds.

In high winds, boats from Grand Cayman employed a "weatherboard," set well outboard on the windward side, upon which one of the crew could perch to add leverage and prevent the craft from capsizing. Sailors from Cayman Brac considered this apparatus a lubberly piece of gear and almost never used it (Doran 1953:314). A deep keel or centerboard was not considered necessary, since the boat operated only part of the time under sail. Sometimes Brackers would put an additional false keel on their boats, with a bolt or screw fore and aft so that it could come off easily. The false keel helped to sail closer to the wind and not drift away.

Rather than a tiller that would become fouled by turtle nets, the catboat's rudder head was fitted with a yoke bar with steering lines that could be dropped quickly in the bilge. This arrangement also allowed the rudder to be unshipped and stored out of the way. Thus, with sailing rig and rudder removed, the double-ended catboat became a highly maneuverable platform under oars for netting turtles and catching fish.

Each catboat was equipped with two to four oars that were 8–10 ft in length, and sometimes with a spare. Called "sculls" by the turtlers, the oars were composite-built: the loom of Spanish elm and the blade of white pine. On Grand Cayman, oars sometimes were made of strawberry and red mangrove. Instead of oarlocks, a small piece of wood called a "pallet" or "oarblock" was fastened to the caprail. Two holes bored through the pallet were threaded with thatch palm rope, which held the oars to the gunwale. As with an oarlock, this arrangement was a highly functional one, for the boat quickly could be backed, changing directions as the turtle chase required. A long paddle, sometimes over seven feet in length with a curious notched grip, or "horn," for the upper hand, also was carried for use as a steering oar when the rudder was unshipped.

The catboat's complement of equipment invariably included a "waterglass," an open wooden box with a pane of glass in the bottom to view the seabed, and a "calabash" bailer, made from the woody shell of fruit from the gourd tree, to eject water from the hull. Often, a long pole, for propulsion in shallow water, doubled as a harpoon with the fixing of barbs at one end. In addition, a wooden fish club, or "muntle," often was carried along with the usual hooks and line.

Caymanian catboats invariably were painted a distinctive bright blue color. The same traditional color, called "catboat blue," also was applied to those articles of boat gear that were used in the water, such as the oar blades, paddle, pole, and the water glass. The blue paint was a functional part of the turtling lore because it acted as an effective marine camouflage, blending with the surrounding water and allowing the fishermen to approach their prey undetected. When not in use, catboats were unrigged, pulled up on the shore, and stored under low, thatch-roofed shelters that were open to the sea breeze. In some cases they were suspended from davits along the rocky shoreline.

For some 200 years, turtling, shipbuilding, and its related activities of woodcutting, rope making, and sail making were among the principle forms of livelihood in Cayman. The Caymanian shipbuilding industry climaxed between about 1903 and the late 1930s. The Cayman Islands had 30 schooners on the water by 1907, averaging 67 tons. By 1930, Grand Cayman's fleet consisted of 23 schooners and about 300 catboats. Regularly scheduled regattas and races were inaugurated in 1935 and were usually held at the end of January, between the shark and turtling seasons (Smith 2000:131). The days of the regatta, when more than 100 catboats raced against each other, were declared public holidays and the western coast of Grand Cayman was filled with cheering spectators and picnic feasts.

Sadly, the Caymanian catboat's twilight came when the marine resource it was specially designed to capture dwindled, and restrictive legislation made turtling unprofitable. On the islands, roads were developed and motorized vehicles were imported. On the sea the arrival of the outboard motor and the development of fiberglass as a boat-building material hastened the catboat's demise.

Fortunately, a surviving member of the catboat fleets was found stored in a shed behind a house on the north shore of Cayman Brac (Fig. 3.3). Named *Ajax*, the boat

Fig. 3.3 Catboat *Ajax* built on Cayman Brac (photo by KC Smith)

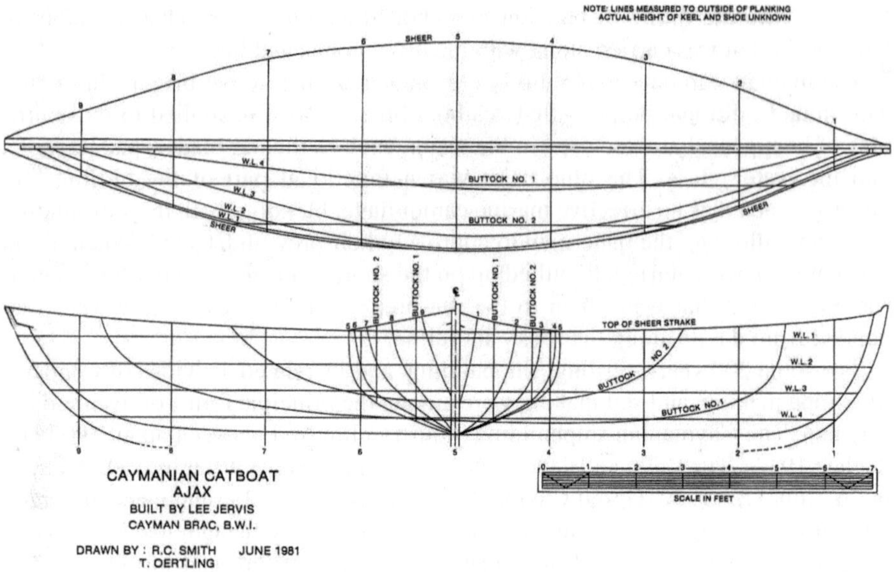

Fig. 3.4 Lines of *Ajax* (Drawing by the author and Thomas Oertling)

was a veteran of sun-bleached shoals and countless struggling turtles, but its well-preserved hull and its associated equipment allowed the entire vessel and gear to be recorded in 1979. The half-model from which the boat had been constructed in 1934 kindly was loaned for examination by Lee Jervis, the vessel's builder. A series of offset measurements was taken from both the model and the boat to establish the catboat's hull shape and characteristics (Smith 1985:335).

One can see that the bow sections (Fig. 3.4 on the right) are fuller than the stern. This feature helped the boat to "float" over the waves. The bow is fuller so it has more buoyancy and floats up, but the stern is not so buoyant and does not force the bow down. With more deadrise, the stern will essentially sink down into the passing sea, and the boat just seems to mold itself right into a wave, and go over it smoothly. The negative side of this feature is that many catboats were known for being swamped from the stern by the same seas.

The seaworthiness of Caymanian catboats is proverbial. A beautiful low-sheared vessel with light displacement and an almost flamboyant sail form, the catboat is very tender and not very efficient sailing to windward. It was a no-nonsense pickup truck of the Islands, serving as a primary means of transportation and a mainstay of employment for much of their history.

In recent years, interest in the catboat's heritage has been revived in the Cayman Islands. Publication in 1999 of a photographic essay, *Love's Dance—the Catboat of the Caymanes*, by the Cayman Islands Seafarers' Association, led to a joint sponsorship with the Cayman Maritime Heritage Foundation to inaugurate a Catboat Club and to reorganize an annual Regatta on Grand Cayman (Ross 1999). With more than a dozen catboats competing in traditional boating tasks, the heritage of this distinctive Caribbean craft is being preserved.

An intrepid island society that developed on the frontier of the sea, the people of the Cayman Islands depended on their watercraft and nautical skills for communication and transportation, not only between the islands, but also with the outside world. Specialized craft evolved in form and function to meet their maritime needs, from local fishing activities along fringing reefs to organized voyages hundreds of miles from home. Blessed with native hardwoods for timber and palms for rope, the outward-looking islanders used the sea as an avenue of subsistence, rather than viewing it as a barrier. They learned its seasonal moods, both good and bad, and how to construct by hand vehicles that were swift, efficient, and durable in the offshore elements. Passing these nautical secrets down through generations, Caymanians forged a unique legacy that carried them not only throughout the Caribbean, but also eventually around the globe.

References

Burton, F. J. (1997). *Wild trees in the Cayman islands*. Grand Cayman, British West Indies: National Trust of the Cayman Islands.

Considine, J. L., & Winberry, J. J. (1978). The green Sea turtle of the Cayman islands. *Oceanus, 21*(3), 50–55.

Craton, M. (2003). *Founded upon the seas. A history of the Cayman islands and their people*. Kingston, Jamaica: Ian Randle.

Doran, E. B. (1953). *A Physical and Cultural Geography of the Cayman Islands*. Doctoral dissertation, Department of Geography, University of California, Berkeley, CA.

Duncan, D. D. (1943). Capturing giant turtles in the Caribbean. *National Geographic, 84*, 177–190.

Hirst, G. S. S. (1910). *Notes on the history of the Cayman islands*. Kingston, Jamaica: P. A. Benjamin.

Langley, W. W, Jr. (1964). *Capturing Green Turtles off Nicaragua*. Master's thesis, School of Public Communication, Boston University, Boston, MA.

Matthieson, P. (1975). *Far Tortuga*. New York: Random House.

McLaughlin, H. R. (Ed.). (1994). *The '32 storm. Eyewitness accounts and official reports of the worst natural disaster in the history of the Cayman islands*. George Town, Grand Cayman, British West Indies: Cayman Islands National Archive.

Morison, S. E. (1963). *Journals and other documents in the life and voyages of Christopher Columbus*. New York: The Heritage Press.

Rebel, T. B. (1974). *Sea turtles and the turtle fishery of the West Indies, Florida and the Gulf of Mexico*. Coral Gables, FL: University of Miami Press.

Ross, H. E. (Ed.). (1999). *Love's dance. The catboat of the Caymanes*. Georgetown, Grand Cayman, British West Indies: Cayman Islands Seafarers' Association.

Smith, R. C. (1985). The Caymanian catboat: A west Indian maritime legacy. *World Archaeology, 16*(3), 329–336.

Smith, R. C. (2000). *The maritime heritage of the Cayman islands*. Gainesville, FL: University of Florida Press.

Williams, N. (1970). *A History of the Cayman Islands*. Grand Cayman, British West Indies: Government of the Cayman Islands. (Reprinted 1992).

Chapter 4
The *Bergantín*, a Little Known Craft from the Early Spanish Period in the New World

Samuel P. Turner

Introduction

The Spanish exploration and colonization of the New World in the sixteenth century had a profound impact on both sides of the Atlantic as ideas and commodities flowed in both directions. Much has been written on colonization processes and vernacular technologies, with vernacular defined as domestic or indigenous, not imported or copied. Vernacular may also emphasize function over form. The *bergantín* is a specific vessel type that played an important role in the success of the Spanish in the early sixteenth century. Unlike vernacular boats and technologies documented in later centuries, the *bergantín* represents the transition from European to American. Prefabricated in Europe and finished with New World timber, the *bergantín* bridges the Atlantic and represented the foundation of vernacular vessel types to come.

The *bergantín* has its origins in the Mediterranean Sea where it was one of the smallest of a variety of water craft that fell under the category of galley. Galleys could be rowed or sailed but derived most of their advantages from their oars. While under oar power, *bergantines* and galleys could maintain a heading regardless of wind direction and were agile and nimble since they could maneuver with oars independently of the wind. They were also capable of considerable bursts of speed. Galleys had been in use in the Mediterranean since ancient times and during the sixteenth century played a major role in naval warfare there. Galleys, in the form of the *bergantín*, also played an important role in the exploration, development, and conquest of the New World.

The Mediterranean *bergantín* used by the Spanish and Italian city states was a slim and fast craft that was fitted with between 10 and 15 pairs of oars. Each of

S.P. Turner, Ph.D. (✉)
Lighthouse Archaeological Maritime Program (LAMP), St. Augustine Lighthouse & Maritime Museum, 81 Lighthouse Ave., St. Augustine, FL 32080, USA
e-mail: sturner@staugustinelighthouse.org

A.M. Evans (ed.), *The Archaeology of Vernacular Watercraft*, When the Land Meets the Sea, DOI 10.1007/978-1-4939-3563-5_4 these oars was pulled by a single oarsmen, thus a vessel's complement would consist of between 20 and 30 oarsmen and approximately 20 soldiers and were often armed with a number of swivel guns in the bow as a forward battery (Konstam 2002:20).

In spite of their small size, or perhaps because of it, the *bergantín* was often used in amphibious landings, for raiding coastal settlements and landing bodies of troops in support of land actions ashore, as well as for moving troops from one ship to another during battle. In some of the larger naval battles, *bergantines* shuttled reserves of troops to galleys under attack or reinforced threatened or weakened sections of the galley line (Konstam 2002:20).

Vessel types used by the Spanish during the earliest period of exploration and colonial expansion in the New World have attracted scholarly attention for many years. Famous among these are the *naos* and *caravelas* used by Columbus and his contemporaries to push the boundaries of discovery. Documentary sources and historical accounts permit a more in-depth examination of *bergantines*, an evolving New World vernacular watercraft which, like the contemporary, *naos* and *caravelas*, saw extensive service in the New World in the earliest period of European settlement and throughout the sixteenth century.

Bergantines, like most galleys, were fore-and-aft rigged. They utilized a lateen rig which resulted in the vessels being swift and capable of sailing close to the wind. Their shallow draft and oars made them the attack craft of preference and they were used by such men as Miguel Diáz de Aux, Juan Ceron, and Juan Ponce de León during the Carib wars in the waters adjacent to San Juan Bautista, today known as Puerto Rico. *Bergantines* were also used as scouting and inshore vessels during exploratory and settlement voyages such as those discussed below.

The earliest references to *bergantines* in the New World would appear to be those utilized by the two Tierra Firme expeditions which departed Española in 1509. Juan de la Cosa and Alonso Ojeda, partners in one of the expeditions, provisioned a *nao* and one or two *bergantines* in Spain with 200 men and as many supplies as possible. Diego de Nicuesa, leading the other expedition, brought four large vessels and two *bergantines* with supplies and personnel from Spain for his operation (Las Casas 1986:II 377). The interpretation of Las Casas related to the provisioning of these fleets is that these *bergantines* sailed from Spain to the New World in 1508 or 1509. This, however, was not the case. As shall be discussed below, evidence suggests *bergantines* were all built in the New World.

In an extensive study of the trade between Seville and the New World between the years 1504–1650, Huguette and Chaunu (1955:6–121) documented hundreds of sailings by *caravelas*, *naos*, and the occasional *barco*. In the first two decades covered by the study, 1504–1520, not one *bergantín* was documented as making a transatlantic voyage. Furthermore, most *bergantines* were small craft most likely undecked or only partly decked, and it was unlikely they were capable of carrying sufficient supplies to sustain a crew on an Atlantic crossing. How can one account for the discrepancy between Las Casas' assertion that the *bergantines* came from Spain and their complete absence in the transatlantic sailing records as published by the Chaunus?

For clarification of this discrepancy, we turn to factors' accounts of King Ferdinand V of Castile and León (1452–1516) from the Island of Puerto Rico. During an audience with the king in 1511, Miguel Díaz and Juan Ceron, Diego Columbus's appointed officials to the island of Puerto Rico, were dispatched to their posts in the Indies with the additional task of making war on the Carib Indians on Puerto Rico and the adjacent islands (Marte 1981:105–106). To this end the king sent a consignment of weapons to the Indies including the materials for two *bergantines* which were shipped from Spain in parts, to be assembled upon arrival in the Indies. This is undoubtedly how the *bergantines* of the Nicuesa and the La Cosa and Ojeda armadas arrived from Spain. Las Casas saw both fleets with their accompanying *bergantines* at Española depart for Tierra Firme in 1509 and had assumed they had originally sailed from Spain with the larger supply ships.

The Architecture of the *Bergantín*

In order to better understand the following section on *bergantín* architecture, a discussion of its close relatives is offered in order to place this evolving New World vernacular watercraft in its proper context. Details of the *bergantín's* form can be traced from its European origins. The *bergantín* was one of the three smallest types of galleys, along with the *galiot* and the *fusta*, but its design was not simply a smaller version of larger galleys (Konstam 2002:20).

The *galiot* had 16–20 banks, or pairs, of oars and these were rowed "*alla scaloccio*" with two men per oar. Larger galleys were also rowed *alla scaloccio* with between five to eight men per oar depending on the size and type of galley. A typical *galiot* could be 27 m (88.59 ft) long with a beam of three meters (9.84 ft) for a length-to-beam ratio of 1:9 and a draft of less than 2 m (6.56 ft). A 20 banked *galiot* would be manned by 80 oarsmen and approximately 60 soldiers and a handful of gunners and officers. *Galiots* carried a single lateen-rigged sail on a single mast. Typically they were armed with only one centerline 16 to 24-pounder gun in the bow supported by a small bank of swivel guns. These vessels were considered "undecked"; that is, they had no complex poop deck structure.

The *fusta* was a smaller vessel than the *galiot*. It typically had between 10 and 15 banks of oars, and like the *galiot*, was rowed *alla scaloccio* with two men per oar. A 15 banked *fusta* would have had 60 oarsmen who were supported by 30–40 soldiers and a few gunners and officers. They carried a single lateen-rigged sail on a single mast. Typically *fustas* were armed with a single centerline gun ranging between a 12 and 18-pounder supported by a few swivel guns. Like the *galiot*, the *fusta* was considered undecked. No precise dimensions for these vessels have come to light but Konstam (2002:20) estimates their length to have been 21.5 m (70.54 ft) long.

As mentioned earlier, the *bergantín* was a slim and fast craft fitted with between 10 and 15 pairs of oars. Unlike the *galiot* and *fusta* however, the *bergantín* was not rowed in the *alla scaloccio* fashion but rather *alla sensile* where each oar was pulled by a single oarsmen. For this reason a vessel's complement would be made up of between 20 and 30 oarsmen, approximately 20 soldiers, with possibly a gunner and some officers. As *bergantines* were the smallest of the galleys they did not carry any great ordnance but instead were often armed with a number of swivel guns in the bow as a forward battery. A Venetian *bergantín* was reported as being 16 m (52.5 ft) long with a beam of 2 m, or 6.56 ft (Konstam 2002:20). This craft was long and narrow with a length-to-beam ratio of 1:8.

When attempting to determine what these craft looked like it is important to remember their origins and their galley-like characteristics. One of the things that made galleys different from other small, rowed craft such as *chalupas* or *bateles* was that they utilized outriggers to hold tholepins to which the oars were attached further outboard. This permitted the interior of the vessel to be kept clear of oars, making room for supplies and a gangway for the fast movement of fighting men. The outriggers of a *bergantín* or any galley were substantial, heavy structures that had to be carefully balanced so as not to have too destabilizing an effect on the hull (Figs. 4.1 and 4.2).

In addition to supporting the tholepins, the outriggers also supported the weight of outer walkways, stanchions, and rails. All this weight required that the outrigger be attached to the hull through substantial timbers that were likely attached to either thwarts or deck beams that ran athwart ships. Historic drawings exist that

Fig. 4.1 A Venetian *galliot* or *fusta* from a sketch attributed to Raphael (1483–1520). Note the extensive outriggers and supports and the method of stowing oars up and outboard while under sail. Also shown are the stanchions, beams, and screens that shielded oarsmen, to a degree, from projectile fire in combat situations. Illustration adapted from Konstam (2002:6)

Fig. 4.2 A Venetian *galliot* or *fusta* by André Zysberg from a watercolor by Raphael. This shows the same vessel as Fig. 4.1 from the stern and under sail. In this illustration the screens are not deployed, exposing oarsmen to view. Illustration adapted from Konstam (2002:45)

show a plan view and section of a galley (Fig. 4.3). The vessel in the drawing has a length-to-beam ratio approaching 1:8 that was typical of most galleys. The long oval shows the shape of the hull with the somewhat triangular ram on the right side of the drawing denoting the bow of the vessel.

A semi-circular foredeck is illustrated in the bow and an aft quarter deck at the stern of the vessel. The rectangular shape overlying the hull represents the outrigger with the various banks of oars denoted by points along the two sides of the vessel. In this case the galley has 27 banks of oars and so represents a sizeable galley.

The section drawing shows the keel, the shape of the hull, the shape of the outriggers, and the central gangway. Though the drawing represents a larger galley, the proportions depicted are correct for a *bergantín*. The drawing thus provides a good indication of the general appearance of these craft in plan view.

Like most shipbuilding at this time, the construction of a *bergantín* was generally managed through a contract. These were fairly formulaic legalistic documents. The study of two such contracts drawn up in the city of Genoa, Italy provides useful information on *bergantín* construction. Vessels could be described in terms of the number of benches or rowing thwarts that it had, as is the case in the first contract

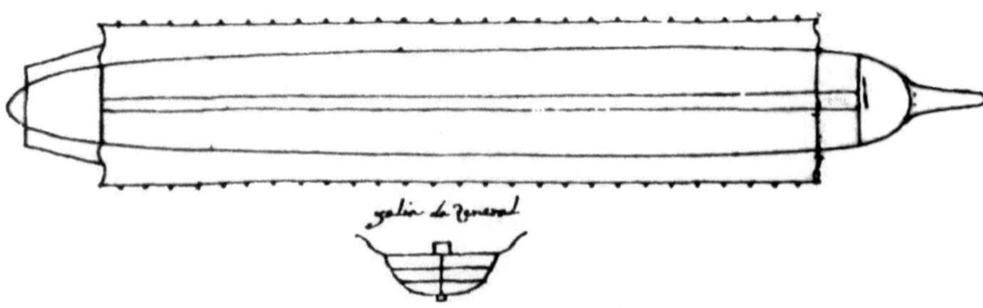

Fig. 4.3 The deck plan and section of a Venetian galley showing clearly the hull and outrigger outlines as well as the central gangway and fore and aft deck structures that were the essential components of *bergantines* and all other galley craft. Illustration adapted from a late fifteenth- or sixteenth-century manuscript held at the *Biblioteca Nazionale Marciana* and found in Konstam (2002:7)

reviewed for this study. In this case, the builder was permitted to determine the general dimensions of the hull given the desired number of benches. The number of benches required was 15 and the time allotted for the work was 1 month. The anticipated quick execution time for the contract was based on the fact that the builder owned a substantial and well-established shipyard and had all of the necessary contacts to obtain the required resources in short order.

The contract stipulated that the vessel was to be delivered without blemish and ready to row. The price stipulated was 65 gold pieces with 20 paid upfront at the time the contract was signed to begin building. The contract was signed on November 9, 1523, and provided no other information with regard to vessel dimensions. Given the number of benches however, this was a larger-sized *bergantín* (Bresc et al. 1975).

The second *bergantín* contract was dated Genoa, November 8, 1564, and had two dimensions listed. The length of the vessel was to be 19 *goa*, or 14.12 m (46.33 ft), and three and a half palms, or 0.88 m (2.88 ft), in height amidships. This height presumably refers to the depth of hold. The other proportions of the *bergantín* were to be worked out based on best practice utilizing the stipulated dimensions as a starting point. The cost of this vessel was set at 60 pieces of the gold of Italy judged sound, that is to say checked for signs of counterfeiting or clipping. As in the contract discussed above, the price of 20 gold pieces was paid upfront to initiate construction. According to the terms of the contract this vessel was to be delivered as soon as possible and, again, it had to be well constructed, labored over, and finished (Bresc et al. 1975:70–71).

Building Two *Bergantines* in Puerto Rico

The above contracts provide a good idea of some of the formalities, pricing, sizes, and delivery times for *bergantines* in sixteenth-century Genoa but little information regarding the details of how the vessels were actually constructed or the materials that went into them. Documents in King Ferdinand's factors' accounts for the island of Puerto Rico discuss the materials that went into making *bergantines*. These *bergantines* are also discussed in correspondence between the king of Spain and his officials concerning their dispatch to the Indies.

Juan Ponce de León wrote King Ferdinand in 1510 or 1511 stating that he had written and requested a *bergantín* be built and sent to Puerto Rico from the island of Española. The vessel was to be used to defend the island from the frequent attacks by Carib Indians who were based on neighboring islands. On September 9, 1511, the king informed his officials in Puerto Rico by letter that he was ordering his officials of the *Casa de Contratación* in Seville to send one or two *bergantines* (Murga Sanz 1960:55, 57). Four months later in January of the following year, the king wrote his officials in Seville on the matter once more ordering them to send two more *bergantines* in addition to those already sent. The king wanted them to have four *bergantines* on the island for their protection and he ordered that they be well maintained and cared for (Murga Sanz 1960:67).

Our search for information relating to *bergantines* resumes in the written records of Puerto Rico among the documents created by the *teniente de factor* Francisco de Cardona from 1510 to 1513 (*Cargo hecho a teniente de factor Francisco de Cardona* 1510–1512:ff. 4–5) and those of factor Miguel Díaz de Aux in 1513 (Tanodi 1971:35). The factor was essentially the king's business manager who handled the king's businesses and accounts. These records provide information on *bergantines* and how they were shipped. Cardona's records show the lists of materials dispatched from Spain to construct the two *bergantines* and the records of Miguel Díaz show what was left of those supplies when the office of factor passed to him from Cardona in October 1513.

By the time Díaz took office, a considerable quantity of naval stores was no longer on the books which suggests they had already been used in ship construction. Still in the inventory however was rudder hardware for two *bergantines* (Tanodi 1971:35). The *bergantines* were not yet at the stage for rudders to be installed, indicating that a naval building program was underway at the time Miguel Díaz took office in October 1513. It seems that sometime before October 1513, the Spanish organized a carpenter shipwright and assistants to build the two *bergantines*. The factor's accounts show that only two *bergantines* were shipped to Puerto Rico, rather than the four that were desired by the king.

The complete list of materials shipped from Spain for the *bergantines* appears in Cardona's accounts and are mixed with farming tools and clothing. Besides these two categories of items, however, all of the remainder of the material was used in naval construction and, in many cases, is described specifically as being for the *bergantines*. Other items not specifically designated for the vessels are likely to have been for them as well since they are all functionally related.

The specific document begins with a brief introduction and description of the inventoried material, specifically rigging, ship's equipment, and fasteners (*Cargo hecho a teniente de factor Francisco de Cardona* 1510–1512:ff. 4–5). No reference is made to the *bergantines* in this introduction. The first items listed are hoes (*azadas*), picks (*azadones*), pruning hooks (*podones*), and axes (*hachas*) which are associated with agriculture and timber work. It is possible that the pruning hooks (*podones*) and axes (*hachas vizcaínas*) could have been used in the processing of trees into ship timber but the numbers, fifty of each, suggest a broader sylvan application.

The next category of material on the list is fasteners, or nails (*clavazón*). Six distinct types of fasteners are listed. Each had a specific function in the construction process that was determined by its length and to a lesser extent its diameter. Of the six types, two are specific to shipbuilding. These are the *clavos estoperoles* and the *clavos de ligazón*. *Clavos estoperoles* (1000 in number) are sheathing nails with wide flat heads used for tacking on lead sheathing or to nail strips of lead over caulked seams to keep the caulking in and shipworm (*Teredo navalis*) out (Arnold and Weddle 1978:236). In this case, the latter is the more likely application since newly built vessels would not likely require the use of lead sheathing, which was usually used to patch leaks. *Clavos de ligazón* (1000 in number) were the nails that were used in framing the vessels, that is to say, for fastening together floors and futtocks which made up the frames. Their numbers, to a certain extent, demonstrate that they have a specific use that precludes them being the principal fastener type of the vessels.

The remaining four nail types included on the inventory are *clavos entrecaravíes*, *clavos de torno*, *clavos de media talla*, and *clavos cabriales*. *Clavos cabriales* (400 in number) derive their name from their use in house carpentry. A *cabrio* is a roof rafter and these nails were used to fasten them to tie beams, king posts, wall plates, ridge beams, and the like. In a ship the natural corollary to roof rafters are deck beams. Of all categories of fastener this type appears least with only four hundred in total. They were used less than half as much as *clavos de ligazón*, or framing nails. It is possible that such nails might also have been used in certain applications in the construction of the outriggers of the *bergantines*.

A third category of ship fastener, the *clavo de torno* (1000 in number), is more difficult to figure out. *Torno* is the Spanish term for a lathe, while the verb "*tornar*" means to go around. Therefore it is possible that this name may actually refer to the fastener's shank as round as opposed to the usual square-shaped fastener shanks that were more common in this period. Archaeological evidence for fasteners with round shanks (variety 1) have been found on the Molasses Reef Wreck, an early sixteenth-century Spanish vessel thought to have been lost sometime before 1513, which places it temporally very close to these documents (Keith 1987:111). If this is the case, these nails would have been used in applications that called for nails with round shanks.

The *Clavo entrecaraví* (5800 in number) belongs to the largest category of nail. Of unknown linguistic origins, this nail type was the most numerous and was probably used in planking the vessels. Other applications for this nail might have included their use in the installation of ceiling planking and decking among sundry other purposes. The last remaining fastener type, the *clavo de media talla* (600 in number), refers to the nail's size or stature (*Cargo hecho a teniente de factor Francisco de Cardona* 1510–1512:ff. 4–5). These were half-sized nails whose application was one that called for a mid-range nail. In terms of quantities of shipped nails, the *clavos de media talla* constituted the second smallest number. Given this, it is likely they probably had a relatively special and targeted application.

The next item in Cardona's accounts following the ship fasteners is of great interest and the first item mentioned specifically for the *bergantines*. The reference is to "*dos guarniciones de timones para bergantines*" or "two sets of rudder hardware for *bergantines*" (Fig. 4.4). This no doubt included the pintles and gudgeons, the most complicated and substantial of all the hardware associated with rudders. It may also have included the fasteners for joining the various timbers that made up the rudders. These last, however, may have been made on the island of Puerto Rico with some of the iron stock that forms the next category of cargo in this inventory.

Three kinds of iron bar stock were included in Cardona's accounts destined for the blacksmith's shop. Such a tradesman was an essential part of any shipbuilding team and this would have particularly been the case in early colonial Puerto Rico. The first type of stock was a group of six large "*palancas*" of iron. These large bars would have been used to forge some of the largest objects in the blacksmith's shop. These were followed by 24 "*barras de hierro medianas*" or medium iron bars. The last group of bar stock consisted of 24 "*barras de hierro pequeñas*" or small iron bars. These, together along with the previously mentioned bar stock, weighed 720 lb.

This weight of bar stock seems consistent with the needs of two *bergantines* beyond the nails that were discussed above. The raw bar stock may have been made into bolts for fastening frames to keels and keelsons, and to make up any short-fall in any of the nail categories included for the *bergantines*. Additionally, any shipbuilding tools that were lacking could be fabricated by the blacksmith for the building project.

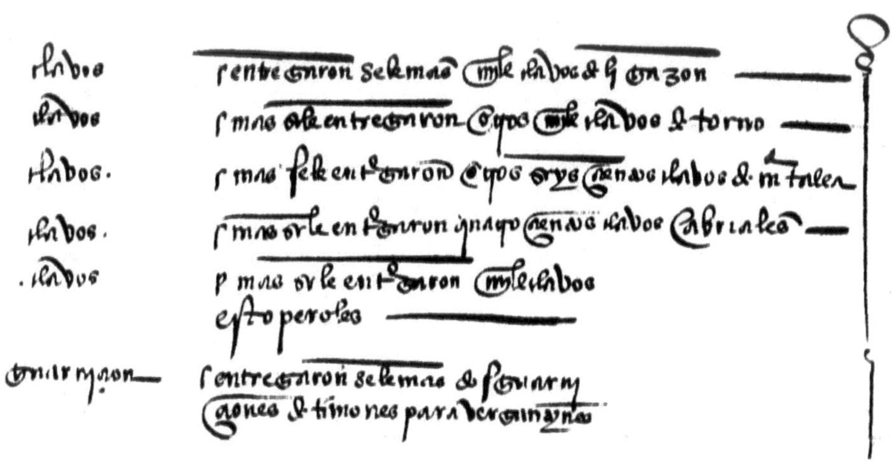

Fig. 4.4 Portion of the materials inventory of the two *bergantines* assembled in Puerto Rico from the records of Francisco de Cardona. The first five entries are for various types of nails and the sixth is for two sets of rudder hardware for *bergantines*. Image is a section of folio four *verso* from *Contaduría* 1071, found in the *Archivo General de Indias*, Seville

The next item listed was 267 lb of tallow (*sebo*) shipped in two crates. This could have been used to make candles or soap, but given its association with the materials for the *bergantines* was likely used in a maritime application such as caulking or coating the vessels themselves, as was the custom during this period (Pontillo 1975:402).

Cardona's accounts give a detailed picture of the variety of materials that go into the construction of a *bergantín* permitting a greater understanding of the construction process and also an appreciation of all the different industries and trades that were of critical importance to it. Another lot of iron stock in these accounts, *cabos de hierro*, took the form of 21 iron off-cuts. Thirteen of these were from plate (*plancha*) and the remainder from round stock (*vergajón*). These seem to have been kept apart from the bar stock previously discussed, possibly because they were incomplete bars. These, like the bar stock above, would have been used for whatever need came to hand.

The next item in the inventory was oakum. Oakum was used in the caulking of the hulls and decks of ships. It was often made by tearing apart and breaking down old lines and rigging into their constituent fibers, usually hemp or jute. In this case 150 lb of oakum (*estopa*) was specified, and described as *torcida*, or twisted. This means that the oakum fiber was rolled between hands and turned into strands that could be used as it was, or with some preparation turned into the necessary strand width for the job. This was an essential material for all shipbuilding and it makes sense that it would be included in an assemblage of supplies to be used in the construction of *bergantines*. The oakum was followed by 600 lb of pitch or tar, called *pez* in Spanish. Like the *sebo* (tallow) and *estopa* (oakum), pitch and tar were also used in caulking the hull as well as in tarring the rigging.

Inventoried materials also included seven dozen large pine planks (84 in total) for the craft. These would have been shipped on transatlantic vessels in case it was difficult to find a readily available local supply of large tall straight trees. It is important to note that these materials were shipped during the initial period of exploration and colonization since local access to specialized materials was severely curtailed by limited knowledge of New World resources. Shipping the ready cut planks would have accelerated the building process considerably. Next in the list is an oak timber (*viga*) for the two *bergantines*. In the left side column, where the lots are itemized, the plural term *vigas* is used suggesting that there may have been more than one (*Cargo hecho a teniente de factor Francisco de Cardona* 1510–1512:ff. 4–5). It is possible that this was used for keels or any other number of applications that called for a piece or pieces of straight oak timber; the shipwrights engaged in building the vessels would have sawn the timber however they needed it.

This revealing list of supplies necessary for constructing the two *bergantines* demonstrates that materials for rigging the vessels were included as well. The inventory shows that certain pulleys and blocks (*poleas y motones*) for the running rigging (*jarcia*) of the *bergantines* were also shipped, as were six manila (*medriñaque*) canvases that were to be used to make the sails. This was followed by six lengths of hemp rigging for halyards and moorings for the vessels. One hundred baskets of esparto grass or fibers (*esparto*) were also included, which may have been used to fabricate additional specialized cordage or may have been used for caulking in some capacity.

The rigging equipment appearing in Cardona's accounts was followed by a number of items including one hundred dibbles (*almocafres*), that are clearly intended for agricultural use as well as 24 locks and latches (*cerraduras con sus cerrojos*) with two keys each that may have had a purpose on board the *bergantines* to lock up tools or supplies. There were also 50 iron wedges (*cuñas de hierro*), which were, as in the case of the *cabos de hierro*, possibly a form of off-cut. There were also one hundred pounds of steel (*acero*). Both of these last items may have been part of the materials for the *bergantines*.

The final two items to appear in the inventory are clearly associated with the *bergantines*. These are six grapnel anchors (*rezones*), and 48 oars (*remos*) for a total of 24 oars per vessel at twelve oars to a side (*Cargo hecho a teniente de factor Francisco de Cardona* 1510–1512:ff. 4–5). This fits nicely with the 10–15 pairs of oars that are the usual range for a *bergantín* (Konstam 2002:20). The *rezones*, or grapnel anchors, are an interesting item in helping us build a picture of *bergantín* equipment and outfitting. Six were shipped and accounted for indicating that each vessel likely carried three of these anchors. These were lighter anchors than the traditional ship's anchor of the time. These, however, had the advantage of holding better in the varied types of seabed it was thought the *bergantines* were likely to encounter closer in shore. The anchors were stockless and had multiple arms so that, no matter what the grapnel's orientation on the seabed, one or more arms and flukes would dig into the seabed or catch on protruding rocks or coral (Fig. 4.5).

Further information on the *rezón* anchor comes from Pontillo's *Nautical Terms in Sixteenth Century American Spanish* (1975:68, 399). In this work, most of the examples cited come from documents referring to a *fragata* or *galera*. This suggests that galley-like craft utilized the *rezón* anchor type as a general rule. *Rezones* of varying sizes were also used in smaller craft such as *chalupas* and *bateles*. Diego García de Palacio (1993:392), a naval architect living in Mexico writing in

Fig. 4.5 A grapnel anchor from site VO 7484 found off Ponce Inlet, Florida. The site is tentatively dated to the seventeenth or early eighteenth century. This anchor is a good example of what one of the *rezones* for the *bergantines* may have looked like. Image courtesy of the Florida Division of Historical Resources

1587 describes *rezones* as having four *uñas*, or flukes, which refers to the number of arms on the grapnel anchor.

Not documented but necessary for the building of these *bergantines* is the timber required for the framing, knees, outriggers, and possibly decking. This would have been obtained locally in Puerto Rico. Compass timbers for the framing would have been especially awkward to stow and ship, as opposed to specialized straight planks of significant length. Also not mentioned are treenails that were common to Iberian shipbuilding methods of the day. These may or may not have been used in the construction of these vessels. If they were used then it is to be assumed they were made from locally available wood on Puerto Rico since they do not feature in the inventory.

In short, the inventory in Francisco de Cardona's accounts provides an extensive parts list for a *bergantín* and is proof positive that *bergantines* were not built in Spain and then sailed to the New World but rather dispatched in parts in the holds of larger transatlantic *naos* and *caravelas*.

Bergantines in the Carib Wars

Having established, to a degree, what went into building a *bergantín*, we can now turn to how crucial *bergantines* were during the Spaniards' wars with the Carib Indians. Their shallow draft, their ease of maneuverability, and their independence of wind power made them perfect for speedy and violent action.

The *bergantines* discussed above were completed and launched some time in 1514 and used in naval raids against Carib Indian encampments on islands in waters adjacent to Puerto Rico, as had been intended by the king. The Taino Indian rebellion of 1511 on Puerto Rico resulted in an unprecedented alliance between the Tainos and their traditional enemies, the Carib. The rebellion was countered by Spanish attacks led by Juan Ponce de León, but the Carib Indians continued their assaults against Spanish settlements and ranches on Puerto Rico. At least one Carib raid was documented in 1512 on the west coast of Puerto Rico where San Germán was burned to the ground. In 1513 the Carib burned Juan Ponce de León's settlement of Villa Caparra in the eastern part of the island.

It seems likely that this 1513 Caparra raid gave impetus to the building of the two *bergantines*, whose parts inventory we just reviewed above, since the dates in the factor's accounts suggest a *bergantín* building project was underway around this time. The following year, 1514, saw a total of five seaborne Carib raids against Puerto Rico. The majority of these raids were lead by Cacimar de Vieques who was killed during the last documented assault on Puerto Rico in 1514 (Negroni 1992:211).

It is interesting to note that the *bergantines*, which were under construction during this period of intense seaborne raids, were not destroyed. This suggests they were built in a remote location away from the principal Spanish landings and settlements targeted by the Carib. Following their completion later that year the governor of Puerto Rico, Cristóbal de Mendoza, launched a retaliatory raid against the Carib base of operation on the neighboring island of Viequez where Jaureybo ruled following his brother Cacimar's death. There is no direct evidence that proves that the two *bergantines* completed in 1514 are the same *bergantines* employed in these retaliatory Spanish naval raids, but the circumstantial evidence is strong. Whether or not it was specifically these two *bergantines* which were employed in the raids, it is clear from documentary evidence that *bergantines* played a critical role in these raids.

In 1514 Cristóbal de Mendoza attacked the island of Vieques to the east of Puerto Rico with 150 men utilizing two *bergantines*, and a caravel with two boats, presumably to land or recover troops and spoils of war. The Spanish utilized the *bergantines* effectively, achieving complete surprise. Some 120 Carib Indians were killed and another 100 captured. These captives were shipped to Puerto Rico in the accompanying caravel where they were sold at public auction as slaves. Immediately following this raid, Mendoza launched another attack against Carib settlements and encampments in the Virgin Islands, and followed it with a third raid in 1514 under Juan Gil (Negroni 1992:211–212). It is possible Miguel Díaz de Aux, one of the factors involved with the *bergantines* discussed above and who was tasked with finishing and launching them, may also have been involved in these expeditions. He had been directly charged by the king to make war on rebellious Taino and Carib Indians. His being given the command of a *bergantín* during the conquest of Mexico in 1521 also supports this supposition (Gardiner 1956:134).

Two Carib raids against Puerto Rico followed in 1515 but sources indicate that few or no attacks occurred between 1515 and 1520, possibly as a result of the retaliatory raids by the Spanish (Negroni 1992:211–212). Two *bergantines*, possibly those discussed above were eventually sold by the royal officials of Puerto Rico some time in 1519 or shortly thereafter. The total cost of the used *bergantines* in gold currency then in use in the New World was 316 *pesos*, 6 *tomines*, and 6 *granos* of gold (Tanodi 1971:40).

The *bergantines* were apparently purchased by Juan Ceron, one of the island's highest-ranking officials, and Juan Martín Peña since both men are described as owing money on the purchase of the *bergantines* (Tanodi 1971:61). Juan Ceron may have used his *bergantín* for communication as well as for moving comestibles and supplies, though no documentary evidence for this has so far come to light. Juan Martín Peña has left a few traces in the notorial records. These indicate he was involved in shipping and naval matters. On one occasion he owed freight to the king's factor for a consignment of goods that was not described. On another occasion he owed taxes for pine pitch and canvas and later he owed the Crown for the cost of a breast plate and helmet that were handed out during the wars. Later still he was owed money for freight on three boatloads of bread he brought to Puerto Viejo from the *bergantín* of Juan Bono de Quejo (Tanodi 1971:25, 35, 36, 53, 298).

Peña's involvement in shipping and trade make it likely he used his purchased *bergantín* for the same. His acquisition of armor, as described in the records indicates he was, in some capacity, involved in fighting the Tainos and possibly the Caribs. His purchase of one of the *bergantines* and his involvement in maritime matters suggest the possibility he was involved in some of the naval raids that took place in 1514, possibly as crew on one of the *bergantines*.

The *Bergantín* as Transport

Bergantines were used for tasks beyond that of exploratory armadas and raiding. During the second decade of the sixteenth century there was a good deal of inter-island trade between the islands of Española and Puerto Rico for which customs documentation has survived. The vessels conducting this trade were *caravelas*, *naos*, and *barcos*. Bergantines were not as a rule used for this type of transportation. There is only a single recorded instance in 1516 of a *bergantín* arriving from Española at the port of San Germán in Puerto Rico. The vessel, belonging to the king's officials and apparently being transferred from that island to the Carib theater of operations, arrived with a cargo of 250 oranges (Tanodi 1971:279). The waterproof nature and small size of the cargo suggests that this *bergantín* may have been undecked or partly decked and so did not carry the more traditional and delicate cargoes of cazabe bread, *tocino*, or salt pork, livestock, and textiles which were very susceptible to damage by water. The crossing of the Mona Passage, a very rough patch of water between the islands of Española and Puerto Rico, is a testament to the seaworthiness of *bergantines*. The single instance of the use of a *bergantín* in the inter-island trade points to the unsuitability of this craft for trade involving delicate goods in rough waters.

This is not to say it was completely unsuited to any sort of trade. Juan Bono de Quejo, formally an employee of Miguel Díaz and Francisco de Garay, owned a *bergantín* on the island of Puerto Rico during 1517 and possibly in the years preceding and following. During this time he operated his *bergantín* in the coasting trade carrying cazabe bread, *tocino*, fish, and tools to the mining region of Luquillo in eastern Puerto Rico and on occasion transferred cargoes to other craft such as that operated by Juan Martín Peña (Tanodi 1971:50–51, 53). The records for these voyages are found in the king's factor's accounts but it is very likely de Quejo was in addition employed in the coasting trade by other private concerns as well. Apparently *bergantines* were more suited to carrying cargoes in relatively calm coastal waters where their exposed cargoes were less likely to suffer water damage.

The *Bergantín* in the Discovery and Reconnaissance of Nueva España

During the period of time the two *bergantines* mentioned above were being utilized for war and Juan Bono de Quejo was engaged in his coastal trade, exploration to the west of Cuba had begun in earnest. Bernal Díaz del Castillo (1956:405) and Bartolomé de Las Casas (1986:III 156) have left us excellent accounts of the use of a *bergantín* employed during the voyage of Francisco Hernández de Córdoba to Campache in 1517.

Diego Velázquez, the Governor of Cuba, was a wealthy and powerful man. He had taken on to himself certain privileges that were normally the right of the king, specifically, that of licensing voyages of discovery. Francisco Hernández de Córdoba received such a license. Cordoba's armada of two *caravelas* and a *bergantín* set sail from Axaruco on the north coast of Cuba. During the voyage the *bergantín* was used to land troops en masse along with the other ship's boats. The shallow draft of *bergantines*, together with their oar-powered capability, and relatively large size compared to ships' boats, made them the ideal landing craft for offloading a body of troops on shore as quickly as possible. The armada also used the *bergantín* as their primary watering vessel. Its ample carrying capacity, masts, and yards made it useful for this purpose, with the yards possibly utilized to hoist filled casks directly from the seashore if the water depth permitted.

The *bergantín* played a crucial role in the removal of troops. In general the *bergantín* was probably used very close in shore and was beached only under suitable circumstances. Beaching may have occurred to land troops but would not have been to take on large cargoes or troops, as the weight would have made it difficult to push off. This is especially true when a large and aggressive Indian army was immediately behind retreating troops. Their utility in close, inshore, waters for troop removal is well illustrated by the rescue of troops during the Spanish rout at Champotón in Nueva España. The ships' boats, which were supposed to be used for troop removal, were flooded and awash as a result of panicked boarding by troops. The majority of the expedition's survivors were only saved by the timely arrival of the *bergantín* which moved in shore to evacuate the troops, possibly covering their retreat with *verso*, or swivel gun fire. The retreating Spaniards clung to their boats half swimming until they reached the *bergantín*.

This is an opportune moment to address the issue of whether or not these vessels had transom sterns, an element of design which would have had a significant influence on the *bergantín*'s efficacy as an inshore vessel of war. It seems likely a *bergantín* built for combat or potential combat situations would have been designed to make speed both forward and astern. In the case of landing or removing troops from a beach head and lightering supplies and materiel to and from the shore, a sharp stern would have been considerably superior to a transom. Such a vessel would have been able to cut the water equally well with both bow and stern. The *bergantín* was classed as a galley and the majority of artistic depictions of galleys make it clear that these vessels had sharp sterns. Some, however, displayed considerable roundness and many had quarterdeck structures that might have been confused with a transom.

A vessel built with a transom stern would not have had the same hydrodynamic quality as a sharp stern. It would have been more at a disadvantage in the case of a moderate shore break or swell. A transom would have provided more surface area for waves to work against. This in turn would have require a vessel keeping station off the beach to work its oars harder in order to prevent beaching or running aground. Once beached, a *bergantín* with a transom would have had considerably more difficulty shoving off.

Following the battle at Champotón, the Spanish were short on manpower for sail-handling on the *caravelas* and so decided to abandon the *bergantín* removing her useful gear and setting her afire (Díaz del Castillo 1956:24, Las Casas 1986:III 165, Smith 1992:20). The expedition returned to Cuba via Florida where the principal expedition pilot, Antón de Alaminos, knew of a good watering site.

The Francisco Hernández de Córdoba expedition was followed by a second under the command of Juan de Grijalva which consisted of four *caravelas* and a *bergantín*. The account of this *bergantín*'s participation in the expedition reveals some of the limitations of the vessel type's design, namely their inability to carry large amounts of stores and keep them dry at sea during long voyages. The four *caravelas* sailed from Santiago de Cuba on January 25, 1518, and proceeded at a leisurely pace up Cuba's north shore where they gathered men and supplies at the harbors of Boyúcar, Matanzas, and Puerto de Carenas. The *bergantín*, which had gone ahead to the rendezvous point at Cabo San Antón found herself having exhausted her supplies long before the other ships finally made their arrival on May 1. Unable to resupply in this remote location, the *bergantín*'s crew was forced to abandon the expedition and returned to port without rendezvousing with the remainder of the fleet. Given the critical role of the *bergantín* in raids and expeditions as discussed above, the *bergantín*'s absence would have been keenly felt (Weddle 1985:67).

Bergantines were capable of traveling alone and carrying their own supplies. Their stores and cargo capacities, however, were limited due to the small size of the vessels, as well as their being undecked, or only partly decked. As a consequence their cruising range was limited. As we see in the case of the Grijalva expedition, on longer voyages, *bergantines* were dependent on supplies being provided periodically by larger vessels sailing in their company.

Bergantines in the Conquest of Nueva España

The use of *bergantines* was instrumental in the conquest of *Nueva España*, today called Mexico. As news of the discovery of new lands to the west spread, a large expedition was outfitted in Cuba as a reconnaissance in force to explore the coasts and islands to the west as well as to search for the Grijalva expedition that had been considered overdue (Weddle 1985:81–82). The expedition commander, Hernán Cortés, had invested all his money in the expedition and, with a number of other investors, was well along in their preparations when Grijalva returned. The personal relationship between Hernán Cortés and Governor Diego Velázquez had deteriorated. The souring of their relationship, in no small part, caused Hernán Cortés to break away from Velázquez's authority and establish his autonomy once in Mexico.

Cortés's fleet of 11 vessels set sail from Cuba in February 1519, and one large vessel bearing critical supplies later joined them in Mexico. The flagship was a *nao* of 100 tons called *Nuestra Señora de la Concepción*. Also present were three *caravelas* that fell within the 70–80 ton range. The rest of the craft were *bergantines* and smaller craft of an indeterminate nature (Díaz del Castillo 1956:41, Gardiner 1956:17–19, Weddle 1985:84).

Having settled on an anchorage Cortés established the settlement of *Villa Rica de Vera Cruz*. Here, to preclude any possibility of a revolt by troops loyal to Velázquez and an attempted return to Cuba, Cortés ordered the ships dismantled. He had all useful equipment removed such as sails, rigging, hardware, and anything else that might serve in naval construction at a later date (Weddle 1985:91). The hulls were burned and scuttled, or in some cases drawn up on the beach. Given that a number of his ships were *bergantines*, which were designed to be beached, it is likely that these were some of the vessels drawn ashore for storage. After having all their equipment and gear removed, including most likely their rudders, they were in no danger of being put to sea without significant time for re-rigging. The equipment and rigging removed from these vessels were to play a vital role in the conquest of Mexico.

The story of the conquest of Mexico is both fascinating and involved. There is neither space nor time to tell it here. The discussion of the role of the *bergantín* in the conquest of Nueva España revolves around the construction and use of two small fleets of *bergantines* on Lake Texcoco. The Aztec capital of Tenochtitlán lay on an island connected to the mainland by a number of causeways, and control of Lake Texcoco was to prove to be indispensable in gaining control of the city (Fig. 4.6).

During their first stay in Tenochtitlán, before the Aztecs had become hostile to the Spanish, a fleet of four *bergantines* was constructed, ostensibly for fishing and pleasure on the lake. Their true purpose, however, was to project Spanish naval power. This first display of Spanish shipbuilding in Aztec territory necessitated the identification of the necessary talent amongst Cortés's men. Martín López was

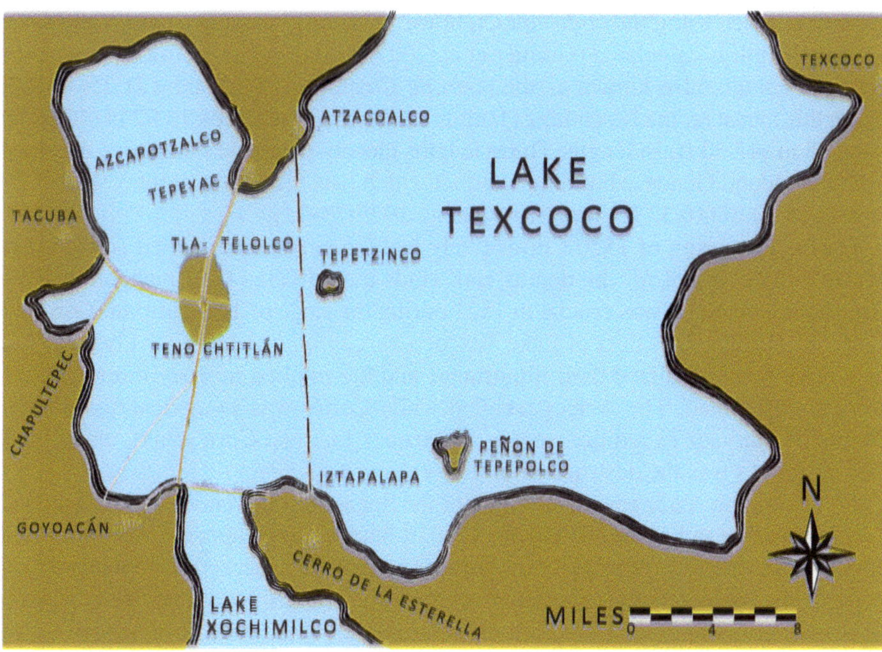

Fig. 4.6 Map showing Lake Texcoco, its littoral communities, the island capital of Tenochtitlán, and the critical causeways that connected the island capital to the mainland. Dominance of the lake and the causeways was of critical importance to the Spanish military campaign. Illustration adapted from Gardiner (1956:36–37)

chosen as shipwright and he had two servants skilled in carpentry, which gave him the necessary skilled labor pool to undertake the job.

We may assume that Martín López was an experienced shipwright or boat carpenter when he arrived in Mexico. Victory and, indeed, survival hinged on the success of a second fleet of *bergantines*. When building this second fleet of 13 *bergantines* López was assisted by Diego Ramírez who described himself as a ship carpenter. Martín López was assisted by a blacksmith named Hernán Martín, who manufactured the necessary tools for the job. Martín was assisted by another smith named Pedro Hernández who may have been involved in the manufacture of ship fasteners. Sawyers and caulkers were also critical parts of the team and have been documented as working on the first fleet of four vessels (Gardiner 1956:64). The materials list of Francisco de Cardona discussed above for the two *bergantines* built in Puerto Rico gives us an excellent idea of the types of materials and resources that went into the construction of these four *bergantines* which were assembled under Cortés.

Martín López eventually took Hernán Cortés to court in 1528 in order to obtain payment for the construction of these *bergantines*, which he stated were built at his cost. Court documents from this case provide critical information concerning the *bergantines*' construction. In his testimony, Martín López reported that the *bergantines* were between 25 and 26 *codos*, or cubits, in length. While doing his calculations to convert into feet, Gardiner used the English cubit of 18 in. (45.7 cm). This comes to a measurement of between 37.5 and 39 ft for length of the vessels (Gardiner 1956:68). When using the *codo común*, however, used in most of Spain at the time, a longer length is calculated (Grenier et al. 2007:IV–319). The metric equivalent of the *codo común*, also known as the *codo de Castilla*, is 55.71 cm (21 15/16 in.). Using this measure, the *bergantines* built in 1519–1520 measured 13.92 m (45.67 ft) to 14.48 m (47.50 ft) in length. These lengths closely resemble that above obtained from the second Genoese building contract which lists the length of the vessel as 19 goa, or 14.12 m (46.33 ft), long (Bresc et al. 1975:70–71).

During the spring of 1520 Cortés likely had the *bergantines* sailed across Lake Texcoco to get a feel for the depths and winds that affected their navigation. This information no doubt was crucial in 1521 during the siege of Tenochtitlán (Gardiner 1956:68, 71). The first fleet of Lake Texcoco *bergantines* had a relatively short life span. The Aztecs realized their importance and the implied military threat that the vessels represented. The Aztecs rebelled while Cortés was away dealing with the Pánfilo de Narváez expedition sent by Velázquez to arrest Cortés, and one of their first objectives was the destruction of the *bergantines*.

Following the Aztec revolt and flight of the Spanish from Tenochtitlán during "*la noche triste*," or sad night of June 30, 1520, it became more apparent than ever that naval power would be of primary importance in order to successfully isolate and lay siege to the Aztec capital. Hernán Cortés regrouped his army and with Spanish reinforcements from Cuba, Jamaica, and elsewhere, joined with thousands of allied Indian troops to undertake a number of campaigns to regain the upper hand, to rebuild morale among his Spanish troops, and restore confidence in his Indian allies.

The losses suffered during *la noche triste* made the Spanish keenly aware of the vulnerability of troops on the causeways without naval support. Consequently, Cortés ordered Martín López to build 13 *bergantines* in order to achieve naval supremacy on Lake Texcoco. This would allow the Spanish to cut Tenochtitlán off from supplies brought in by canoe as well as to cover the advance of Spanish troops along the causeways which were the only approaches to the city (Gardiner 1956:89).

The timbers to be used in the construction of the *bergantines* were cut and shaped far out of Aztec striking distance in the allied Indian city of Tlaxcala. The wood used was specified as oak (probably a white oak variety), evergreen oak (most likely a variety of live oak), and pine.

The timber was felled and shaped into floors, futtocks, planks, and all other necessary timbers in Tlaxcala. The plan called for the prefabricated *bergantines* to be ready so that when the time to lay siege to Tenochtitlán came, the many parts of the vessels could be transported to the final assembly point near Lake Texcoco and quickly assembled and launched. When all was ready, thousands of Tlaxcalan Indian porters were used to transport the materials, including the remaining stored naval equipment and rigging, from Tlaxcala to the city of Texcoco which had been taken and occupied by Spanish and allied troops. These porters were accompanied and protected by thousands of Tlaxcalan warriors (Gardiner 1956:115).

As previously done for the first four *bergantines*, Gardiner used the English cubit of 18 in. to calculate the lengths of this second fleet of 13 vessels. According to his calculations, twelve of these vessels were about 42 ft long, or 28 English cubits. The flagship was somewhat larger at 48 ft or 32 English cubits. Again, applying the *codo común* to the conversion results in different figures. The 12 *bergantines* of 28 cubits are calculated to be 15.59 m (51.15 ft) long. The Venetian *bergantín*, discussed previously and reported as being 16 m (52.49 ft) long, would seem to be a close correlate for the *bergantines* in Cortés's second fleet. The *bergantín* flagship was longer, measuring 32 cubits measures 17.82 m (58.46 ft) long when applying the sixteenth-century *codo común*.

Once arrived in Texcoco, a site for the assembly of the prefabricated parts was selected that was inland from the water so as to distance the craft as much as possible from Aztec canoe raids. The bank of a dry creek bed was selected. While the vessels were being assembled, allied Indian laborers dug the creek bed into a canal from the shipyard to the lake in order to move the fleet out once it was completed. The fleet was finally launched on April 28, 1521, and made its way to Lake Texcoco via the canal that was approximately 12 ft wide and 12 ft deep (Gardiner 1956:126–127). The siege of Tenochtitlán had begun.

Some sixteen men have been documented as commanding the *bergantines* on Lake Texcoco. One of these men was Miguel Díaz de Aux, who had been instrumental in the building of the two *bergantines* earlier that decade on the island of Puerto Rico and who saw subsequent action in the Carib wars. It is likely that his previous experience in *bergantines* made him a natural choice for a command on Lake Texcoco (Gardiner 1956:134).

Bernal Díaz claims that the number of mariners and troops assigned to the *bergantines* was 325, not including additional artillerymen. This was approximately one-third of the total of 928 Spanish soldiers present at the siege of Tenochtitlán (Díaz del Castillo 1956:358). The remaining two-thirds of the Spanish troops were divided amongst three principal bodies of men that each assailed one of the principal causeways. This made the fleet the largest contingent amongst the four principal units taking part in the fighting. These numbers are approximate since Cortés himself gives slightly different numbers but the proportions are consistent (Gardiner 1956:155).

Bernal Díaz gives a clear picture as to how the *bergantines* were manned. Each vessel he says was manned by 12 crossbow men and musketeers who were strictly fighting men and not required to row. Additionally there were 12 oarsman, six to a side, as well as a captain. There were also a number of artillerymen on board who operated the small brass guns mounted in the bows of the *bergantines* as well as a number of swivel guns. Here we come to some differences between Cortés's *bergantines* and those of standard Mediterranean design. Cortés's vessels contained fewer oarsmen and had brass artillery mounted in their bows in addition to the standard swivel guns. These are some of the ways this developing New World vernacular type differed from the standard Mediterranean *bergantín* which had more oarsmen and no mounted artillery beyond small swivel guns in its bows.

A helmsman may also have been assigned to the *bergantines* on Lake Texcoco if this task was not carried out in person by the Captain. The regular complement for one of these vessels probably approached 30 men with a slightly larger number on the *bergantín* flagship (Díaz del Castillo 1956:358). Of significance is Bernal Díaz's statement that each vessel was rowed by twelve men, a clear example of a New World adaptation to a specific situation. In these circumstances, it was decided to reduce the number of rowers from the traditional Mediterranean and New World *bergantín* configuration that had between 10 and 15 oars per side, such as had been the case with the earlier examples built in Puerto Rico. This may have been the result of a lack of sufficient manpower or a reduced need for speed or simply because they eliminated oarsmen in this instance to make room for additional gunners operating the brass artillery.

The flagship of the fleet, which was the largest vessel at 32 cubits in length, was armed with two cannon in her bow as well as swivel guns. She was likely manned by a larger crew with more oarsmen, a fact that may have slipped Bernal Díaz del Castillo's mind when he wrote his account of the conquest of Mexico. This would particularly be the case if the *bergantín* flagship had not seen action in the vicinity of the Tacuba causeway where he had fought.

During the first action on the lake, the Spanish delayed their entrance until a strong and favorable wind rose up which, combined with their rowing, gave them a great deal of momentum and speed. They then bore down on a force of approximately 1000 Aztec canoes, ran them down and fired among them, taking many prisoners and destroying many canoes (Díaz del Castillo 1956:368).

Ramming was a primary tactic from the very beginning of the naval action on the lake. Cortés routinely assigned a number of *bergantines* to each of the three principal units of his army, which were each attacking along an assigned causeway. Not long after the beginning of the naval action, a *bergantín* that seems to have been built undersized was taken out of action and laid up on shore. It was found to be too small to effectively ram and run down canoes. Its crew was divided up and distributed among the other twelve *bergantines* (Díaz del Castillo 1956:370).

The principal role of the *bergantines* was to provide support for troops making their way along the causeway. A strategy used by the Aztecs to counter this close support was to set sharpened stakes underwater in specific areas where they wished to prevent the approach of *bergantines*. The Aztecs also lured *bergantines* onto the stakes to immobilize them. This technique resulted in the capture of at least two *bergantines* (Díaz del Castillo 1956:373).

Once Cortés's men had isolated the city of Tenochtitlán by cutting off its water supply and holding the landside approaches to all the causeways, the Aztecs resorted to supplying the city by canoe. To counter this, two *bergantines* were assigned to a night patrol to interdict this traffic. Ambushes were a regular tactic used by the Aztecs and when the Spanish had intelligence of a planned ambush they would plan one of their own, camouflaging their *bergantines* with branches and muffling their oars for silence. Frequent close quarters fighting along the causeways exposed *bergantín* crews to darts, javelins, stones, and other missiles, the same projectiles which were used on those fighting on the causeways, killing many and wounding most of those serving in the *bergantines*. In this particular military campaign, the screens hung to protect and conceal oarsmen from view would have been of considerable benefit (Figs. 4.1 and 4.2).

Eventually, the Spanish found they could knock down and overrun the submerged stakes by hitting them at speed under oar and sail. Their hulls were sufficiently robust and were not holed by such action. Consequently, they were able to overcome these submerged obstacles and get in very close to support troops wherever necessary. This was an important development in strategy and helped bring about Spanish victory (Fig. 4.7) (Díaz del Castillo 1956:394).

Ultimately, the *bergantín* literally brought the siege to an end when the crew of a vessel under the command of García Holguín captured the Aztec King Guatemuz,

Fig. 4.7 A fragment from the *Codex Florentino* showing a *bergantín* landing Spanish troops during the siege of Tenochtitlán. The vessel and all other illustrations in the codex were drawn by contemporary Mexican artists and show the *bergantín* from a Native American perspective. They focused more on the activity being depicted than detail and accuracy of watercraft. Illustration adapted from Gardiner (1956:192–193)

concluding the 93 day siege. Following the end of hostilities those in the *bergantines* were in the most advantageous position wrote Bernal Díaz, for they were best equipped to sack the houses that were built on stilts out on the lake and search for hidden riches in the reeds where many had been hidden. When Cortés demanded the treasure of Moctezuma, the Aztec captains and Guatemuz himself told him that the crews of the *bergantines* had already stolen most of it (Díaz del Castillo 1956:405).

Conclusions

The *bergantín* was derived from, and closely resembled those used in the Mediterranean Sea and played an essential part during the contact period in the New World. This Spanish vernacular watercraft was successfully adapted to a variety of New World roles in fleets of exploration and discovery as well as cargo transports and fast attack craft in the Carib wars of Puerto Rico. When this vessel type was first introduced into the New World it was shipped in parts in the holds of larger transatlantic vessels and completed with timber and ironwork forged in the New World. Spanish records have yielded a great amount of detail regarding the wide variety of iron fasteners, hardware, rigging, and other materials that went into building these vessels. They have also shed light on the activities of different industries and trades that were of critical importance to shipbuilding and demonstrate the established and active presence of these industries at this early stage of the contact period.

The long and narrow nature of this craft and its independence from the wind made it an ideal warship. Independence from the wind also made it capable of achieving an element of surprise which proved invaluable in the Carib wars. The building contracts examined for this study contain details that could potentially be used for experimental archaeology in terms of building a *bergantín* replica. Combining the length-to-breadth ratio determined for the Venetian *bergantín* (Konstam 2002:20) with the length and depth of hold of the Genoese *bergantín* (Bresc et al. 1975:70–71) combine to make a good starting point for making a hypothetical reconstruction of such a craft.

From a Spanish colonial perspective, there is no doubt that the most important role played by these craft was in the siege of the Aztec capital of Tenochtitlán for which *bergantines* were purpose built out of New World timber and materials. This represented the next step in terms of development of this type of vernacular watercraft. These *bergantines* were not shipped in parts in the holds of vessels. Their design was specific to what was not only required at the time, but what was possible given the number of men and the materials at hand.

The successful building and use of *bergantines* in the conquest of Mexico had a profound impact on world history. Mexico produced vast wealth for the Spanish Crown and Spain that would help make Spain a wealthy and militarily invincible power in Europe for most of the sixteenth century.

The *bergantín* continued in use in both the new and old worlds throughout the sixteenth century. The latest use of this vernacular water craft in the New World that this author has found occurred in the Spanish settlement of St. Augustine, Florida where a *bergantín* named the *San Felipe* was documented in port records dating between 1597 and 1601 (*Cuentas de Bastamientos* 1597–1601). This demonstrates clearly this crafts practical usefulness in a variety of roles throughout the sixteenth century.

Acknowledgements The author wishes to express his thanks to Dr. John Young of the History Department of Flagler College, St. Augustine, for his Latin translations of Italian shipbuilding contracts and to Dr. Francisco Fernández-González of the *Gabinete de Historia de la Ciencia y la Tecnología Navales* in Madrid, Spain, for materials and insights provided during the course of this investigation. A special thanks are due Mr. Brendan Burke of the Lighthouse Archaeological Maritime Program (LAMP) who produced the figures in the text.

References

Archivo General de Indias. 1510-1512. Cargo Hecho a Teniente de Factor Francisco de Cardona desde que se Pobló la isla de Puerto Rico hasta el 8 de Mayo de 1512. Contaduría 1071, Archivo General de Indias, Seville, Spain.

Archivo General de Indias. 1597-1601. Caja de San Agustín de Florida, Cuentas de Bastamientos No. 2, Resultas Contra Personas Particulares. Contaduría 950, Archivo General de Indias, Seville, Spain.

Arnold, B., & Weddle, R. (1978). *The nautical archeology of padre island, the Spanish shipwrecks of 1554*. New York: Academic.

Bresc, H., Gatti, L., Grendi, E., & Borzone, P. (1975). *Studi di Storia Navaei*. Giunti, Italy: Centro per la Storia Tecnica in Italia del Consiglio Nazionale delle Ricerche.

Chaunu, H., & Chaunu, P. (1955). *Séville et l'Atlantique (1504-1650)*. Paris: École Pratique des Hautes Études, Centre de Recherches Historiques, Librairie Armand Colin.

Díaz del Castillo, B. (1956). In A. Idell (Ed. & Trans.), *The Bernal Díaz Chronicles: The true story of the conquest of Mexico*. Garden City, NY: Dolphin Books, Doubleday.

García de Palacio, D. (1993). *Instrucción Náutica*. Madrid, Spain: Editorial Naval, Museo Naval.

Gardiner, C. H. (1956). *Naval power in the conquest of Mexico*. Austin, TX: University of Texas Press.

Grenier, R., Bernier, M.-A., & Stevens, W. (Eds.). (2007). *The underwater archaeology of Red Bay: Basque shipbuilding and whaling in the 16th century (volumes 1-5)*. Altona, Victoria, Canada: Parks Canada.

Huguette, C., & Chaunu, P. (1955). *Séville et l'Atlantique (1504–1650)*. Paris: École Pratique des Hautes Études, Centre de Recherches Historiques, Librairie Armand Colin.

Keith, D. (1987). *The molasses Reef Wreck*. Doctoral dissertation, Department of Geography, Texas A&M University, College Station, TX.

Konstam, A. (2002). *Renaissance War galley 1470-1590*. Oxford, England: New Vanguard 62, Osprey Publishing.

Las Casas, B. d. (1986). *Historia de las Indias (I-III)*. Mexico City, Mexico: Fondo De Cultura Económica.

Marte, R. (1981). *Santo Domingo en los Manuscritos de Juan Bautista Muñoz*. Santo Domingo, Dominican Republic: Fundación García Arévalo.

Murga Sanz, V. (1960). *Puerto Rico en los Manuscritos de Don Juan Bautista Muñoz*. Río Piedras, Puerto Rico: Ediciones de la Universidad de Puerto Rico.

Negroni, H. (1992). *Historia Militar de Puerto Rico*. Madrid, Spain: Sociedad Estatal Quinto Centenario.

Pontillo, J. (1975). *Nautical terms in sixteenth century American Spanish*. Doctoral dissertation, Department of Linguistics, State University of New York at Buffalo, University Microfilms International, Ann Arbor, MI.

Smith, R. (1992). Ships in the exploration of La Florida. *Gulf Coast Historical Review, 8*(1), 18–29.

Tanodi, A. (1971). *Documentos de la Real Hacienda de Puerto Rico* (Vol. 1, pp. 1510–1519). Buenos Aires, Argentina: Centro de Investigaciones Históricas, Universidad de Puerto Rico.

Weddle, R. (1985). *Spanish Sea*. College Station, TX: Texas A&M University Press.

Chapter 5
Expressions of a Dying Tradition: Vernacular Watercraft in Apalachicola, Florida

Della A. Scott-Ireton and Christopher E. Horrell

Introduction

The sleepy community of Apalachicola, Florida, once was among the largest ports on the Gulf of Mexico, serving as the distribution center for goods shipped up and down the Apalachicola River. Fleets of watercraft plied the river and bay transporting cotton and other products from the interior of the southeastern United States and manufactured goods from all over the world to feed the booming regional economy. Many of these ships and boats were the products of local boatyards, and some were simply "homemade," knocked together as needed by local people without the benefit of plans or much experience in boatwrightry. Built by people living along the Apalachicola River and Bay specifically for local needs and conditions, these watercraft are the very definition of "vernacular." In 2002, the Florida Bureau of Archaeological Research recorded the remains of several local boats in the Apalachicola River and its tributaries. This chapter describes these watercraft and their relationship to the economy of a coastal town on the north Florida Gulf Coast (Fig. 5.1).

The watercraft described here are not the "great" vessels of history. They cannot compare to Henry VIII's *Mary Rose*, the Confederate Navy's *H.L. Hunley*, or La Salle's *La Belle* in terms of fame or relation to a specific, significant moment in time. These humble boats were, however, enormously significant to the northern Gulf coastal economy they supported and enabled to grow to regional and even

D.A. Scott-Ireton, Ph.D. (✉)
Florida Public Archaeology Network, 207 E. Main Street, Pensacola, FL 32591, USA
e-mail: dscottireton@uwf.edu

C.E. Horrell, Ph.D.
Bureau of Safety and Environmental Enforcement,
1201 Elmwood Park Blvd, New Orleans, LA 70123-2394, USA
e-mail: christopher.horrell@bsee.gov

A.M. Evans (ed.), *The Archaeology of Vernacular Watercraft*,
When the Land Meets the Sea, DOI 10.1007/978-1-4939-3563-5_5

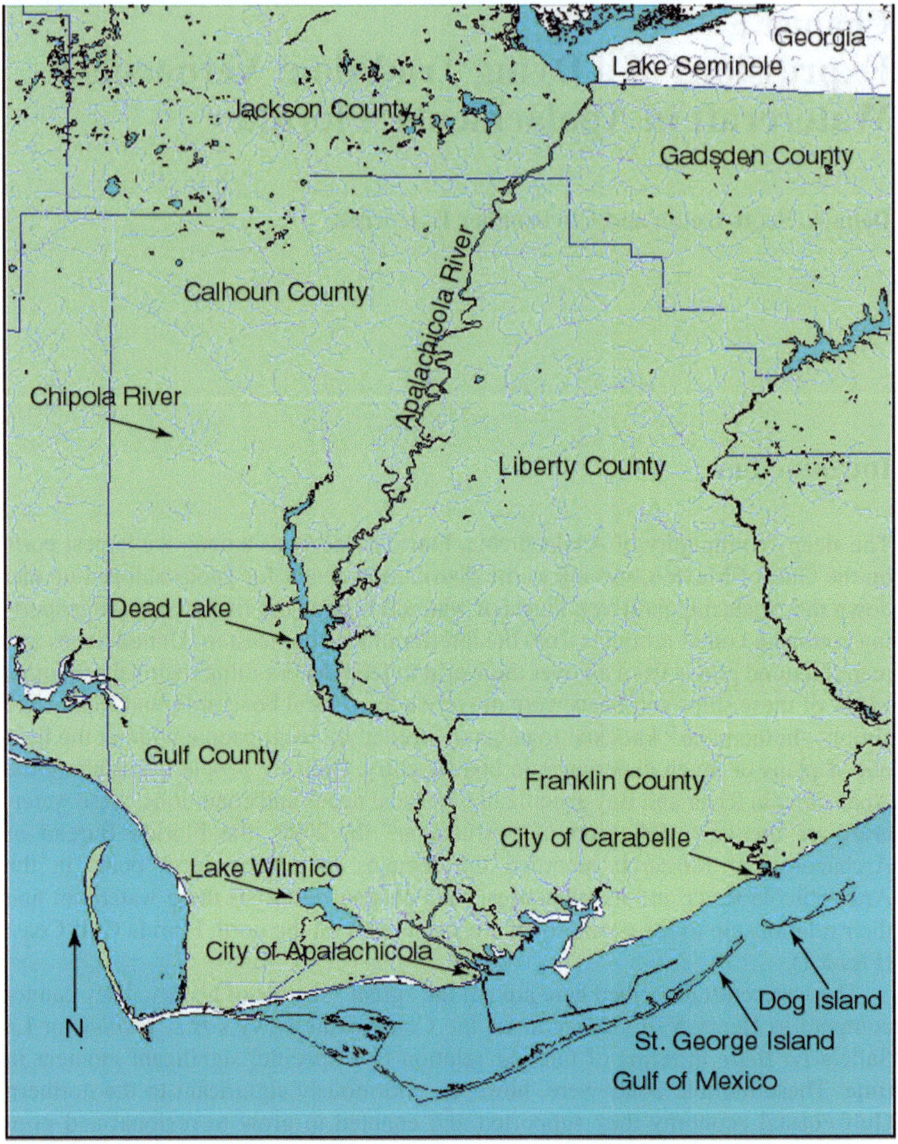

Fig. 5.1 Map showing location of the Apalachicola River in the Florida Panhandle

global importance. The remains of Apalachicola's vernacular watercraft are critical to understanding the boat-building traditions of a particular region and time. Further, they are the physical representation of local labor participating within communal, regional, and global maritime economies extending from the northern coast of the Gulf of Mexico (Horrell 2005).

History of the Northern Gulf Coast and Apalachicola

The Apalachicola River, formed by the confluence of the Flint and Chattahoochee Rivers in extreme southern Georgia, is Florida's largest river in terms of water volume, and is one of the major navigable waterways into the interior of the southeastern United States. The wide, brown river winds for over a hundred miles through the dense pine, hardwood, and scrub forests of north Florida until it spills into the Gulf of Mexico. Sediment issuing from the mouth of the Apalachicola River is deposited a short distance offshore, forming Dog, St. George, and St. Vincent Islands (Donoghue and White 1995:652–653; White et al. 1995:3–4; Damour 2002). These barrier islands, essential for protecting the mainland from storm wind and waves, create a calm estuarine setting where oyster beds and marine nurseries thrive. Like many barrier island environments along the Gulf Coast, the sounds and bays behind Dog and St. George Islands are shallow, making navigation between the Gulf and the Apalachicola River difficult for vessels with deep drafts. As a result, the sheltered waters behind the barrier islands served as natural anchorages where vessels loaded and offloaded cargo onto lighters for transport. Cargo lightered from sea-going ships to the small town of Apalachicola at the river's mouth was exchanged for cotton, lumber, seafood, naval stores, and other locally produced goods shipped down the Apalachicola River (Rogers and Willis 1997:26). This commerce sustained the community and, eventually, made possible its growth into a major port city (Horrell 2005).

The port town of Apalachicola and the river that shares its name were inextricably linked in the manufacture and exchange of goods (Horrell 2005:43). Founded in 1828, Apalachicola served as one of Florida's leading ports during the nineteenth century (Owens 1966:78–79). Initially, the community of Apalachicola rode the fluctuating surge of the cotton exchange which caused the town to become the third-most important port on the Gulf Coast (Owens 1966:222). As the cotton trade blossomed, a seasonal pattern of exchange developed within the region. During winter months the river rose due to increased rainfall, making water depths sufficient for steam-powered vessels to carry goods up and down the river and to transport cotton from the interior to Apalachicola for shipment abroad (Willoughby 1993:12). The population of Apalachicola increased in the winter as transient labor moved south to find work shipping cotton. During the summer growing season, transient workers left Apalachicola to find employment elsewhere, returning again in the winter (Owens 1966:82; Willoughby 1993:12; Horrell 2005:54). This seasonal trade pattern continued throughout the early and middle nineteenth century, coming to a halt at the beginning of the American Civil War in 1861 when hostilities disrupted the cotton industry.

With the end of the Civil War and subsequent expansion of railroads in the south, the transportation of cotton became faster and cheaper than ever before (Rogers 1987:92; Rogers and Willis 1997:44). Steamboats could not compete economically and ceased to be the prime method of commodity transportation, effectively strangling the economic life from Apalachicola (Horrell 2005:57). The town of

Apalachicola turned inward and found new ways to survive economically (Horrell 2005:63). Lumber, naval stores, and seafood industries sprang up shortly after the end of the Civil War in 1865, coinciding with the decline of cotton shipment (Rogers 1987:93). These industries were not seasonal in nature and, as a result, the permanent population of Apalachicola increased. This shift in economic direction had profound effects on local maritime commerce, population, labor pool, and, ultimately, on the construction of vessels for local needs (Horrell 2005:82–83).

At the height of Apalachicola's economic importance, maritime labor consisted of a broad range of occupations including draymen, fishermen, oystermen, spongers, captains, sailors, harbor masters, pilots, and stevedores (Horrell 2005:56). These occupations were critical for maintaining the economy as well as the community of Apalachicola. Continued reliance on maritime labor for the lumber, naval stores, and seafood procurement industries and other related activities provided social stability, especially in times of economic hardship. The economic significance of maritime-related occupations in Apalachicola continued throughout the nineteenth century, growing from 16 % of total jobs in 1850 to 43 % by 1910 (Horrell 2005:81). In addition, census records indicate that many individuals employed in maritime-related occupations hailed from the northeastern United States where boat building and other maritime activities were essential to the local economy. Census data suggest that northern boat building traditions likely were communicated and passed from generation to generation in Apalachicola (Horrell 2005:83). This knowledge, coupled with economic needs and environmental constraints, resulted in the development of vernacular vessels uniquely suited to the needs of Apalachicola's watermen (Horrell 2005:82–83). Several of these local craft, ranging from small river boats to large ocean-going cargo ships, have been recorded from archaeological contexts in the vicinity.

Vernacular Watercraft of Apalachicola

Porter Lake Steamer (8FR916)

This small steam-powered vessel is located in the St. Marks River, a tributary of the Apalachicola River, near Porter Lake about three-quarters of a mile south of the East River cut-off. The vessel lies against the east bank of the river with its bow pointed southward downriver and is overgrown with vines and small trees. The starboard side of the bow is partially exposed, and a section of the gunwale is exposed at the stern. Although most of the engine machinery has been removed, the compound boiler extends above water. A large length of chain is wrapped around the boiler and extends to a tree near the bow. This small steamboat likely aided larger sailing ships as they moved around the sound or into harbor and transported goods up-river, and probably was in use from the mid-nineteenth century into the early twentieth century (Fig. 5.2).

Fig. 5.2 The boiler of the Porter Lake Steamer (photo courtesy Florida Bureau of Archaeological Research)

At the bow, an eroded section of the cutwater extends into sediment. A stempost also is present with dimensions of 10 in. molded and 6.5 in. sided with 34 in. exposed above sediments. Eight frames are exposed along the starboard bow; the first two frames represent single cant frames, the third frame is composed of sister frames, while the fourth is single. The alternating pattern of framing continues below the sediments. The exposed frames on the starboard side facilitated recording of the frame intervals. Frame five represented one of the better-preserved frames exposed on the starboard side and has dimensions of 6 in. molded and 4 in. sided. Planks are fastened to frames with 6 in. long, square-shanked, round-headed planking nails with chiseled points. The average outer planking width is 6 in. with a thickness of 1.5 in. The average inner planking measures 6 in. wide and 2 in. thick. At 18 ft. aft of the stempost, a large iron hatch coaming is partially exposed under the bank. The hatch coaming measures 11 ft. wide; overall length fore and aft was impossible to record due to the buried nature of this feature. The coaming extends at least 6 in. above the deck; it probably extends more but the buried portion could not be measured. The forward decking starboard of the coaming is still present and outboard of the coaming the handrail or gunwale is present.

The vessel's compound boiler is located 25 ft. aft of the stem and is constructed of riveted cast-iron plating. The boiler measures 10 ft. long and extends to 35 ft. aft of the stem. The vertical compound section measures 1.5 in. wide by 2.5 in. high. The vertical section of the compound boiler has an approximate diameter of 3 ft. The horizontal portion of the boiler has a pressure relief valve with two brackets. The approximate diameter of the horizontal cylinder is 6 ft. A heavy chain is wrapped around the vertical cylinder and extends into the water. This vessel, probably a tugboat, was stripped and abandoned after the vessel was used beyond repair. While the wood has suffered some erosion, the majority of the vessel appears to be in sound condition.

Porter Lake Barge (8FR917)

This barge is located in Porter Lake just off the St. Marks River. The barge lies along the east bank of the lake and is partly exposed during low tide. Partially articulated rectangular wooden structure is present, measuring 43 ft. long by 14 ft. wide. The side planking is fastened edge-on-edge with through-bolts extending through the planks. The top plank measures 8 in. wide and 2 in. thick. Both ends of the barge are canted up, indicating double-ended construction. Planking is fastened to frames with bolts and nails. Plank dimensions at the southern end of the vessel measure 2 in. thick by 6 in. wide. The southern end is missing the top strake, although the top strake is present at the north end. The wood is in poor condition.

Gibby's Boat (8GU122)

This wooden boat is located in a little natural lake at the mouth of a small creek off the Saul's Creek Cutoff, a small tributary of the Apalachicola River. It is completely submerged and is buried in thick mud up to the gunwales; the wood is in an excellent state of preservation. The bow of the vessel protrudes out of the entrance to the small lake and points toward Saul's Creek Cutoff. Overall length of the vessel is 35 ft. with a maximum breadth at the mast tabernacle of 12.5 ft., giving the vessel a length-to-beam ratio of 2.7:1. The vessel appears to have had a rounded hull although the heavy sedimentation makes a positive determination impossible.

The stempost is composed of a single piece of wood with dimensions of 8 in. molded by 5 in. sided and is beveled and shaped with five sides. Lower hull planking is fastened to the stempost with iron fasteners. Outer hull planking in the bow measures 4 in. wide by 1.5 in. thick. The stern hull planking measures 3 in. wide by 1 in. thick. Only the upper sections of a few frames are exposed in the bow where spacing is 2 ft., 4 in. Deck planking just aft of the bow measures 4 in. wide but is broken and in disarray. A forward transverse bulkhead is located 14 ft. aft of the bow and has an opening in the center for access into the lazarette. Aft of the lazarette, the deck planking is deteriorated and, on the port side, planking is broken off. The fasteners observed on the deck planking have round heads and shanks. A mast tabernacle is located 21 ft. from the bow. It is constructed of ¾ in. planks, measures 6 in. square, and extends through the deck planking. The 1.5 in. deck planking is deteriorated forward of the tabernacle, especially on the starboard side. A large wooden hatch coaming is present 22 ft. aft of the bow. The coaming measures 1 in. thick and protrudes 6 in. above deck. The hatch measures 5 ft. wide and 6.5 ft. long fore and aft. Maximum beam of the vessel aft of the hatchway measures 10 ft. The square stern, 7.5 ft. wide, is intact. The stern is decked and the rudder is missing. In the center of the stern deck is a square bronze or brass plate, 5 in. by 7 in., with a 2 in. diameter round fitting. The raised bronze flange and socket (perhaps for a flagpole) is set inside a six-sided raised nut. This vessel appears to be a small, locally constructed craft, likely used for fishing as well as cargo transport in the river as well as St. George Sound.

Saul's Creek Boat (8GU121)

This vessel is located in Saul's Creek Cutoff, near the location of Gibby's Boat. The abandoned boat is on the south bank with the bow pointing east. Planks, frames, and a cylindrical metal tank protrude from the water. The vessel is approximately 30 ft. long with an estimated beam of 12 ft., 2 in., giving the vessel an estimated length-to-beam ratio of 2:5. The port side of the boat has collapsed and fallen into the river. The boat has a rounded hull with no evidence of a hard chine. The bow is composed of a stempost and stemson bolted together with a washer and hexagonal nut. A series of frames (floors and futtocks) are set at 1 ft. intervals. Planking is present aft of the stem; hood ends are eroded. Outer planks are 3.5 in. wide and 1 in. thick, and are fastened with square cut nails. The inner planks are fastened with wire nails with round heads. Due to the presence of a thick layer of sediment, the ceiling planking is in good condition. The outer hull planking is fragile and thin, showing evidence of some erosion. Within the hull are two transverse bulkheads. The forward bulkhead is located 13.5 ft. from the bow and is composed of vertical and horizontal members. The aft bulkhead is located approximately 31 ft. from the bow (Figs. 5.3 and 5.4).

Between the two bulkheads is a large tank that appears to be constructed of two oil drums with the bottoms removed and welded together. The tank is oriented along the longitudinal axis of the vessel and is 6 ft., 2 in. in length. The estimated diameter is 2 ft., 8 in. and approximately 7 ft. in circumference. The forward part of the tank is located 16 ft., 3 in. from the bow. The tank is constructed of cast iron and is extremely corroded; no valves or pipefittings were located. Some sections of the tank have several holes and at least one portion of the tank has been cut away. How the tank is fastened to the vessel or its intended purpose is impossible to determine without removing the protective sediment; it likely carried some form of liquid, but whether for cargo, fuel, or other purpose is unknown. Machinery is present near the tank, although its use could not be determined due to the poor condition and thick layer of sediment in the hull. Piping was also observed near the tank, as was the presence of other metal debris that may be associated with a second tank.

The boat's transom is eroded and fallen way. A few pieces of wood are located just aft of the vessel but their purpose or function is unknown. No evidence of a

Fig. 5.3 Structure and the tank of Saul's Creek Boat (photo courtesy Florida Bureau of Archaeological Research)

propeller or shaft was noted. Upper portions of the vessel are deteriorated, but wood protected by sediment is in excellent condition. The function of this vessel remains unclear at this time. The vessel may have been involved in fishing or perhaps served as a personal pleasure craft; the presence of the tank indicates it likely was used to transport some form of liquid. Additional investigations at the site may provide further information about the role of the vessel in the region.

Ingram Creek Steamboat (8GU123)

This vessel is located in Ingram Creek, another small tributary of the Apalachicola River. The abandoned boat lies along the west bank of the creek pointing northward. This is a composite-built vessel with planks, machinery, and a large compound boiler protruding above the water. The vertical compound section is missing. The

Fig. 5.4 The authors recording Saul's Creek Boat (photo courtesy Florida Bureau of Archaeological Research)

remains of the boiler appear to be constructed from cast iron and riveted together. Some of the machinery is present and a few components are exposed above the water, and planking is visible just under the water. A stanchion is visible toward one end of the vessel. The site is in poor condition and little additional information is known about the vessel, although it may represent the remains of a stern-wheel steamboat (Fig. 5.5).

Ballast Cove Wreck A (8FR903)

Fieldwork conducted by students with Florida State University from 2000 to 2003 identified articulated vessel remains located on the bottom of Ballast Cove off Dog Island as the remnants of a carvel-planked, wooden sailing ship fitted with a centerboard slot (Horrell 2005:137–140). Portions of hull structure below the turn of the bilge remain and are buried in the sediment of the cove. Preservation of the remains of the vessel varies throughout the extent of the site. Generally, deeply buried features are in excellent condition while structure that exists above or at the sand surface is eroded and pitted with shipworm (*Teredo navalis*) activity (Horrell 2005:132–135). The ship measures 65 ft. long with an estimated beam of 22 ft.

Fig. 5.5 Steam machinery of the Ingram Creek Steamboat protrudes from the water (photo courtesy Florida Bureau of Archaeological Research)

based on extant starboard hull remains. The vessel's port side seems to have been entirely salvaged and no timbers remain. The material culture recovered from excavations indicates the vessel probably was constructed and used during the mid-to-late nineteenth century. Location of patent sheaves suggests the vessel was rigged as a two-masted schooner. Built entirely of local yellow pine, possibly heart of pine, this shipwreck represents the type of coastal schooner used to move goods through the shallow waters of Apalachicola Sound and along the Gulf Coast. A key component of the local maritime labor, this vessel may have been used as a fishing boat, harbor pilot, coastal trader, or in all these capacities as well as other maritime commerce (Horrell 2005:182–183).

Conclusions

The shipbuilding industry that developed in and around the town of Apalachicola, Florida, was a reflection of the needs and desires of the local commercial situation. Vessels, such as the Ballast Cove Wreck, capable of sailing in the open waters of the Gulf of Mexico and beyond were needed to carry the products of the interior, including cotton, lumber, and naval stores, to distant markets. Shallow-drafted lighters, originally small schooners and later steam-powered smallcraft such as the Ingram Creek Steamboat, were necessary to carry cargo from warehouses in town to ships waiting in the anchorages behind Dog and St. George Islands. Steam tugs and barges similar to the Porter Lake vessels freighted goods up and down the river, taking manufactured items to markets in the interior and bringing raw products downriver for transshipment elsewhere. Small craft like the Saul's Creek Boat probably were intended for personal use including movement of people and small amounts of goods. As products and tools of the local maritime labor, these vessels sustained the small community of Apalachicola throughout the nineteenth and early part of the twentieth century.

Vernacular traditions in boatbuilding are seen in the unusual, idiosyncratic construction of the watercraft recorded in the Apalachicola River. None of the small vessels seem to have been built from standard plans and oddly placed bulkheads and strange fittings are common. The vessels appear to be sturdy and would have proven extremely effective in moving people and products on the river and into the sound. They likely were locally built in the late nineteenth or early twentieth century and were used until they had completely worn out or outlived their usefulness. Taken to out-of-the-way creeks and small lakes away from the traffic of the main river channel, the boats were stripped and abandoned.

Today, Apalachicola is a remote resort community on what is called the Forgotten Coast of Florida. The local economy depends primarily on tourism and fishing. Apalachicolans whose ancestors were sailors and stevedores today make their living as river guides and charter boat captains. Although the vernacular boatbuilding traditions of Apalachicola's bustling past are dying, the remains of this important industry exist in the watercraft abandoned in the river. Generally well preserved by thick river sediments, the boats represent exciting opportunities to learn more about the maritime activities of the northern Gulf Coast.

Acknowledgement The authors would like to thank both Dr. Roger Smith and Mr. Gibby Conrad for their help in locating and identifying these sites.

References

Damour, M. J. (2002). Looking for HMS *Fox* (1799): A model for applying barrier island geomorphology to shipwreck survey. Master's thesis, Department of Anthropology, Florida State University, Tallahassee.

Donoghue, J. F., & White, N. M. (1995). Late Holocene sea-level change and delta migration, Apalachicola River region, Northwest Florida, U.S.A. *Journal of Coastal Research, 11*(3), 651–663.

Horrell, C. E. (2005). *Plying the waters of time: Maritime archaeology and history on the Florida Gulf Coast.* Doctoral dissertation, Department of Anthropology, Florida State University, Tallahassee. University Microfilms International, Ann Arbor, MI.

Owens, H. P. (1966). *Apalachicola before 1861.* Doctoral dissertation, Department of History, Florida State University, Tallahassee. University Microfilms International, Ann Arbor, MI.

Rogers, W. W. (1987). *Outposts on the Gulf.* Gainesville, FL: University of Florida Press.
Rogers, W. W., & Willis, L. (1997). *At the water's edge: a pictorial and narrative history of Apalachicola and Franklin County*. Virginia Beach, VA: Donning.
White, N. M., Ryder, K. D., Grammer, S. M., & Mayo, K. L. (1995). *Archaeological survey of Dog Island Franklin County, Florida.* Report submitted to the Barrier Island Trust, Tallahassee, FL. Department of Anthropology, University of South Florida, Tampa.
Willoughby, L. W. (1993). *Fair to middlin: the antebellum cotton trade of the Apalachicola River Valley*. Tuscaloosa, AL: University of Alabama Press.

Chapter 6
Some Assembly Required: The Analysis and Reassembly of the Larkin Boat, a Vernacular Watercraft Recovered from Gregory Mill Creek in Liberty County, Florida

Chuck Meide

Introduction

Vernacular ship and boatbuilding traditions in the South, and more specifically along Florida's waterways, are poorly understood and documented, and relatively few examples of archaeologically investigated vernacular watercraft exist. This may be somewhat surprising considering the important role that boats and waterways have played in Florida's history and prehistory, and in the lifeways of past residents. Vernacular boatbuilding by definition though, was a backyard operation handed down by oral transmission from generation to generation, and few if any written accounts remain to document even broad evolutionary developments (Wilde-Ramsing and Alford 1990:3). In addition, the lightweight nature of historic small craft makes their abandoned remains highly susceptible to deterioration, and relatively few specimens survive for study, despite the prevalence throughout Florida of wetlands that could potentially preserve such remains (and have preserved many examples of more robust prehistoric dugout canoes; see, for example, Purdy 1991; Meide 1995; Wheeler et al. 2003). The construction of seemingly ordinary wooden boats for fishing, working, travel, and pleasure—once a vital yet commonplace skill—has literally become, since the introduction of plywood, gasoline engines, and the mass production of fiberglass hulls in the second half of the twentieth century, a lost art.

A planked punt known as the Larkin Boat, discovered in 1996 on the border of the Apalachicola National Forest in the Florida Panhandle, is a rare example of

C. Meide (✉)
Lighthouse Archaeological Maritime Program (LAMP), St. Augustine Lighthouse Maritime Museum, Lighthouse Avenue, St. Augustine, FL 32080, USA
e-mail: cmeide@staugustinelighthouse.com

A.M. Evans (ed.), *The Archaeology of Vernacular Watercraft*,
When the Land Meets the Sea, DOI 10.1007/978-1-4939-3563-5_6

historic vernacular watercraft. It was conserved at the State of Florida Research and Conservation Laboratory over the following 2 years, and in 2002, when employed by the US Forest Service, the author oversaw the recording and reassembly of its disarticulated remains (Meide 2002a, b). It appears to be a truly unique vessel, though it bears resemblances to various other punt-type craft in the historical and archaeological records. The history of the Larkin Boat's discovery, recovery, conservation, documentation, and reassembly for display, as well as the results of an archaeological analysis of its structural remains, are presented here. Specific details of the Larkin Boat are used in a discussion that explores comparable examples of similar boats and attempts to better understand design features that appear to be unique to this vessel.

Discovery and Conservation

Sometime before mid-July 1996, Mr. Mitchell Larkin of Liberty County, Florida, discovered the remains of a 5.32 m (17.45 ft.) long plank-built vernacular watercraft submerged in the riverbank mud of Gregory Mill Creek, in the Apalachicola River floodplain (Fig. 6.1). His property, on which the boat was located, is adjacent to the Apalachicola National Forest near the Florida River area. Liberty County, in Florida's Panhandle, is predominantly rural and both the least populous and least-densely populated county in the state. This forested area east of the Apalachicola River is dominated by longleaf and slash pine with an understory of palmetto, gallberry, and wiregrass, and the generally flat terrain is pocked by sinkholes, creeks, and swamps (James 2006:2). The original old-growth forest featured trees as tall as 36.5 m (120 ft.), many of which were 120–250 years old, along with dense groves of younger tall saplings. This old-growth forest was heavily exploited for lumber between 1880 and the early 1900s, and many of the remaining pines were tapped for gum to distill for turpentine and rosin (James 2006:1). Demand for timber resources fueled continued forest exploitation, the only real industry in the area, during and after World War II.

Believing the boat he had found dated to the Spanish colonial period, Mr. Larkin decided to recover it. It unfortunately broke apart as it was pulled from the muddy, aquatic environment that had ensured its preservation since the time of its abandonment, perhaps centuries earlier. If there were any diagnostic artifacts associated with the boat, their presence was not noted by Larkin. He soon became alarmed at the rate at which the waterlogged wood began to deteriorate once removed from its submerged context, and realized that without professional attention the remains of the vessel might be lost. Understanding its historical significance and wishing to preserve it for research and interpretation, he contacted US Forest Service personnel at the Apalachicola National Forest. Staff archaeologist Dr. Andrea Repp felt that the disarticulated remains could be safely stored in water pending the formal donation of the boat to the Forest Service for conservation and curation. The various planks and timbers comprising the boat find were stored in a shallow, water-filled

Fig. 6.1 Map showing the location of the Larkin Boat, discovered submerged in the muddy riverbank at Gregory Mill Creek in Liberty County, adjacent to the Apalachicola National Forest. Illustration by Brendan Burke, satellite image courtesy of Google Earth

trench within a secure area at the Forest Service's Wakulla Work Center in Wakulla County (Colaninno 1996). Mr. Larkin offered in writing to donate the vessel to the Forest Service, which was formally accepted a month later on 10 August 1996 (Larkin 1996; Siderits 1996; Joslin 1996).

By this time the State of Florida's Bureau of Archaeological Research had been notified of the find, and state authorities had offered their assistance with the conservation of the waterlogged timbers (Miller 1996). The boat remains, which included a number of planks, frames, and the remnant of a somewhat crudely fashioned paddle, were therefore transported to the state capital of Tallahassee, to be conserved in the State of Florida Research and Conservation Laboratory. Under the direction of James Levy, the chief conservator of the laboratory at the time, the timbers were successfully stabilized through treatment with polyethylene glycol (PEG), a standard procedure for conserving waterlogged wood (Hamilton 2010). They were treated for over 2 years with PEG of an average molecular weight of 1450 and the process was very successful, due to the relatively thin planks (James Levy, 04 February 2013, elec. comm.). The PEG treatment was finished by April 1999 (Smith 1999).

Recording, Pre-2002

Before its transportation to Tallahassee, the Larkin Boat had only been documented through photography. The first systematic recording of the vessel remains appears to have been conducted by a Sea Scouts organization designated Sea Explorer Ship 141. Correspondence archived in the Florida Master Site File indicates that at the suggestion of State Underwater Archaeologist Dr. Roger C. Smith, Mr. Dan Tonsmeire (1996a, b) contacted both the Bureau Chief Dr. Jim Miller and the US Forest Service archaeologist Dr. Andrea Repp, proposing that the scouts record the vessel and build a replica under the direction of David Wyman (a Panama City naval architect who had participated in the *Defence* excavation and hull analysis). It should be noted that in the archived correspondence the Larkin Boat is sometimes referred to as the "Bristol Boat" or the "Bristol Skiff," a reference to Bristol, Florida, the county seat and only incorporated town in Liberty County, located some 32 km (19.9 miles) north of the Larkin Boat site. In September 1996 the Sea Scouts, under the direction of Wyman and Smith, "measured up all of the parts and pieces of the boat, and made a small cardboard model" (Smith 1996). Unfortunately no report, notes, or drawings from this project appear to have been provided to the Florida Master Site File or the US Forest Service, nor does a full-scale replica boat appear to have ever been built (Smith 1999).

Photographs and a brief description of the Larkin Boat were also sent by Dr. Smith to the North Carolina Department of Cultural Resources, an agency that has had notable experience in the research and management of small historic craft (Wilde-Ramsing 1990; Wilde-Ramsing and Alford 1990). This information was forwarded to Michael Alford, the former Curator of the North Carolina Maritime Museum, who provided preliminary comments to Dr. Smith (Alford 1996).

The 2002 Project: Recording and Reassembly

A more thorough documentation of the Larkin Boat took place in early 2002, when the author was temporarily employed as an archaeologist for the US Forest Service. By this time the disarticulated, conserved remains had been returned to the Apalachicola National Forest's Wakulla Work Center. Between 20 February and 21 March the planks and other component pieces were recorded in detail, and 1:4 scaled drawings were produced of the bottom planks, side planks, ceiling plank, and frames. In late March a system of steel frames were designed and fabricated to hold the component pieces together in their original configuration, and the timbers were successfully reassembled between 25 March and 3 April (Meide 2002b). Further recording was carried out at this stage to document the shape of the reassembled vessel. On 16 April, the reassembled boat was transported using a moving van from the Wakulla Work Station to the Ranger District Office in

Bristol, 32 km (19.9 miles) north of the site on Mitchell Larkin's property where the boat was originally discovered. It has remained at this office where it is on display with interpretive materials for the public.

Recording Methodology

For the most part, the boat remains were documented using practices standard for nautical archaeologists and small craft specialists (Lipke et al. 1993; Steffy 1994). The first step was simply to retrieve the timbers from their storage location and to lay them out flat on a series of foam rubber mats in a large metal outbuilding, with enough interior space to maneuver around the planks and other timbers during the initial inspection and subsequent recording process. As various timbers were identified by function, their layout was repositioned. The two side planks and two bottom planks were arranged from port to starboard to provide an "exploded view" or real-life strake diagram (cf. Greenhill and Morrison 1995:64–66) of the vessel, while a single ceiling plank and the three intact and two partially intact frames were laid out to be recorded separately. The various components were thoroughly documented via digital photography, and on one occasion the entire assemblage was moved outside during a sunny day, laid out in order on a layer of flattened burlap sacks on an asphalt driveway surface, to generate a series of plan view photographs from atop a tall ladder. These were used to generate a series of photomosaics showing the planks by themselves, with the frames and intercostal keelson in place, and with the frames, intercostal keelson, thwart risers, and ceiling plank in place (Fig. 6.2).

Once all of the identifiable members were arranged according to their original position, the next step was to inventory all of the timbers, including a number of pieces whose original function or placement remained a mystery. Most of the timbers were previously tagged with identifying numbers, and in fact two separate numbering systems had been employed, presumably during the time the boat remains were housed at the conservation laboratory in Tallahassee. Most timbers

Fig. 6.2 Photomosaic of the Larkin Boat planks arranged to show an exploded view or strake diagram, with frames, intercostal keelson, thwart risers, and ceiling plank in place. The side and bottom planks are labeled by letter, while the frames are not labeled, they are numbered I through V starting at the stern

had two attached tags, though some had lost one or both. One set of tags were very small and oval shaped, featuring a pre-printed five digit number ranging from 16,143 to 16,166. It is believed that these identifying tags were used by the conservation laboratory to keep track of the timbers during the PEG treatment. These pre-numbered tags appear to have been somewhat arbitrarily assigned to timbers, though hull members of the same type (i.e., frames, centerline nailers, or paddle fragments) tended to feature sequential numbering. During the 2002 study, and in this volume, this set of numbers is referred to as the BAR Lab number. Email queries to the current and former chief conservators at the state lab did not result in any inventory, notes, or other lab documentation which might have further clarified these numerical designations.

The second set of tags differed in that they were large, plastic, hand-cut, and hand-labeled. While several were missing or illegible, they indicated a simple numbering system labeled 1–29, corresponding to the 29 extant timbers. It is presumed that these tags were attached to the timbers during the original recording conducted by the Sea Scouts. This numbering system is hereby referred to as the Timber Number. For the timber inventory conducted during the 2002 analysis, a binomial Hull Member Designation was used combining the Timber Number following a description of the hull member type, i.e., Thwart Riser 10, Intercostal Keelson Pieces 11–15, Paddle Shaft 22. The exceptions to this nomenclature were the frames, which were designated with Roman numerals (i.e., Frames I through V), so that their designation number would correspond to the individual frame's position on the boat ordered from stern to bow, and also the primary planks (most of which were missing their original timber numbers), which were labeled Bottom Planks A and B, Side Planks C and D, and Ceiling Plank E. Table 6.1 is the master timber inventory and scantling list, which includes every extant timber by their binomial Hull Member Designation, Timber Number (large tags), BAR Lab Number (small tags), along with their dimensions and any particular comments.

The dimensions of each timber were measured, using folding rulers, tapes, or calipers as appropriate. In addition, by taking offset measurements from a longitudinal baseline, detailed 1:4 plan views were generated of the interior surface of the bottom planks, the interior (inboard) surface of the side planks, the upper surface of the ceiling plank, and the forward, top, and bottom surfaces of all frames (Figs. 6.3 and 6.4). Positions of fastener holes were noted and plank beveling was recorded, in order to better understand the nature and sequence of construction. Great care was taken in recording every detail, because it was understood that when the various components were reassembled and secured in place, some features would be obscured or even destroyed. Once the planks and frames were successfully reassembled and held in place by a steel framework, interior body sections were recorded using a leveled centerline, tapes, folding rule, and plumb bob at seven points along the length of the vessel (Fig. 6.5). Two stations were placed at the ends of the boat, where the (missing) transom pieces originally joined the foremost and aftermost ends of the bottom planking, with five additional stations spaced evenly at 1 m intervals along the length of the hull, centered at the midships station or widest point of the hull. Finally, a 1:1 tracing of the ceiling plank denoting

Table 6.1 Larkin boat timber list

Member designation	Timber no. (large tag)	BAR lab no. (small tag)	Max dimensions (cm)	Comments
Bottom plank A	?	?	528.2×35.0×1.70	Port side
Bottom plank B	4	16143	522.3×35.1×2.35	Starboard side
Side plank C	?	?	535.2×29.8×2.3	Port side
Side plank D	?	?	533.2×33.5×2.15	Starboard side
Ceiling plank E	26	16165	431.2×18.3×1.95	Starboard side
Frame I	9	16148	32.9×3.35S×4.2M	Deteriorated and incomplete; original length approx. 40 cm
Frame II	5	16144	53.5×3.55S×5.2M	
Frame III	7	16146	60.5×3.4S×5.05M	
Frame IV	6	16145	40.5×3.45S×5.25M	Incomplete; original length approx. 65 cm
Frame V	8	16147	56.0×3.65S×5.1M	
Board 16	16	16155	32.3×17.5×1.58	Boards 16–17 may be one piece and possibly could be transom
Board 17	17	16156	24.9×8.6×1.45	
Thwart riser 10	10	16149	28.8×4.4×2.4	
Thwart riser 18	18	16157	28.3×3.5×2.3	
Thwart riser 19	19	16158	26.7×3.2×2.1	
				All keelson pieces are beveled (convex in cross section)
Intercostal keelson 11	11	16150	62.6×8.8×1.55	9 Nail holes (4 one side, 5 other)
Intercostal keelson 12	12	16151	50.8×8.0×1.48	8 Nail holes (4 one side, 4 other)
Intercostal keelson 13	13	16152	86.1×7.9×1.48	11 Nail holes (6 one side, 5 other)
Intercostal keelson 14	14	16153	87.1×8.0×1.50	11 Nail holes (6 one side, 5 other)
Intercostal keelson 15	15	16154	89.0×8.8×1.15	10 Nail holes (5 one side, 5 other)
Intercostal keelson 25	25	16158	74.6×7.9×1.2	Incomplete and deteriorated. 4 nail holes (2 one side, 2 other)
Unknown fragment 21	21	16160	29.0×5.6×1.2	Flat fragment; 2 nail holes
Unknown fragment 23	23	16162	26.5×2.9×1.6	Could be thwart riser

(continued)

Table 6.1 (continued)

Member designation	Timber no. (large tag)	BAR lab no. (small tag)	Max dimensions (cm)	Comments
Unknown fragment 28	28?	16168	35.2×4.4 (diam.)	Round, burnt, small log; Probably intrusive
Unknown fragment 29	29?	16169	44.5×4.6×2.15	Previously thought to be paddle shaft
Paddle blade 20	20	16159	45.15×13.0×1.23	
Paddle shaft 22	22	16161	27.0×3.9×1.7	Prob. upper end of paddle shaft
Paddle shaft 24	24	16163	15.8×3.5×1.45	24 and 27 fit together and form part of paddle shaft
Paddle shaft 27	27	16166	30.0×3.6×1.3	Shaft section adjacent to blade

S sided, ***M*** molded

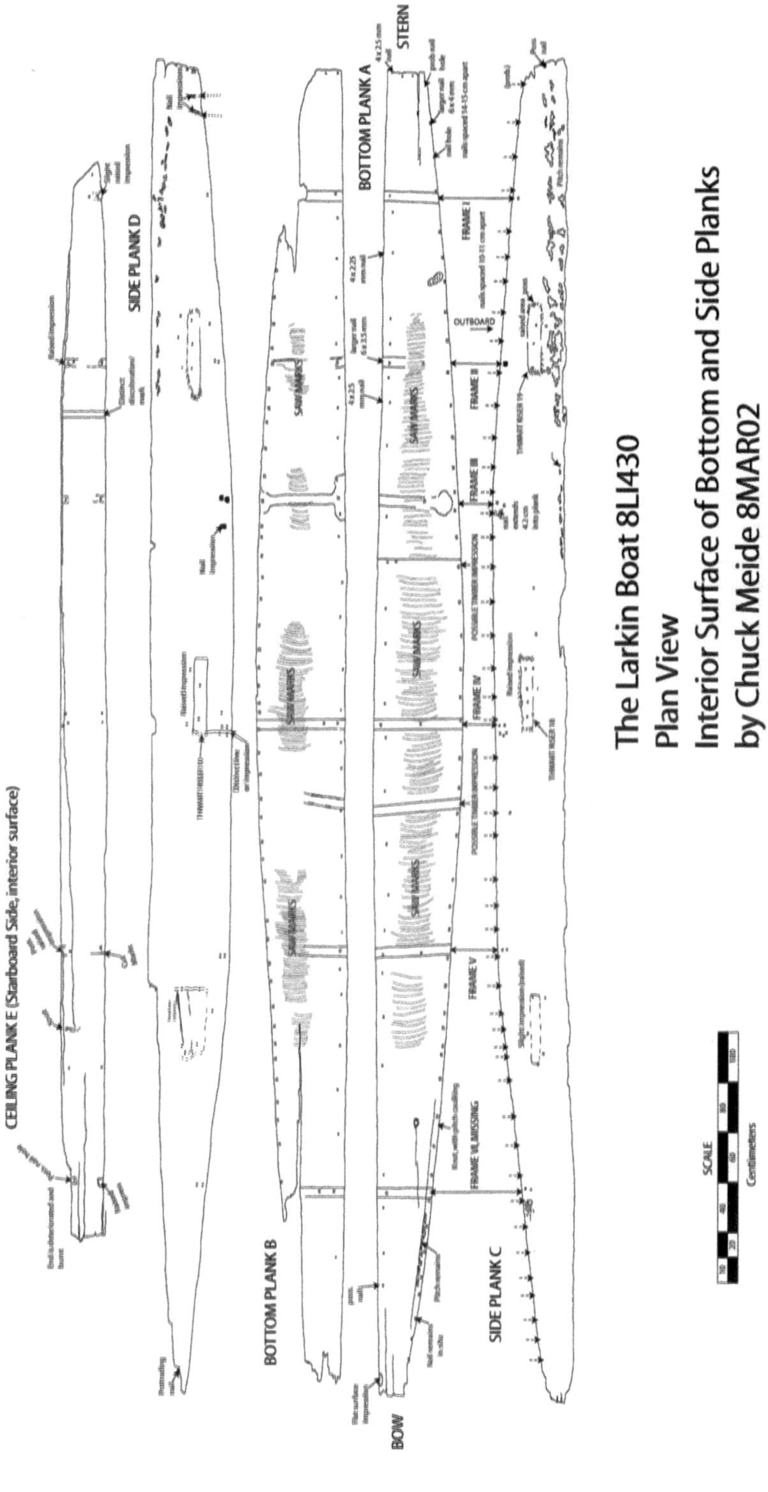

Fig. 6.3 Strake diagram or plan view of the Larkin Boat planks. Recorded and drawn by Chuck Meide, digitized by Tim Jackson

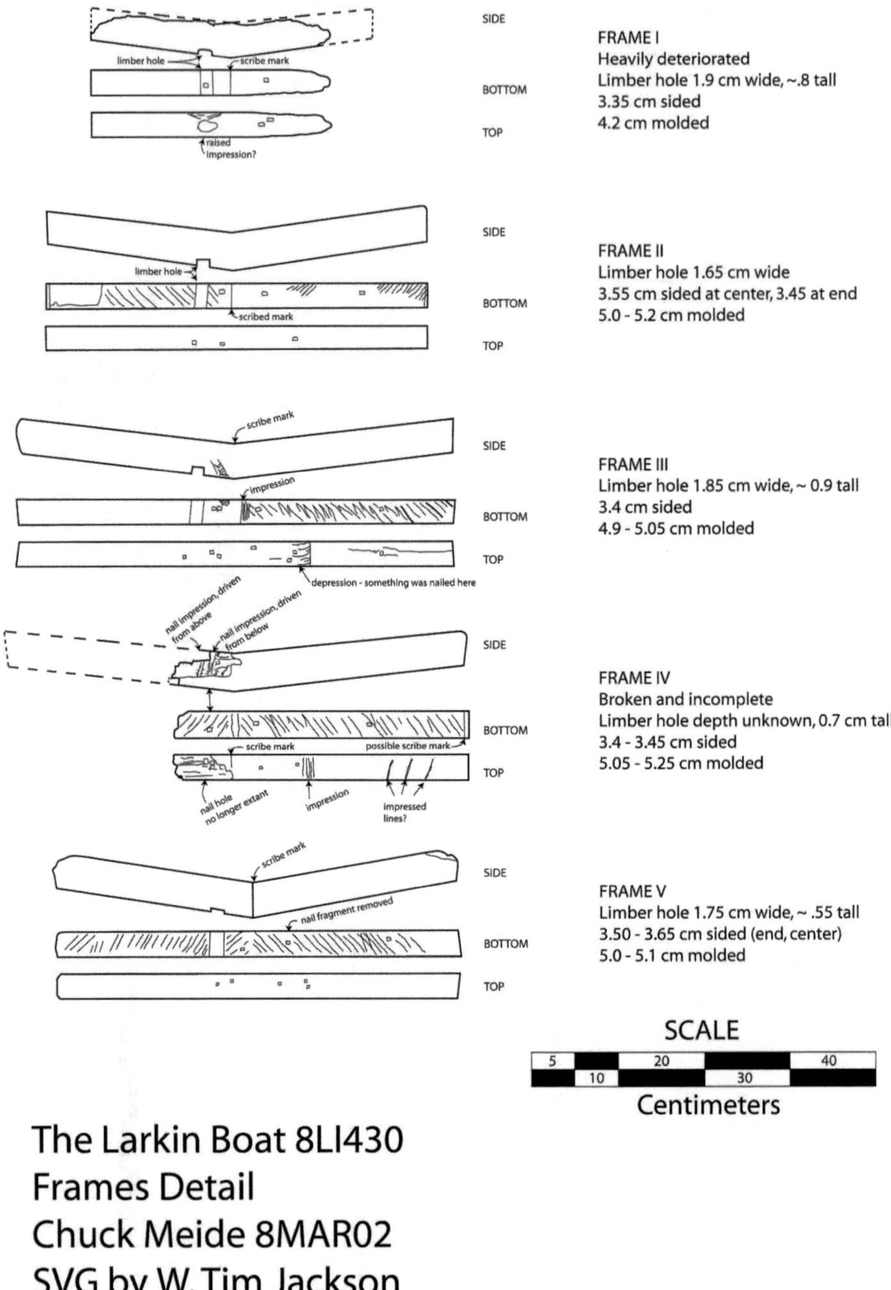

Fig. 6.4 Larkin Boat frames, showing the forward, upper, and lower surfaces. Recorded and drawn by Chuck Meide, digitized by Tim Jackson

Fig. 6.5 The author recording the interior curvature of the hull at the midships body station, or the widest point of the vessel (located 2.46 m forward of the stern, which is in background of photograph)

all its markings and fastener holes was made on clear plastic sheeting, a technique used on both the *Belle* and Red Bay shipwreck excavations (Waddell 1986; Bruseth and Turner 2005; Grenier et al. 2007), to better determine the placement of this member on the reassembled vessel.

Reassembly Methodology

The objective of the 2002 study of the Larkin Boat was not only to thoroughly record its historic fabric but also to design, fabricate, and implement a system to secure the reassembled component pieces so that the intact vessel could be placed on permanent public display (Meide 2002a). There are few examples in North America of a disarticulated small craft recovered from an archaeological context being reassembled for display, the most prominent being the sixteenth-century Basque whale boat (*chalupa*) excavated and reassembled by Parks Canada (Harris 2006; Grenier et al. 2007; Ontario Service Centre of Parks Canada 2009). The *chalupa* was much older, more complex, and larger than the Larkin Boat, not to mention better-funded, and for it a sturdy framework running the length of the vessel was designed that served as both a stand and a cradle for permanent display at the Red Bay National Historic Site in Labrador. Other examples of cradles for supporting historic boats are discussed in Lipke et al. (1993:13–15). Given the lightweight nature of the Larkin Boat, and the limited funding available for the project, a system was devised that was considerably simpler and less expensive than the aforementioned examples, yet which proved to be quite effective.

The author solicited the assistance of his cousin, Nik Hartney, who at the time was an engineering student at Florida State University (FSU), and together a design was developed. The concept involved the use of four pairs of 1 in. (2.54 cm) wide stainless steel straps that were pre-bent to form brackets matching the shape of the hull at four points along its length. The lower bracket of each pair would be placed on the outer hull, with its mate positioned directly over it on the inside of the hull. The two brackets would then be bolted together through a series of drilled holes in both the brackets and the planks, sandwiching the planks firmly in their original positions. When feasible these brackets were to be placed in the exact position of a frame, which would then be attached to the interior of the boat, directly on top of the upper paired bracket, obscuring that portion of the bracket from view. Screws driven up from below, through another series of pre-drilled holes in the straps, would hold the frames firmly in place.

Before the four sets of paired support brackets were built it was necessary to determine exactly where they would be placed. It was most important to secure the ends of the vessel, rather than the middle, as the flattened bottom and side planks needed to be bent back to their original shape, and their curvature was more acute towards the ends. Thus it was decided to place two sets of brackets towards the stern, at the locations of Frame I and Frame II (around 0.48 and 1.68 m forward of the stern, respectively), and two more towards the bow, at the location of Frame V and adjacent to the impression of the missing Frame VI (around 1.72 and 0.79 m aft of the bow, respectively). The space between the two sets of support brackets, encompassing the midships area of the boat, is around 1.76 m.

The brackets were fabricated at the joint FSU-Florida A&M University mechanical engineering laboratory under the direction of Mr. Hartney. The first step in the process was to generate a template based on the exact shape of the hull at the four selected locations. Actual floor timbers from the Larkin Boat were traced to draw 1:1 cross sections of the boat, and the templates were in turn used when bending the steel straps to the shape of the hull. Figure 6.6 shows a completed set of brackets, those to be positioned under Frame III. They have been bent to shape, drilled for fasteners, and include a threaded steel rod in place at their top ends, designed to firmly hold the side planks in place. After the initial stage of reassembly, when the bottom and side planks were secured with all four sets of brackets, it was realized that the thwartship rods were not needed for stability. They were omitted, eliminating the need to drill additional holes into the historic fabric of the boat, and making for a less obvious support framework for display purposes.

All the stainless steel brackets, washers, and fasteners were then painted with several coats of black spray paint, in an attempt to prevent the metal from coming into contact with the PEG-soaked wood, as PEG is known to exacerbate the corrosion of ferrous material (Guilminota et al. 2002). The steel support brackets were installed and the main hull components were assembled over a 4 day period. A

Fig. 6.6 The paired steel brackets to be positioned under Frame III are shown on the paper template. The brackets have been bent to shape, drilled with fastener holes, and include a threaded steel rod in place at their upper ends to more firmly hold the side planks in place. It was later determined that the thwartship rods were not needed for stability and they were omitted

temporary wooden scaffold, padded with burlap, was erected to hold the bottom planks horizontally above waist level, so that it would be easy to access the historic timbers from above or below during the reassembly. Each set of paired brackets were held in place by one worker while holes were drilled through the bottom planks by another, using the pre-drilled holes in the brackets as guides. The brackets were then secured with ¼ in. (0.64 cm) bolts. Washers were used as spacers between the brackets and the planks to minimize contact between PEG and the steel straps. Ratcheted come-along straps were used to temporarily tighten the ends of the planks, holding them together in their original positions until the steel straps were all in place.

Once the four bracket pairs were installed, fixing the bottom planks in their original, v-shaped configuration, first one and then the other side plank were carefully lowered into place. This entailed sliding the side plank in between the paired bracket uprights while cautiously bending the ends of the side plank inwards. It was also necessary to carefully bend the bottom planks upwards at the ends of the boat, so that the outer edges of the bottom planks met the lower edges of the side planks before drilling through the side planks and bolting them in place at the outermost brackets. This was done successfully, though wooden spacers were used within the forwardmost pair of brackets, which had been fabricated slightly off the required specifications. When the planks were securely assembled within the bracket system, the final step was to attach the frames and intercostal keelson pieces. The frames were secured by screws driven through the bottom planking from below, and the centerline nailers or intercostal keelson pieces were then positioned in their original places between the frames and secured by two screws each, driven from above through existing nail holes (Fig. 6.7). The remaining thwart risers were also attached to their original locations on the side planks using screws driven through historical nail holes.

The ceiling plank was not placed until after the boat's final move from the Wakulla Work Station to the Ranger District Office in Bristol. Instead of permanently attaching it to the hull with modern fasteners, it was simply placed in the boat in its original position.

Fig. 6.7 View of the reassembled Larkin Boat from the stern looking the length of the boat towards the bow. Note the deadrise at the stern and the frames and intercostal keelson pieces in place between frames. The thwart risers and ceiling plank are not yet installed

Analysis

Bottom Planks

The bottom of the Larkin Boat consists of two longitudinal planks, designated Bottom Plank A (port side) and B (starboard side) (Fig. 6.3). The surviving overall length of Plank A is 5.28 m (17.3 ft.) and that of Plank B is 5.22 m (17.1 ft.). The maximum breadth of Planks A and B are 35.0 and 35.1 cm (13.78 and 13.82 in.), respectively. Their inner edges were fit together forming a seam along the centerline of the vessel. The two bottom planks show a slight deadrise form, meaning that they are angled upwards from the centerline towards the sides of the boat, forming a slight v-shape to the hull. This angle of deadrise is readily apparent when observing the frames, and can be seen in Figs. 6.4 and 6.7. The divergence of the disassembled seam between bottom planks at the bow and stern ends of the boat, visible in Figs. 6.2 and 6.3, indicates a fair degree of deadrise is carried to the ends, a feature which was confirmed upon reassembly. The outer edges of the bottom planks are beveled so as to be flush with the side planks, which flair outwards and were originally attached to the upper face of the bottom planks by nails. The inner edges of the bottom planks sometimes appear to display a bevel, to accommodate the join between the two planks and their angle of deadrise, but this bevel is less distinct and not always readily observable. The inner edge of the bottom planks tends to be narrower, ranging between 1.4 and 1.95 cm (0.55 and 0.77 in.) thick, than the outer edge, which measures between 1.9 and 2.35 cm (0.75 and 0.93 in.) thick. Plank A is somewhat thinner than Plank B, with maximum thicknesses of 1.7 cm (0.67) compared to 2.35 cm (0.93 in.).

Nail holes line the outer perimeter of the bottom planks, where they were hammered from the bottom surface of the bottom planks into the lower edge of the side planks. These nail holes are somewhat regularly spaced, usually ranging between 6 and 13 cm (2.36 and 5.12 in.) apart. In some areas of this nailed join with the side planks there are remnants of pitch remains. Nail holes are also present along the inner edges of the planks, marking where the intercostal keelson pieces were nailed down along the centerline, in between frames. Additional nail holes, and impressions on the interior surface of the bottom planks, mark the original locations of six frames. The stern end of the planks is relatively intact (Fig. 6.7), while the bow end shows significant degradation, precluding a reliable understanding of their original shape and configuration at their forward terminus.

Frames

The Larkin Boat originally featured six floor timbers, designated Frames I–VI in order from stern to bow. The forwardmost of these, Frame VI, was missing when the boat was recorded in 2002 and was probably never recovered when the boat remains were originally pulled from their archaeological context. Its impression on the bottom planks, along with nail holes on the bottom and side planks, indicate its original placement within the hull. The other frames have survived, usually in an excellent state of preservation, though two are incomplete. Frame I is highly eroded and missing the upper portion of its starboard wronghead and much of its port arm, and Frame IV is broken near the centerline and missing almost its entire starboard arm. Figure 6.4 is the scaled drawing of the upper (sided), forward (molded), and lower (sided) faces of the frames. Figure 6.8 shows the aft molded faces of the surviving frames. Fastener holes and saw marks were visible on the bottom sided surfaces of the frames. Each frame features a small limber hole notched into its lower surface, just to the starboard of the centerline (Fig. 6.9). In addition, all five frames feature a scribe mark, either on the forward molded surface (Frames III and V), the lower sided surface (Frames I and II), or on the top or upper sided surface (Frame IV). These marks, such as the one visible in Fig. 6.9, denote the centerline of the vessel and were used during the design and manufacturing of the frames.

As the frames span the inboard breadth of the boat those in the middle are longer than those towards the ends. The shortest frame (Frame I) is deteriorated but was probably around 40 cm (15.75 in.) long, while the longest frame (Frame IV) is broken but was probably around 65 cm (25.6 in.) long. Their sided dimensions range from 3.35 to 3.65 cm (1.32–1.44 in.) and their molded dimensions range from 4.2 to 5.25 cm (1.65–2.07 in.). The angle of deadrise on all frames is consistent at 7–8°.

The frames were fastened to the bottom planks by nails, usually two in each arm of the frame. Nails were both driven up from below and down from above, and the latter were likely to attach ceiling planking or flooring rather than to fasten the frame to the bottom of the boat. In addition, two nails were hammered transversely through the side planks into the outer ends or wrongheads of each frame, to attach them to the side planks of the boat.

Fig. 6.8 View of the aft molded surface of the five surviving frames

Fig. 6.9 Close-up view of Frame V (forward molded surface) showing the limber hole and the scribe mark denoting the centerline

The frames are somewhat irregularly spaced. It is 48.8 cm (1.60 ft.) from the aftermost edge of the bottom planking to Frame I, 61.2 cm (2.00 ft.) between Frame I and Frame II, 51.6 cm (1.69 ft.) between Frame II and Frame III, 86.8 cm (2.85 ft.) between Frame III and Frame IV, 88.4 cm (2.90 ft.) between Frame IV and Frame V, 92 cm (3.02 ft.) between Frame V and the missing Frame VI, and 80.4 (2.64 ft.) between the missing Frame VI and the forward edge of the bottom planking.

Intercostal Keelson

In between the frames along the centerline were laid a series of thin boards, covering the inner seam between bottom planking (Fig. 6.7). These had been nailed in place using between 4 and 6 nails along each side of the board. When initially recorded during the 2002 study they were referred to as "bottom nailers" or "centerline nailers," but they are more appropriately described as intercostal keelson pieces. They served to help hold the bottom planking together, and to protect the seam between bottom planks. Butted against the frames, they also helped strengthen the hull longitudinally. By matching nail holes in the bottom planking with those present in the intercostal keelson pieces, the original location of each of the surviving six pieces was determined. One was placed between each of five spaces between frames, and between the foremost frame (Frame VI) and the bow transom. There do not appear to be any fastener holes along the bottom plank seams aft of the aftermost frame (Frame I), so it is assumed that there was no keelson piece between Frame I and the stern transom.

The intercostal keelson pieces range in length from 50.8 to 89.0 cm (1.67–2.92 ft.), in width from 7.9 to 8.8 cm (3.11–3.46 in.), and in thickness from 1.2 to 1.55 cm (0.47–0.61 in.). They are rounded or convex in profile, particularly on their upper surfaces. Due to their lighter coloration, they appear to be fashioned from a different type of wood than the other hull members, though species identification has not yet been completed for any samples. All of the intercostal keelson pieces have survived in a great state of preservation, except for the forwardmost (Keelson Piece 25) which is severely degraded, along with the bottom and side planking at the bow.

Ceiling Plank

One ceiling plank survived on the vessel (Ceiling Plank E). This is another longitudinal plank some 4.31 m (14.15 ft.) in length, which was placed on top of the frames on the starboard side of the centerline (Fig. 6.2). Presumably a second ceiling plank once existed and was situated on the port side of the vessel. These would have acted as a floor in the vessel to protect the frames from wear and to provide further longitudinal strength. The plank is relatively uniform in width at around 18.3 cm (7.2 in.) but it does narrow towards its aft end, which is cut diagonally. It is 1.95 cm (0.77 in.) thick. The inner surface of this plank was recorded (Fig. 6.3); the drawing shows frame impressions and fastener holes which helped researchers position it in the boat during reassembly. As mentioned previously, a 1:1 depiction of this timber was traced onto clear plastic sheeting, which was also used to help fix its original location within the boat.

Side Planks

The sides of the Larkin Boat were formed by attaching a single plank on each side of the bottom planking (Figs. 6.2, 6.3, 6.5, and 6.7). These two planks have been designated Side Planks C (port side) and D (starboard side). Each has been shaped so that in profile they are rounded on the bottom, following the bottom contour of the boat with its rockered ends, and are flat at the top edge. The upper portion of Side Plank C is degraded so that its upper edge is no longer extant, though the original upper edge on Side Plank D survives intact from the stern to a point about 1.7 m from the bow end. Nail holes indicate the side planks were positioned on the edge of the bottom planks, and were nailed from outboard through the outer edge of the bottom plank into the lower edge of the side plank. The side planks were angled outwards from the bottom planks by about 10–15°.

Nails were hammered from the outside of the side planks to secure the frame wrongheads. Visible on the interior surface of the side planks were the impressions where seat supports or thwart risers were once nailed. These riser impressions appear as raised areas on the interior surface of the side plank, and when initially encountered it was assumed these raised areas had been achieved by adzing the surrounding plank. After further inspection it is now believed that the raised areas are the result of wear on the interior surface of the side plank outside the area covered and protected by the risers and outboard seat edges. Also present on the interior face of the side planks, particularly towards the stern end, are the remains of pitch or a similar waterproofing material.

The sides of the vessel were originally around 33.5 cm (13.19 in.) high. They are of similar thickness to the bottom planking, ranging from 2.15 to 2.33 cm (0.85–0.92 in.) thick. As with the bottom planks, the forward extent of the sides display severe erosion, but at least one portion of the original forward edge on Side Plank C has survived, providing a total length of 5.35 m (17.56 ft.).

Thwart Risers and Seats

The Larkin Boat once had three seats running thwartships across the hull. The seats are all missing, but three of the seat supports, or thwart risers, do remain, and the impressions of both these and the three missing risers are present in the interior surface of the side planks, as mentioned in the previous section. The aft seat was positioned about 1.02 m (3.35 ft.) forward from the stern, the midships seat was about 1.10 m (3.61 ft.) forward from the aft seat, and the forward seat was about 1.04 m (3.41 ft.) forward from the midships seat, and about 1.38 m (4.53 ft.) back from the bow. Fastener patterns suggest that the thwart risers were nailed into their positions from the inside of the boat usually using three nails, and that the seats resting on top of them were secured with an additional 2–3 nails, driven in from the outside. One riser impression (associated with Thwart Riser 18) featuring six nail holes was probably due to the replacement of a broken riser.

The surviving risers, Thwart Risers 10, 18, and 19, measure between 26.7 and 28.8 cm (10.5 and 11.33 in.) in length. These dimensions are probably close to the original seat diameters. They measure between 3.2 and 4.4 cm (1.26 and 1.73 in.) in height and 2.1 and 2.4 cm (0.83 and 0.94 in.) in thickness. The interior sides are beveled so that the thickest part of the riser is directly under the seat it supports.

Transoms

The Larkin Boat appears to be flat-ended, with both a transom stern and bow. Neither of these end pieces has survived, though two unidentified boards (Boards 16 and 17) may represent a transom piece. The stern of the boat is better preserved, and the shape of its transom may be surmised from the deadrise of the bottom planks and the angle of the side planks (Fig. 6.7). There was a rake to the transom piece of about 15°, derived from the angle of the aftermost ends of the side planks. The transom piece was originally set in place so that its bottom rested on the aftermost edge of the bottom planks, with nails driven upwards through the bottom planks to secure it. Additional nails were driven through the aftermost edge of the side planks into the sides of the transom. While the bow end of the vessel is significantly deteriorated, preventing an exact understanding of its original structure, it appears to have featured a transom piece attached in the same manner.

Tool Marks

Saw marks were clearly visible on the interior surface of the bottom planks, and also on the lower surfaces of some of the frames. The marks on the bottom planking were particularly distinctive (Fig. 6.10), with saw kerf marks vertical across the wood, parallel, and regularly spaced. This appears to be the signature of a machine-operated mechanical pit saw (InspectAPedia 2012). The saw marks have a very slight curve to them, which probably reflects "slop" in the machine between up- and down-strokes (Brendan Burke 2013, pers. comm.). This curvature would be noticeably more pronounced if the plank had been cut on a circular saw mill. Mechanically operated pit saws were in use as early as 1840 in industrial areas such as New York, and would likely date to somewhat later in more remote locations such as Florida's Panhandle (InspectAPedia 2012).

The saw marks visible on the lower sided surface of the frames were also clearly visible, and were made not by machine but by hand-sawing. They are characterized by irregularly spaced straight saw kerf cut lines that display intersecting angles marking the up- and down-strokes.

Fig. 6.10 Detail of saw marks visible on interior surface of Bottom Plank B at midships area (*top* of image is outboard). These are believed to have been made by a mechanical pit saw, which would date to sometime after ca. 1840

Fasteners

Hundreds of iron nails were used in the construction of the Larkin Boat. Most are no longer extant, though there are numerous partially preserved fragments still within their original holes. In general, the many visible fastener holes are similar in size and form, tending to be rectangular or ovoid in shape. The few extant nails that were readily observable were identifiable as machine cut nails. Two specimens were extracted from Frame V during the 2002 study. The shorter was removed from the bottom (sided) surface, on the port side from the closest fastener hole to the center of the frame. The nail is incomplete, weighing 0.99 g with a surviving length of 1.9 cm. The dimensions of its shaft, taken near the missing head, are 0.48 by 0.35 cm. It is more or less rectangular in cross section, though the longer sides are slightly convex and the shorter sides are concave. The other specimen collected was almost complete, though it too was missing its head. It was pulled from the port wronghead. It weighs 2.16 g, its surviving length is 4.9 cm, and its shaft measures 0.38 by 0.33 cm near the missing head and 0.31 by 0.29 cm at its tip. Two of its sides are parallel while the other two display a taper from the missing head to the square tip.

The lack of an extant head on either specimen complicates precise dating, despite an extensive understanding of the technical development of cut nails in the nineteenth century. Nail cutting machines in the US were rapidly developed in the late eighteenth century, and cut nails had almost completely displaced hand-wrought nails by around 1815. Cut nails were the most common form of fastener in general use in the United States throughout the nineteenth century until the late 1890s when they were largely supplanted by wire nails (Edwards and Wells 1993:14, 17; Wells 1998; Adams 2002; McCarthy 2005).

In his book *Practical Boat-Building for Amateurs*, first published in England in 1880, Neison includes a chapter describing how to make a punt that is similar in size and design to the Larkin Boat. He states that the best fasteners to use, other than a few iron screws, are either "galvanized iron hammered nails" or the more expensive copper boat nails, with the latter being preferable (Neison et al. 1929:42). He does note that "common iron cut nails, if heated to a dull red heat in the fire, and then allowed to cool gradually, will answer very well. These, however, are not to be recommended when better can be had" (Neison et al. 1929:40).

Paddle

Recovered along with the remains of the boat itself were the remnants of a somewhat primitive paddle (Fig. 6.11). It was in four pieces by the 2002 study, and it is not known if it was similarly broken at the time of its initial discovery or not. While the proximal or upper end of the paddle appears intact, it seems likely that a broken segment of the shaft is missing, as all fragments fit together seamlessly except for the proximal section. The paddle was rather crudely constructed from a flat plank, with little if any rounding done to the shaft. The proximal or upper end, while slightly deteriorated, is flat-edged

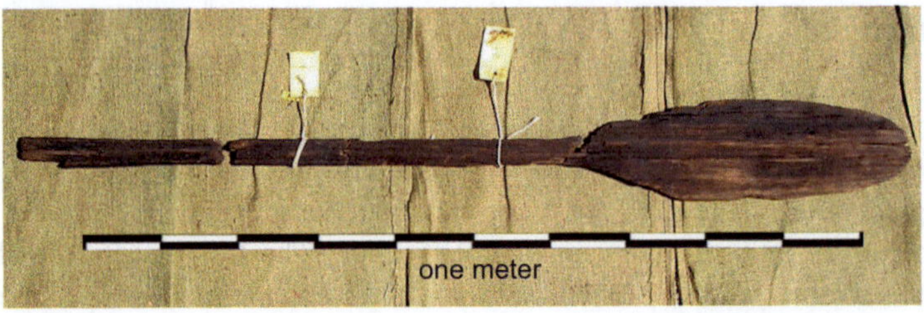

Fig. 6.11 The remains of the paddle recovered with the Larkin Boat. Four pieces have survived, though a segment of the shaft appears to be missing below the uppermost fragment, suggesting that the original paddle was somewhat longer than its surviving length of 117.95 cm (3.87 ft.)

and seems to have been slightly wider than the rest of the shaft, so that it would have fit easily in the hand with the fingers doubled over the top. The overall surviving length (the sum of the extant length of all four paddle fragments) is 117.95 cm (3.87 ft.), the width of the shaft ranges between 3.5 and 3.9 cm (1.4–1.5 in.) and its thickness between 1.45 and 1.7 cm (0.57–0.67 in.), the length of the blade is around 45 cm (17.7 in.), and the width of the blade is 13 cm (5.1 in.). As mentioned above a segment of the shaft is likely missing, so that the original length of the paddle was somewhat longer. According to Stephen's 1884 treatise *Canoe and Boat Building*, a typical paddle was around 5½ ft. (167.64 cm) long, with a blade 5 in. (12.7 cm) across, virtually identical to the blade width of the Larkin Boat paddle (Stephens 1884:68).

Wood Samples

In May 2002 three wood samples collected from the hull of the Larkin Boat were sent to FSU, however that species identification was not completed (Colaninno 2002). The samples were returned to the author at the Lighthouse Archaeological Maritime Program in St. Augustine in 2009. In 2011 the samples were transported to University of West Florida. Due to a back log of such samples, and the lack of funding to prioritize them, they have not yet been analyzed at the time of this publication.

The three wood samples are from a plank (Sample FSU-11), an intercostal keelson piece (Sample FSU-12), and a frame wronghead (FSU-13). The specific hull members these samples originated from remains uncertain.

Discussion and Conclusion

The Larkin Boat is a fascinating example of a unique Florida vernacular watercraft. It is a finely shaped and rather slender punt (Fig. 6.12). With an overall length of 5.32 m (about 17 ft. 5½ in.) and a maximum breadth of 78 cm (about 2 ft. 6¾ in.), this shapely little vessel has a particularly extreme length-to-beam ratio of 1:6.8. This would have been considered quite narrow, given the rule of thumb provided by the 1880 book *Practical Boat-Building for Amateurs*: "Length four and a half times the beam is a very good proportion for a fishing punt" (Neison et al. 1929:41–42).

Upon first glance the boat exemplifies a rather simple design with each side consisting of a single plank, bent in plan view to give a pronounced curve towards the vessel ends. This side plank curvature is such that the boat is about half as wide at the stern as it is midships, and even narrower at the bow. In profile there appears to be just the slightest amount of sheer (this is hard to confirm given the deterioration towards the bow), and the bottom is given considerable rocker fore and aft. While the boat is square-ended, featuring transoms at both ends, it is not symmetrical. The stern transom is wider than that of the bow by about a third. The widest part of the boat is located a bit aft of the actual center of the boat, by about 20 cm. The sides and transoms flare outwards so that the bottom of the boat throughout is narrower than the top. The bottom of the boat, like the sides, is also made up of two planks, but they are angled in a modest deadrise form carried to both ends. This sophisticated hull shape is rarely if ever seen in common punts, and somewhat at odds with the rather primitive paddle associated with the boat. The finely crafted frames equipped with limber holes and the ceiling plank flooring are also features not usually associated with simple flatboats.

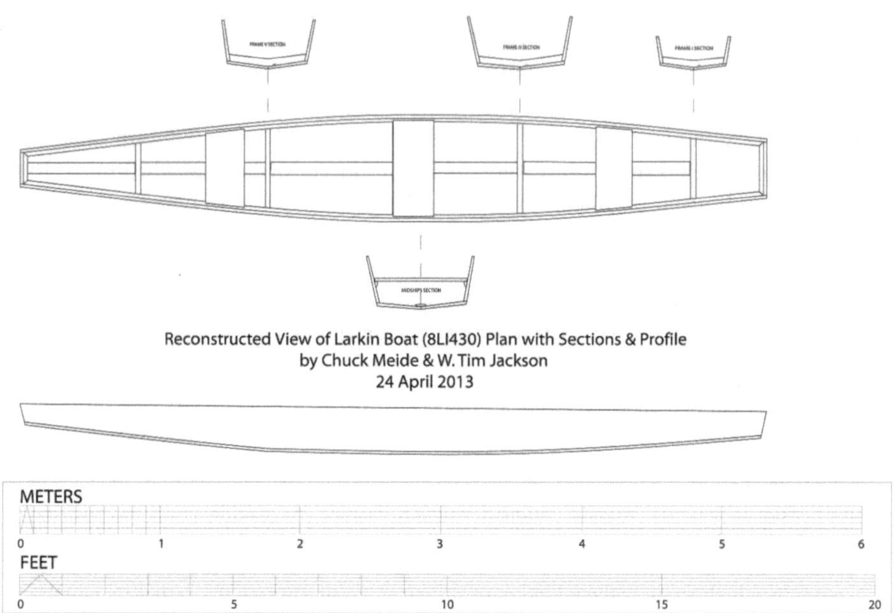

Fig. 6.12 Reconstructed drawings of the Larkin Boat. The ceiling planking has been omitted to show interior details. Stern is to right

It appears that the Larkin Boat pre-dates the introduction of outboard motors, an assessment of its structure that is supported by the two temporally diagnostic features encountered. The mechanized pit saw marks date to sometime after 1840, and the cut iron nails likely date to between the 1820s and the 1890s. It is therefore quite likely that the Larkin Boat dates to the second half of the nineteenth century. "Pirogue" and "bateau" are two terms used during this period across the South for small, manually propelled boats (Alford 1996). The traditional definitions of these boats vary from the punt design seen in the Larkin Boat, however. Both pirogues and bateaux feature flatbottom construction (except for earlier versions of the pirogue which were dugout logboats) instead of a deadrise hull (see also Damour 2016). Pirogues were usually double-ended, and the earliest versions of bateaux were as well, though traditional bateaux or skiffs usually featured a square stern and pointed bow (Stephens 1889:256–257; Fleetwood 1995:146). While the Larkin Boat doesn't fit these criteria it may have been referred to by one of these common names under colloquial usage (Alford 1996). Examples of such variation in vernacular boat nomenclature include the "diamond-bottom" bateau, a deadrise rather than flat bottomed pleasure boat manufactured by a Savannah builder in 1879 (Fleetwood 1995:148), and the Charleston bateau described further below, which has a square rather than pointed bow.

Other than the deadrise bottom, the Larkin Boat displays characteristics most commonly seen in punts. The punt was a variant of the flatboat, a type sometimes called a scow or often simply a flat, which is exactly what it sounds like: a flat-bottomed, square-ended, barge-like hull built for maximum carrying capacity, stability and shallow draft. Flats were the workhorses of the colonial and territorial periods and were regular sights along Florida's waterfronts and inland networks of rivers, marshes, and creeks. Typically, a flatboat would be poled along the shallows or maneuvered by sweeps (oars) while drifting with currents. While their shape made them unsuitable for offshore navigation, flats could be rigged with sails, such as a schooner rig, and sometimes were decked, to increase their seaworthiness. Flats were commonly used on plantations as working vessels and to carry supplies and agricultural products to and from town or waiting ships in the harbor (Leech et al. 1994:41–44; Fleetwood 1995:102, 311–315).

The term punt indicates a rowboat-sized, flat-bottomed vessel, as opposed to a scow or barge which can be used to describe significantly large vessels. The classic punt was a simple flatboat with:

> square ends and straight sides, giving it a rectangular shape in plan view. Its sides are vertical, and its bottom is quite flat athwartships, although fore and aft the bottom will show rocker, the amount varying from boat to boat. It has no sheer curve, being perfectly straight on top. In short, the shape is that of a rectangular box, except for the longitudinal curve of the bottom, which lifts the ends, cutting down on resistance and making it easier to push the punt through the water. Some punts with well-proportioned sled-form bottoms offer surprisingly little-resistance when moving, and, if lightly built, can be brought up to planing speeds by a high-speed outboard motor of sufficient power. (In fact, the extremely popular Boston Whaler is virtually a punt in form, although with secondary modifications which improve its appearance as well as its performance with an outboard motor.) (Gardner 1977:11).

Gardner (1977:12–13) provides the plans for just such a simple punt measuring 14 ft. by 2 ft. 8 in. (4.27 by 0.81 m) that was known as a "Charleston bateau." This working boat was associated with the lumber trade on the Ashley and Cooper Rivers in South Carolina at the turn of the century. Unlike many flatboats which were cross-planked (Stephens 1889:251–253; Leech et al. 1994:43–44; Fleetwood 1995:312; Simmons and Duff 1996:45–54) this example featured lengthwise planking, which offers advantages such as less resistance when sliding over muddy shallows and fewer seams to damage and leak (Gardner 1977:13). A close variant of this vessel representing an early Spanish *barca chata* (literally "flat boat") was built at the LAMP Boatworks at the St. Augustine Lighthouse & Maritime Museum. The author conducted sea trials on this vessel in March 2008 (Fig. 6.13). It was found to be extremely stable (with two in the boat one passenger could literally stand on a side plank without tipping it!) and it paddled and maneuvered much more easily than had been anticipated, and with greater speed. While with its rockered ends it looks very similar to the Larkin Boat when viewed from the side, and is only about 2 in. (5.08 cm) wider, the *barca chata* is somewhat shorter and displays a 1:5.1 length-to-beam ratio. It also has vertical instead of angled or flared sides and transoms, and a flat bottom athwartships with no seats and simple, straight frames without limber holes.

Fig. 6.13 Sea trail of the *bacca chata* (Spanish for "flat boat"), a reconstruction of the Charleston bateau, bulit in 2008 by the LAMP Boatworks in St. Augustime, Florida. Courtesy of the Lighthouse Archaeological Maritime Program and the St. Augustime Lighthouse & Maritime Museum

There are other examples of punts from the historical record that are closer in form to the Larkin Boat. One of these appears in Stephens' 1889 edition of *Canoe and Boat Building*:

> The punt, as it is commonly called, is a scow of rather better design than the one described above, but the operations of building are similar. These boats are often used for fishing on rivers and ponds, as they are roomy, stiff and safe from any danger of capsizing, and the occupants can sit all day in comfort, or move about freely, which cannot be done in a round-bottomed boat of similar size. Such a boat may be 14 to 16 ft. long, 4 ft. beam at gunwale, 3 ft. 4 in. at bottom, and the sides 14 in. deep (Stephens 1889:253).

Stephens' punt is a more shapely craft than the boxy Charleston bateau (Fig. 6.14). Its sides in the plan view are bent inwards, though not to the same degree as the Larkin Boat, as its bow and stern are only about three-quarters the midships beam compared to one half or less. Unlike the Larkin Boat, Stephens' punt is symmetrical, with both bow and stern ends being the same size. Its sides are angled outwards so the top is wider than the bottom, like the Larkin Boat, though the transom pieces are vertical. Viewed from the side, the boat displays a bit of sheer, and its ends sweep upward similarly to the Larkin Boat. While somewhat shorter than the Larkin Boat, it is more than a foot and a half (1.57 ft. or 47.85 cm) wider, resulting in a more conventional length-to-beam ratio of about 1:4. It contains two seats, which along with a series of transverse deck planks at the bow and stern, serve to stiffen the hull. This is critical as there are no frame timbers in this punt. A continuous keelson plank is positioned along the centerline. This is thinner towards the ends to aid in bending it in place, a feature unnecessary with the interspaced keelson pieces on the Larkin Boat (Stephens 1889:253–255). A major difference between the Stephens punt and both the Larkin Boat and Charleston bateau, along with the other examples discussed below, is that the Stephens boat is cross-planked instead of longitudinally planked.

Another similar punt design is presented in a book first published in 1880 in England by Adrian Neison titled *Practical Boat-Building for Amateurs* (Neison et al. 1929:40–53). Neison's punt is just slightly longer than the Larkin Boat and somewhat longer than the Stephens punt at 18 ft. (5.49 m). It measures 4 ft. (1.22 m) in breadth amidships and 3 ft. (0.91 m) at the ends, providing a length-to-beam ratio of 1:4.5, comparable to the Stephens punt. Neison's punt is similar in other ways to Stephens', with rockered ends, symmetrical body, gracefully narrowed ends in plan view, and decking at each end (Fig. 6.14). The most noticeable difference between the two designs is that the British example is equipped with a series of composite frames to provide the "foundation and strength of the punt" (Neison et al. 1929:44). Each frame consists of a bottom frame or floor, and two side frames or knees that are cut to fit the chine of the boat. There are ten composite frames regularly and tightly spaced 15 in. (38.1 cm) apart, a contrast to the six irregularly spaced frames in the Larkin Boat. Neison's punt also featured two longitudinal stringers running along the join between side and bottom planks, to further strengthen the hull. Neison does comment on the robust build of the vessel, noting that it is heavily built by design for its intended purpose:

> In designing a punt, the first thing to decide on is the purpose it is intended for, which, in this case, I will suppose to be for fishing on a quiet river or lake. For shooting in very shallow flats and marshes, a much lighter punt than the following may be designed. A punt for the purposes mentioned here may be heavy, for it is not required to carry it about, comfort, durability, and strength being the first considerations (Neison et al. 1929:40).

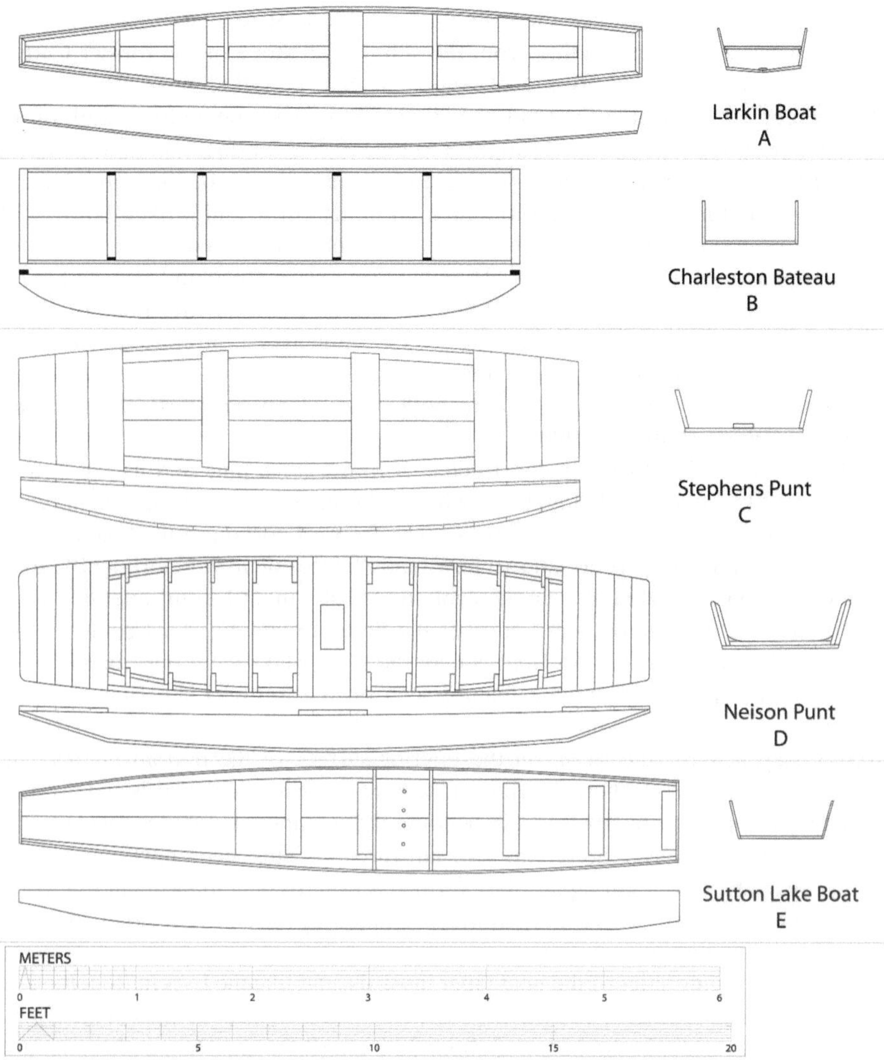

Fig. 6.14 Scaled comparative drawings of the Larkin Boat and other historical punts described in the text. Adapted from Gardner 1977:13; Stephans 1889:254; Neison et al. 1929:41; Lawrence 1984: Figure 3. Digitized by Tim Jackson

The British boat differs from the American longitudinally planked punts in that it does not feature two wide bottom planks, but four narrower ones; this may reflect the scarcity of old growth lumber in England at the time.

There is also one archaeological example of a comparable vernacular punt, the Sutton Lake Boat from North Carolina (Alford 1996). This site was discovered by then-state underwater archaeologist Leslie Bright and recorded by Bright and his colleague Richard Lawrence the following week on 29 August 1984 (Lawrence 1984). A vessel was protruding from the submerged slope of the former Catfish Creek bed in 12 ft. (3.66 m) of water in Sutton Lake, near Wilmington. Archaeologists excavated the muddy sediment to expose the entire vessel, disassembled it on the bottom, and brought it to the surface for documentation. A pole with a hand hewn paddle blade on one end, measuring 11 ft. long and 2 in. in diameter was found nearby. Additionally, an intact bottle and broken bottle with diagnostic embossings were found within the boat, and dated to the late nineteenth or early twentieth centuries. The boat was reassembled and returned to the site afterwards (Lawrence 1984). When the North Carolina Inventory of Small Craft Remains was developed in 1990 by state managers, the Sutton Lake Boat was the only example of its type, categorized as: Plank (slab and post) skiff-built, Flat bottom skiff, Swim head (square both ends), Transom stern, Longitudinal planking (Wilde-Ramsing 1990:13).

The Sutton Lake Boat was constructed of rough sawn pine and its flat bottom was built of two planks held together by internal frames or floors spaced at approximately 2 ft. (61 cm) intervals (Fig. 6.14). The forward portion of the boat had suffered deterioration so it is not known how many frames there were originally. Each side of the boat consisted of a single plank, nailed at 4 in. (10.2 cm) intervals along its lower edge into the outboard edge of the bottom planking. This configuration differs from that of the Larkin Boat, where the side planks were positioned on top of the bottom planks. A unique feature seen on the Sutton Lake Boat was two athwartships sawcuts, about 3/8 in. (0.95 cm) deep towards the bow and 1/4 in. (0.64) deep towards the stern, on the interior surface of the bottom planks. The sawcuts are positioned where the ends of the boat begin to rise upwards and presumably they assisted the boatbuilders in bending these planks (Lawrence 1984).

Another interesting feature was two athwartships bulkheads positioned amidships. Spaced 2 ft. (61 cm) apart, they formed a watertight compartment that could be filled to store fish or live bait, and then drained by way of four holes that contained wooden plugs.

In plan view, the sides of the boat taper inwards, and like the Larkin Boat the bow is narrower than the stern. The Sutton Lake Boat measures about 5.66 m (18.58 ft.) in length, with a maximum beam of about 88.8 cm (2.91 ft.). The resulting length-to-beam ratio is 1:6.4, which is the closest to the Larkin Boat of any of the comparative examples discussed above. Indeed, the two share a very similar look in plan view, though the Larkin Boat features a proportionally narrower stern.

It can be seen that the Larkin Boat bears similarities to many of the examples noted above (Table 6.2). Yet it is also quite distinctive in that it is the only one of these punts to feature a V-shaped rather than a flat bottom. It is the only example with raked transoms, and it displays the greatest narrowing of the hull towards its ends. The Larkin Boat is almost more like a planked canoe than a flat-bottomed punt. No parallel of a square-ended, rockered boat with a v-shaped hull has been identified to date by the author. The deadrise cross section along with the particularly narrow hull suggests that this boat was built, perhaps in part, for speed and maneuverability. These criteria may seem out of place for a vernacular boat found

Table 6.2 Comparison of various punts

	Larkin boat		Charleston bateau		Stephens punt		Neison punt		Sutton lake boat[a]	
Approximate date	ca. 1850s–1890s		ca. 1900		ca. 1889		ca. 1880		Late nineteenth/early twentieth centuries	
Location	Florida		South Carolina		USA		England		North Carolina	
Context	Archaeological		Oral history		Historical		Historical		Archaeological	
Length overall	5.32 m	17.45 ft	4.27 m	14 ft.	4.88	16 ft.	5.49 m	18 ft.	5.66 m	18.58 ft.
Max breadth	0.78 m	2.56 ft	0.81 m	2.67 ft	1.22 m	4 ft.	1.22 m	4 ft.	0.89 m	2.91 ft.
Length to beam	1:6.8		1:5.2		1:4		1:4.5		1:6.4	
Max depth	0.385 m	1.26 m	0.381 m	1.25 ft.	0.36 m	1.17 ft.	0.3048 m	1 ft.	0.34 m	1.10 ft.
Narrowed breadth at ends?	Yes		No		Yes		Yes		Yes	
Symmetrical ends?	No		Yes		Yes		Yes		No	
Stern breadth	0.42 m	1.38 ft.	0.81 m	2.67 ft	0.91 m	3 ft.	0.91 m	3 ft.	0.696 m	2.28 ft.
Bow breadth	0.28 m	0.92 ft.	0.81 m	2.67 ft	0.91 m	3 ft.	0.91 m	3 ft.	0.437 m	1.43 ft.
Breadth ratios, Midships:Stern:Bow	1:1.9:2.8		1:1:1		1:1.3:1.3		1:1.3:1.3		1:1.3:2	
Frames	6 Floors		4 Composite		None		10 Composite		>6 Straight floors	
Bottom shape	Deadrise		Flat bottom		Flat bottom		Flat bottom		Flat bottom	
Bottom planking	Longitudinal		Longitudinal		Cross-planked		Longitudinal		Longitudinal	
Sides angled out	Yes		No		Yes		Yes		Yes	
Sides positioned on top of bottom, or outboard of bottom	Sides on top		Sides outboard		Sides on top		Sides on top		Sides outboard	
Rockered ends	Yes		Yes		Yes		Yes		Yes	
Decked ends	No		No		Yes		Yes		No	
Raked ends	Yes		No		No		Yes		No	

(continued)

Table 6.2 (continued)

	Larkin boat	Charleston bateau	Stephens punt	Neison punt	Sutton lake boat[a]
Seats	3	None	2 Plus decks	1 Plus decks	None
Reference		Gardner 1977:12–13	Stephens 1889:253–255	Neison et al. 1929:40–53	Lawrence 1984

[a]Measurements of the Sutton Lake Boat were measured by the author from a scaled drawing in Lawrence 1984

in a remote, rural area of swamps and sluggish creeks. If this craft had been intended for use as a work boat in the predominant industry of the area, the lumber and naval stores trade, one would expect to see a wider hull in relation to its length, to increase stability. This was the case with the Charleston bateau, which was built for that same industry in South Carolina. On the other hand, just because a boat doesn't conform to the ideal proportions for, say, a fishing punt, as prescribed by a boatbuilding treatise published in London (Neison et al. 1929:41–42), doesn't mean it couldn't be used for fishing. The Sutton Lake Boat, with its hull proportions close to that of the Larkin Boat, is a perfect example; with a live well, it was certainly used for fishing, just as long and narrow aluminum canoes are today.

The quote included earlier from Neison et al. (1929:40) suggesting that a lightly built punt might be perfect for hunting across "very shallow flats and marshes" where one might be frequently moving the boat from one body of water to another, or dragging it across muddy shallows, could be relevant to this discussion. The significantly narrowed ends of the Larkin Boat, which reduced both weight and drag in or out of the water, would lend itself to this kind of activity. The narrow proportions of the boat are also somewhat reminiscent of a craft that appeared in the sixth edition of H.C. Folkard's *The Sailing Boat*, published in 1906. This was a "gunning-punt" built "after the author's invention, suitable both for sailing and paddling in pursuit of wild-fowl" (Gardner 1977:12). Round-bottomed and double-ended, it was not really a punt in the traditional sense, but with its elongated, narrow body it might have handled like the Larkin Boat, which would certainly be suitable for hunting in the environment in which it was found.

We will likely never know to what exact purpose, or purposes, the Larkin Boat was used. It certainly could have been used along creeks and rivers and in wetlands for hunting or fishing. It could also have served as a work boat, though its gracile form was not particularly suited for cargo capacity or stability. It very likely was used for a variety of pursuits on the forested and wetland landscapes surrounding the site of its discovery, as the need arose. Regardless, this is a unique example of Florida's vernacular boatbuilding, and one that does not seem to have a parallel in or outside the state. Even the examples that come closest to it in form are rare in the archaeological record, with the closest, the Sutton Lake Boat from North Carolina, being the only known of its type in that entire state. In this way the Larkin Boat reflects our greater understanding of Florida's historical watercraft: we know they were used extensively, but we have very few examples remaining, and we have only a limited understanding overall of their use, form, regional variation, and general evolution. Piecing together the Larkin Boat, in a small way, has moved us forward in piecing together our understanding of vernacular watercraft and the role they played along Florida waterways over the last 500 years.

Acknowledgments My warmest thanks goes out to US Forest Service archaeologist Dr. Andrea Repp, who was a pleasure to work for, and without whom this unique historical watercraft would still be in pieces, or even lost entirely. Her support for this project was enthusiastic and greatly appreciated. I must also thank Rhonda Kimbrough and all of the other Forest Service staff members who I greatly enjoyed working with, and who helped with various aspects of the recording, reassembly, and transportation of the Larkin Vessel. Mitchell Larkin himself, though I never had a chance to meet him, deserves our gratitude for realizing that this wooden boat was precious and worthy of saving, and for being willing to donate it to the Forest Service with no strings attached. Thanks to my cousin Nik Hartney for his engineering expertise and to the joint FSU and Florida A&M University engineering program for the use of their shop and equipment. Much appreciation goes to Amy Mitchell-Cook for agreeing to conduct the analysis of wood samples from the boat, and to Dr. Cheryl Ward, formerly at FSU, for her initial offer to do so. Thank you to the former conservators at the Florida Bureau of Archaeological Research, Jamie Levy and Claire Tindal, for providing information related to the conservation of the boat. LAMP archaeologist Brian McNamara assisted by scanning numerous drawings so that they could be digitized, which was skillfully accomplished by LAMP volunteer Tim Jackson. LAMP archaeologist Brendan Burke also provided assistance for this publication, by producing the location maps and by lending insight into tool marks and wooden boat construction. Thank you to my family, Brenna Ryan and the Buds, for their patience with me when I spend hours writing. Finally, I must sincerely thank my friend and colleague Amanda Evans, the editor of this volume, for her years of assistance with my first submission to this volume, which was eventually deemed not to fit geographically with the rest of the book, and for her encouragement both to publish that piece elsewhere and to continue my longstanding participation with this book by submitting the present work. It has been many years since our symposium on the archaeology of vernacular watercraft, but thanks to your hard work this book is surely worth the wait.

References

Adams, W. H. (2002). Machine cut nails and wire nails: American production and use for dating 19th-century and early-20th-century sites. *Historical Archaeology, 36*(4), 66–88.

Alford, M. B. (1996). *Letter to Roger Smith, 30 September. Larkin Boat (8LI430)*. Tallahassee, FL: Florida Master Site File.

Bruseth, J. E., & Turner, T. S. (2005). *From a watery grave: The discovery and excavation of La Salle's shipwreck, La Belle*. College Station: Texas A&M University Press.

Colaninno, A. (1996). *Letter to Karl P. Siderits, 30 July. U.S. Forest Service File Code 6400. Larkin Boat (8LI430)*. Tallahassee, FL: Florida Master Site File.

Colaninno, A. (2002). *Letter to Cheryl Ward, 2 May. Larkin Boat (8LI430), Meide papers*. St. Augustine, FL: Lighthouse Archaeological Maritime Program.

Damour, M. (2016). Born on the bayou: Louisiana's vernacular constructed watercraft. In A. M. Evans (Ed.), *The archaeology of vernacular watercraft* (pp. xx–xx). New York: Springer.

Edwards, J. D., & Wells, T. (1993). *Historic Louisiana nails: Aids to the dating of old buildings*. Baton Rouge, LA: Geoscience Publications, Department of Geography and Anthropology, Louisiana State University.

Fleetwood, W. C., Jr. (1995). *Tidecraft: The boats of South Carolina, Georgia and Northeastern Florida, 1550–1950*. Tybee Island, GA: WBG Marine Press.

Gardner, J. (1977). *Building classic small craft*. Camden, ME: International Marine.

Greenhill, B., & Morrison, J. (1995). *The archaeology of boats and ships: An introduction*. Annapolis, MD: Naval Institute Press.

Grenier, R., Marc, A. B., & Willis, S. (Eds.). (2007). *The underwater archaeology of red bay: Basque shipbuilding and whaling in the 16th century* (Vol. 5). Ottawa, Canada: Parks Canada.

Guilminota, E., Dalarda, F., & Degrignyb, C. (2002). Mechanism of iron corrosion in water–polyethylene glycol (PEG 400) mixtures. *Corrosion Science, 44*(10), 2199–2208.

Hamilton, D. L. (2010, March 1). *Methods of conserving archaeological material from underwater sites*. Conservation files: ANTH 605, Conservation of Cultural Resources I. Texas A&M University, Nautical Archaeology Program, Conservation Research Laboratory. Retrieved April 30, 2010, from http://nautarch.tamu.edu/crl/conservationmanual/.

Harris, R. (2006). *Chalupa No. 2: A comparative study of 16th century basque whaleboat construction from examples excavated at Red Bay, Labrador.* Unpublished Masters Thesis, Program in Maritime Studies, East Carolina University, Greenville, NC.

InspectAPedia. (2012). *How adze, axe, or saw cut marks on lumber indicate building age and wood cutting methods.* Poughkeepsie, NY: InspectAPedia. Retrieved from http://www.inspectapedia.com/structure/Lumber_Cutting_Methods.php.

James, F. C. (2006). An introduction to the history, ecology, and management of the Apalachicola national forest. *Friends of the Apalachicola National Forest, 1*(1), 1–6. Retrieved from http://bio.fsu.edu/FANF/Fact%20Sheet%201%220-%220ANF%220basics.pdf.

Joslin, R. C. (1996). *Letter to Karl P. Siderits, 10 August. U.S. Forest Service File Code 6400. Larkin Boat (8LI430).* Tallahassee, FL: Florida Master Site File.

Larkin, M. (1996). *Letter to Karl Siderits, 10 July. Larkin Boat (8LI430).* Tallahassee, FL: Florida Master Site File.

Lawrence, R. W. (1984, September 4). *Sutton lake site. Memorandum to underwater archaeology unit files.* North Carolina Underwater Archaeology Unit, Division of Archives and History, Department of Cultural Resources, Kure Beach, NC.

Leech, R. W., Jr., Wood, J. L., Cook, G., Bowling, K., & Pevny, T. (1994). *Archival research, archaeological survey, and site monitoring: Black river, Chatham county, Georgia, and Jasper county, South Carolina.* Savannah, GA: U.S. Army Corps of Engineers.

Lipke, P., Spectre, P., & Fuller, B. A. G. (Eds.). (1993). *Boats: A manual for their documentation.* Nashville, TN: Museum of Small Craft Association, American Association for State and Local History.

McCarthy, M. (2005). *Ships' fastenings: From sewn boat to steamship.* College Station: Texas A&M University Press.

Meide, C. (1995). *The dugout canoe in the Americas: An archaeological, ethnohistorical, and structural overview.* Tallahassee: Program in Underwater Archaeology, Florida State University.

Meide, C. (2002a). *Proposal for recording and reassembling small boat for display, 29 January, manuscript on file at the Wakulla work station.* Wakulla, FL: Apalachicola National Forest.

Meide, C. (2002b, May 3–4). *Some assembly required: The discovery, conservation, and reassembly of the probable 19th century Larkin boat (8LI430), Liberty County, Apalachicola National Forest, Florida.* Paper presented at the 54th Annual Meeting of the Florida Anthropological Society, St. Petersburg, FL.

Miller, J. J. (1996). *Email to Roger C. Smith, 14 July. Larkin Boat (8LI430).* Tallahassee, FL: Florida Master Site File.

Neison, A., Kemp, D., & Cowper, F. (1929). *Practical boat-building for amateurs: Containing full illustrations for designing and building punts, skiffs, canoes, sailing boats, etc.* (4th ed.). London: The Bazaar, Exchange, & Mart.

Ontario Service Centre of Parks Canada. (2009). *Archaeology of a sixteenth-century Basque whaling boat, Ontario, Canada.* Retrieved from http://www.pc.gc.ca/lhn-nhs/nl/redbay/natcul/basque.aspx.

Purdy, B. A. (1991). Florida canoes: A maritime heritage from the past. In B. A. Purdy (Ed.), *The art and archaeology of Florida's wetlands* (pp. 265–284). Boca Raton, FL: CRC Press.

Siderits, K. P. (1996). *Letter to Robert C. Joslin, 1 August. U.S. Forest Service File Code 6400. Larkin Boat (8LI430).* Tallahassee, FL: Florida Master Site File.

Simmons, J. J. III., & Duff, J. A. (1996). *Phase II archaeological data recovery, area 4, Fig island channel site, Savannah harbor, Savannah, Georgia.* Report prepared by Panamerican Consultants, Inc., and submitted to the U.S. Army Corps of Engineers, Savannah District, and Gulf Engineers & Consultants.

Smith, R. C. (1996). *Letter to Michael B. Alford, 28 October. Larkin Boat (8LI430).* Tallahassee, FL: Florida Master Site File.

Smith, R. C. (1999). *Letter to David Wyman, 31 March. Larkin Boat (8LI430).* Tallahassee, FL: Florida Master Site File.

Steffy, J. R. (1994). *Wooden ship building and the interpretation of shipwrecks.* College Station: Texas A&M University Press.

Stephens, W. P. (1884). *Canoe and boat building.* New York: Forest and Stream.

Stephens, W. P. (1889). *Canoe and boat building: A complete manual for amateurs* (4th ed.). New York: Forest and Stream.
Tonsmeire, D. (1996a). *Letter to James Miller, 22 August. Larkin Boat (8LI430)*. Tallahassee, FL: Florida Master Site File.
Tonsmeire, D. (1996b). *Letter to Andrea Repp, 22 August. Larkin Boat (8LI430)*. Tallahassee, FL: Florida Master Site File.
Waddell, P. J. A. (1986). The disassembly of a 16th century galleon. *International Journal of Nautical Archaeology, 15*(2), 137–148.
Wells, T. (1998). Nail chronology: The use of technologically derived features. *Historical Archaeology, 32*(2), 78–99.
Wheeler, R. J., Miller, J. J., McGee, R. M., Ruhl, D., Swann, B., & Memory, M. (2003). Archaic period canoes from Newnans Lake, Florida. *American Antiquity, 68*(3), 533–551.
Wilde-Ramsing, M. U. (1990). *Management of small craft (notes from the North Carolina Maritime Workshops I-VI)*. North Carolina Underwater Archaeology Unit, Division of Archives and History, Department of Cultural Resources, Kure Beach, NC.
Wilde-Ramsing, M. U., & Alford, M. B. (1990). *North Carolina small craft historical context: An underwater archaeology unit management plan*. North Carolina Underwater Archaeology Unit, Division of Archives and History, Department of Cultural Resources, Kure Beach, NC.

Chapter 7
Born on the Bayou: Louisiana's Vernacular Constructed Watercraft

Melanie Damour

Introduction

In his 1985 article "Folk Boats of Louisiana," Comeaux (1985:162) stated that, "changes are molded by people to fit their ideals, needs, and knowledge, and in these changes there is continuity, as the new ways are superimposed on the old. In this manner some boat types are abandoned while others evolve and change." Changes in vernacular constructed watercraft are often prompted by cultural influences, the introduction of new technologies and techniques, the physical environment, available construction materials, and exploitation of various resources. Over a span of three centuries, Native American, French, Spanish, English, Italian, Cajun, *Isleño* (Canary Islanders), and German traditions of boatbuilding influenced the evolution of Louisiana's vernacular constructed watercraft (Saltus 1988:31; Brassieur 1989; Brassieur 2003d, k). Later influences have been traced to other regional American and even Croatian origins. With the influx of foreign immigrants, new techniques for boatbuilding or design can be introduced. New technologies also play a role in the evolution of boatbuilding traditions, from hollowing out logs or fastening wooden planks on an intricate framework to the use of marine plywood and steel hulls.

The physical environment of southern Louisiana, exemplified by its network of narrow streams, bayous, rivers, lakes, marshes, and shallow coastal bays, also influenced the evolution of boatbuilding. Flat-bottomed or shallow drafted boats are better suited to this environment than the deeper drafted, oceangoing ships. Availability of local resources and cost can affect the design and construction of

M. Damour (✉)
Independent Researcher, Mandeville, LA 70448, USA
e-mail: melaniedamour@netscape.net

A.M. Evans (ed.), *The Archaeology of Vernacular Watercraft*,
When the Land Meets the Sea, DOI 10.1007/978-1-4939-3563-5_7

desired watercraft forms, perhaps leading to the abandonment of one form of vessel for another. There are more than 150 species of trees native to Louisiana though the types that are relevant to boatbuilding include live oak, pine, and cypress, which typically grow in lowland swamp areas (Davis 1971:10). Cypress was especially valued for boatbuilding as it was abundant and resisted rot. Louisiana has historically offered a wide range of other resources for commercial exploitation; these included furs, hides, foodstuffs, tobacco, lumber, sugar, coffee, cotton, seafood, and, in the twentieth century, oil and natural gas (Davis 1971). In order to participate in the emerging economy, vernacular watercraft were often modified to suit the needs of a particular industry, such as structural modifications to a barge for lumber transport or the forward placement of a cabin to create more deck space for collecting and transporting seafood.

While historical records recounting the loss of ships before the nineteenth century are usually obtainable, documentation of the loss of vernacular constructed watercraft rarely exists as they were often unregistered. These vessels were deposited into the archaeological record through abandonment, accidental loss, or foundering though few were ever reported. Due to the lack of available historical documentation describing the loss of vernacular watercraft, archaeologists must turn to historical accounts, contemporaneous paintings and sketches, oral traditions, and the archaeological record in order to examine Louisiana's vernacular watercraft heritage.

Watercraft typologies are difficult to create, especially for vernacular constructed watercraft due to the nature of changes in form over time, local or regional variations or use of a name, and scarcity of archaeological examples. Regional variations of a single form can lead to multiple names for nearly identical boats. Conversely, the same name can be used in two different regions to describe two different forms of watercraft. Typically, archaeologists categorize boat types by form, function, and propulsion. As Saltus (1987, 1988) described, forms of vernacular watercraft include canoes, rafts, keelboats, and flatboats; those classified by function include coal barges, cattle barges, ferries, packets, and mail boats; and boats classified by propulsion include schooners, sloops, steamboats, and rowboats. These "types" of vernacular craft can evolve over time within one or more of the three categories thereby making it difficult to develop a rigid taxonomy. For the purpose of this article to describe general characterizations of vernacular constructed watercraft in Louisiana, a broad classification based on form will be used. The primary forms of Louisiana's vernacular constructed watercraft include pirogues and canoes, rafts, flat-bottomed craft, keeled craft, sailing craft, and motorized craft (Table 7.1).

Table 7.1 Louisiana's vernacular constructed watercraft forms, hulls shapes, waterways used, general function, and propulsion

Form	Hull shape	Waterway used	General function	Propulsion
Pirogue/canoe				
Pirogue	Round bottom	Inland	Personal/hunting/fishing	Paddle/row
Pirogue	Flat bottom	Inland	Small cargo	Paddle/row
Planked pirogue	Flat bottom	Inland	Small cargo	Paddle/pole
Canoe	Flat bottom	Inland	Small cargo	Paddle/sail
Raft/skin craft				
Raft	Flat bottom	Inland	Cargo/logs/sold at destination	Pole/sail
Bundle craft (cajeux)	Flat bottom	Inland	Small cargo/personal possessions	Swimmer
Keelless craft				
Bateau	Flat bottom	Inland	Cargo	Row/pole/sail
Chaland	Flat bottom	Inland	Ferry people/cargo	Paddle
Radeau	Flat bottom	Inland	Bulk cargo	Pole
Scow	Flat bottom	Inland	Ferry people/cargo	Pole/paddle/row/drift
Flatboat	Flat bottom	Inland	Cargo/sold at destination	Paddle/row/drift
Barge	Flat bottom	Inland	Bulk cargo	Tow
Skiff	Flat bottom	Inland	Cargo	Row/sail/cordelling
Yawl	Flat bottom	Inland	Service boat	Row/sail
Keeled craft				
Keelboat	Keeled	Inland	Bulk cargo	Row/pole/sail/drift/cordelling
Barge	Keeled	Inland	Cargo	Sail/pole/cordelling
Longboat	Keeled	Inland/coastal	Service boat	Sail/row
Launch	Keeled	Inland/coastal	Service boat	Row/sail
Chaloup	Keeled	Coastal	Cargo/service boat	Sail/row
Pinnace	Keeled	Coastal	Cargo/service boat	Sail/row
Yawl	Keeled	Coastal	Pilot boat/service boat	Sail/row
Cutter	Keeled	Coastal	Cargo/multipurpose	Sail
Felouque	Keeled	Coastal	Service boat	Row/sail
Sailing/motorized craft				
Schooner	Keeled	Inland/coastal/gulf	Cargo/fishing/multipurpose	Sail/motor
Sloop	Keeled	Inland/coastal/gulf	Multipurpose	Sail/motor
Lugger	Keeled	Coastal/gulf	Fishing/shrimping/oystering/workboat	Sail/motor
Trawler	Keeled	Coastal/gulf	Fishing/shrimping	Motor
Lafitte/Atchafalaya skiff	Keeled	Coastal/gulf	Fishing/shrimping/recreation	Motor

Pirogues and Canoes

Pirogue

Perhaps the earliest watercraft used by prehistoric inhabitants and utilized into the modern era is the pirogue, or dugout canoe (Fig. 7.1). The pirogue is a long, narrow, double-ended craft that can be paddled or rowed. Pirogues exhibiting a round bottom were intended for rapid transport while flat-bottomed versions provided greater stability and capacity for carrying cargo (Bremer 1907:12 in Saltus 1987:60). Other names used to describe pirogues include *perogues, periagua, bacassa, kouliala, couliala, barrackas,* and *kanoa* (Table 7.2) (Saltus 1988:38; Edwards and Pecquet du Bellay de Verton 2004:222–223).

Upon felling a tree, usually cypress, fire and hand tools, such as shells, stone scrapers, or adzes, were used to hollow out the log. Seventeenth-century French and Spanish explorers recognized the importance of these vessels to local Native American groups and adopted their own modified forms (Saltus 1988:38). The process of building a dugout was described by Andre Penicaut, a ship's carpenter accompanying the French Canadian explorer Pierre Le Moyne de Iberville's expedition in 1699. Penicaut observed:

> To make these they kept a fire burning at the foot of a tree called cypress until the fire burned through the trunk and the tree fell; next, they put fire on top of the fallen tree at the length they wished to make their boat. When the tree had burned down to the thickness they

Fig. 7.1 Malcolm Comeaux poses next to a pirogue or dugout canoe (Brassieur 2003l)

Table 7.2 Louisiana's vernacular constructed watercraft forms and alternate names

Form	Alternate names
Pirogue/canoe	
Pirogue	Dugout canoe, *perogues, periagua, bacassa, kouliala, couliala, barrackas, kanoa*
Planked pirogue	*Pirogue en planche, penuche*
Canoe	Bark canoe
Raft/skin craft	
Raft	
Bundle craft	*Cajeux*
Keelless craft	
Bateau	Skiff, john boat, joe boat, launch, put–put, gas boat
Chaland	Barge, plank boat, punt, scow, *chaland a boeufs*
Radeau	Raft, flatboat, *bateau plat*
Scow	Flatboat, scow barge, scow boat, scow schooner, *chaland, radeau,* barge, flat, pirogue
Flatboat	Ark, barge, broadhorn, *bateau de cent, bateau plat,* ferry, wharf boat, quarterboat, New Orleans boat, *chaland*
Barge	Flatboat
Skiff	*Esquiff, peniche, chaloupe, galere,* Creole skiff, Mississippi skiff, lake skiff, Lafitte skiff, Atchafalaya skiff, *canotte*
Yawl	Skiff
Keeled craft	
Keelboat	*Bercha,* berge, barge, *berchita*
Barge	*Bercha, lanchon,* wherry
Longboat	*Bateau plat, chaloup*
Launch	*Chaloup, lancha*
Chaloup	Launch, shallop
Pinnace	
Yawl	
Cutter	
Felouque	*Felucca, falua*
Sailing/motorized craft	
Schooner	*Goelette, goleta,* pilot schooner, trading schooner, fishing schooner, packet schooner, scow schooners, barge schooner, pungy schooner, file bottom schooner, ram schooner, centerboard schooner
Sloop	Sloop-of-war, ship-sloop, brig-sloop, corvette
Lugger	*Canot, canotte*
Trawler	South Atlantic trawler, shrimp trawler, Florida-type trawlers, *floridiane,* South Lafourche trawler, trawling skiff
Skiff	Lafitte skiff, Atchafalaya skiff

wanted for the depth of the boat, they put out the fire with thick mud; then they scraped the tree with big cockle shells as thick as a mans [sic] finger; afterward, they washed it with water. Then they cleared it out as smooth as we could have made it with our tools. These boats may be twenty-five feet long. The savages make them of various lengths, some much smaller than others. With these they go hunting and fishing with their families and go to war or wherever they want to go (Pearson et al. 1989:72 quoting McWilliams 1953:8–9).

While indigenous pirogues exhibited blunt ends and thick hulls, French boatbuilders fashioned theirs with pointed ends and thinner hulls to reduce their weight for portage and paddling. A dugout could be split in half lengthwise and a plank inserted between them to increase its beam for greater cargo carrying capacity (Saltus 1988:38). The method of construction and tools used for building pirogues varied little over the centuries; however, size did change with early pirogues exhibiting lengths of 40 ft. (12.2 m) or more with a wide beam to later pirogues that were generally shorter with a narrower beam (Comeaux 1985:164). Pearson et al. (1989:71) reproduced a segment of Garcilaso de la Vega's 1543 account of an attack by a flotilla of "Indian canoes," which described dugout pirogues with a capacity for holding 75–80 passengers. After the early twentieth century, the number of pirogues considerably declined, appearing only in the southern coastal areas and swamps. Their decline in numbers is attributable to a reduction in popularity for transportation of bulk goods in a growing economy and a decreasing supply due to the value of available cypress logs for the lumber industry (Brassieur 2003l). The dugout pirogue was supplanted by technological innovations in boatbuilding which allowed for boats constructed of wooden planks, and later, marine plywood, fiberglass, or aluminum (Brassieur 2003m).

Planked Pirogue

A variation of the dugout pirogue is the planked pirogue, also known to Cajuns as *pirogue en planche* or *péniche* (Pearson and Saltus 1991:100; Edwards and Pecquet du Bellay de Verton 2004:222). Usually constructed of cypress planks, the planked pirogue has a wider bow than stern, flat bottom, slightly flared gunwales that curve toward the front and rear, small fillets to reinforce the bow and stern, and two thwarts, one of which also functions as a seat (Comeaux 1985:166). Of a similar size to the later dugout pirogues, planked pirogues are propelled by paddle or pole throughout the swamps and marshlands of Louisiana and could even be outfitted with a small sail (Brassieur 1989). Appearing in the early nineteenth century, planked pirogues are still used to this day by fishermen, swampers, and hunters, and the form has remained relatively unchanged for centuries (Pearson and Saltus 1991:100; Brassieur 2003f). The term "pirogue" has also been used interchangeably in the vernacular to describe other hull shapes such as scows, punts, and shallops (Saltus 1987:60).

Pirogues have been documented in the archaeological record. A section of a flat-bottomed pirogue or punt was recorded in the Natalbany River in Livingston Parish (Fig. 7.2). This craft measured 13.2 ft. (4.02 m) in length by 3.1 ft. (0.94 m) in beam with a 1-ft. (30.5 cm) depth and was estimated to have an original length of 17.75 ft. (5.41 m) (Saltus 1987:64). Another pirogue, a modern variety constructed of plywood and brass screws and clenching nails, was discovered in Bedico Creek near the Tangipahoa River in Tangipahoa Parish (Fig. 7.2). This craft measured 11.75 ft. (3.58 m) in length, 2.3 ft. (70 cm) in beam, and 10.5 in. (24.4 cm) in depth and was fitted with a small maststep (Saltus 1988:124). Pearson et al. (1989) recorded a 27-ft. (8.23 m) portion of a dugout found partially exposed in marshland on the north shore of Lake Salvadore in St. Charles Parish (Fig. 7.3). The vessel had a 2-ft.

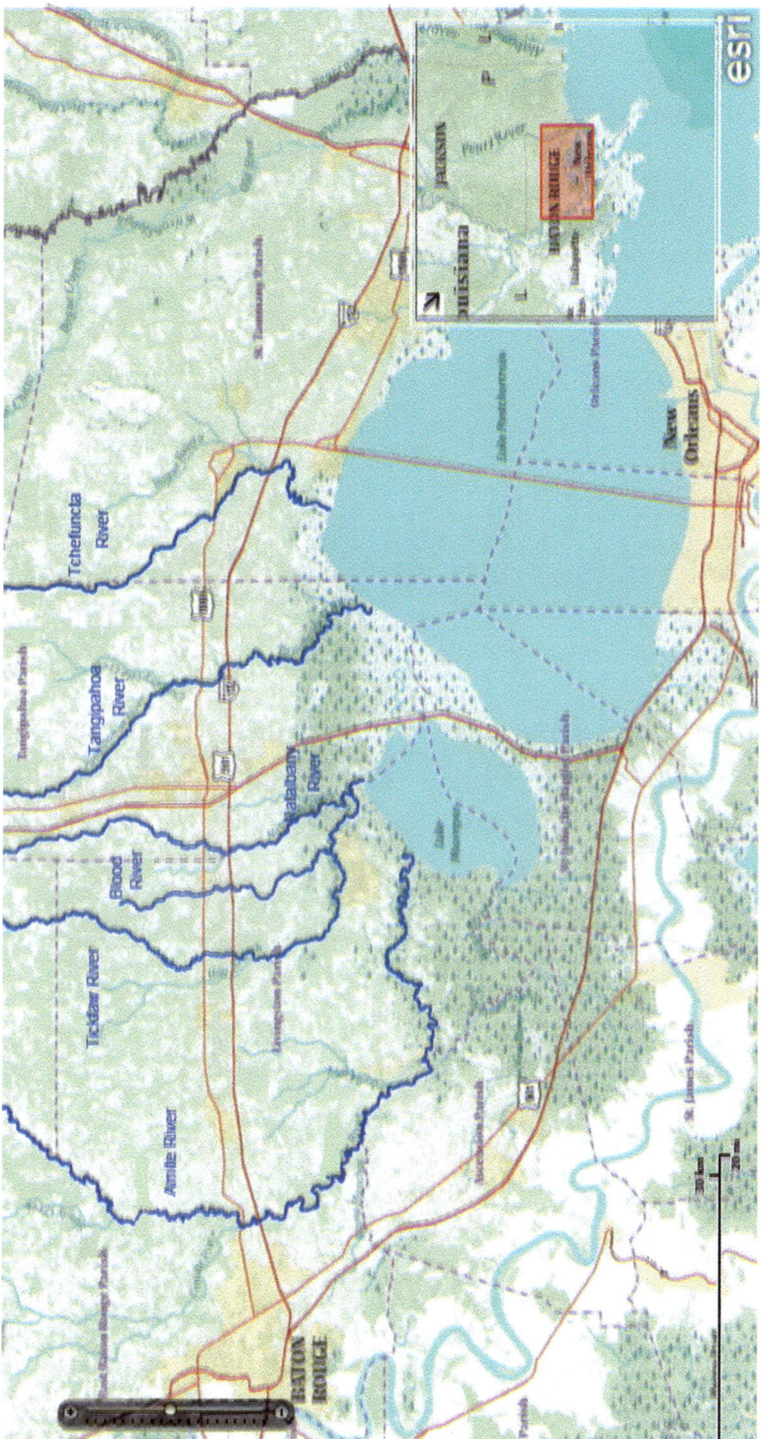

Fig. 7.2 Rivers in the Lake Pontchartrain and Lake Maurepas basins, southeast Louisiana

Fig. 7.3 Atchafalaya River, Atchafalaya basin, and Maurepas basin in southeast Louisiana

beam and a flat bottom. Due to the lack of an intact sample, the overall length of the vessel could not be determined (Pearson et al. 1989:72). Another dugout pirogue, measuring 12.5 ft. (3.81 m) long, 18 in. (45.7 cm) wide, with a 7-in. (17.8 cm) depth of hold was recorded at Fluker's Buff in the Amite River, St. Helena Parish (Fig. 7.2). Pearson et al. reported that a radiocarbon sample from this vessel returned a date of "A.D. 1222" (1989:72). These craft were discovered in the waterways of southeastern Louisiana and demonstrate the longevity of the pirogue form into the twentieth century.

Canoe

Though the term "canoe" can also be used to describe pirogues (e.g., dugout canoe), here the term refers to a type of skin craft with a concave bottom as opposed to the flat-bottomed pirogue used by Native Americans. This form of watercraft was the first to be used by Europeans in the seventeenth century and proved advantageous when exploring Louisiana's circuitous network of waterways (Saltus 1988:41). Saltus described skin craft construction, in general, as consisting of a wooden framework covered with either bark or animal skins (Saltus 1988:41). Bark canoes and other skin craft were paddled, had no rudder, and could be sailed using a small sail crafted from birch bark and upright poles (Surrey 1916). In the Maurepas and Pontchartrain basins, bark canoes represent the only type of skin craft employed during early European exploration of southern Louisiana (Fig. 7.2) (Saltus 1988:40). Iberville's journals from 1698 to 1702 refer to the use of pirogues and canoes during their explorations of the Louisiana interior (McWilliams 1981). Due to the abbreviated period of time that skin craft appear to have been used in this region, they have a low probability of discovery in the archaeological record.

Rafts

Raft

Rafts were simply constructed floating platforms of cane, logs, or reeds fastened together (Comeaux 1985:172). Rafts were generally built for temporary use, such as crossing waterways or transporting a small cargo. They could also be constructed of the logs that were transported downriver and intended for sale (Saltus 1988:42). Usually propelled by poling, rafts could also be fitted with a small lean-to for shelter or a sail when winds were favorable. Due to their often short-term existence and dismantling upon arrival at their intended destination, rafts are not commonly found in the archaeological record. Saltus, however, reported an archaeological investigation of a log raft found on the Natalbany River in Springfield, Louisiana (Fig. 7.2). This craft was constructed with cypress pins, called wooden dogs, driven into sycamore timbers laid across the logs (Saltus 1988:42). Later modifications to raft construction employed metal chain dogs or ring dogs strung together with cable in place of wooden dogs for fastening the timbers to the logs (Saltus 1988:42). The use of wooden dogs generally dates to the first half of the nineteenth century while chain dogs date to the second half of the nineteenth century (Saltus 1987:187).

Bundle Craft

Bundle craft, or *cajeux*, are a smaller variety of raft and were constructed using bound cane (Saltus 1988:41; Edwards and Pecquet du Bellay de Verton 2004:223). These vessels were very small and were used to transport small cargo or personal belongings rather than passengers across waterways. In 1700, the French Canadian explorer and brother of Iberville, Jean-Baptiste Le Moyne de Bienville, described the construction and use of *cajeux* to carry their baggage while they swam behind the craft, pushing it to the other side of the Mississippi River (Pearson et al. 1989:81). Even though they were simple to make, bundle craft never became a popular mode of transportation and, as such, are unlikely to be found in the archaeological record.

Keelless, Flat-Bottomed Craft

Flat-bottomed boats are ideally suited for navigation of Louisiana's shallow and narrow waterways. The early varieties developed alongside colonial-built pirogues and became dependable watercraft forms for movement of goods and passengers. Many terms are used to describe these watercraft such as skiff, *bateau*, flatboat, and barge, and one term can be used to describe markedly different forms of other watercraft (Damour et al. 2005:86). The term "flatboat" is used in general to describe any flat-bottomed craft though it also refers to a specific hull form. Other forms of flat-bottomed craft include the *chaland*, *radeau*, scow, and yawl (Saltus 1988; Edwards and Pecquet du Bellay de Verton 2004:222–223).

Bateau

Bateau, French for "boat," is a Cajun vernacular term applied to more than one form of vessel. The term has been used interchangeably with "skiff" to describe any small, cargo-carrying, flat-bottomed craft, leading to much difficulty when attempting to define vessel types for archaeological purposes (Pearson et al. 1989:90; Edwards and Pecquet du Bellay de Verton 2004:222; Damour et al. 2005:88). The eighteenth-century variety of *bateau* appeared as a flat-bottomed, double-ended craft with a sharply tapered bow and stern, ranging in length from 12 to 80 ft.

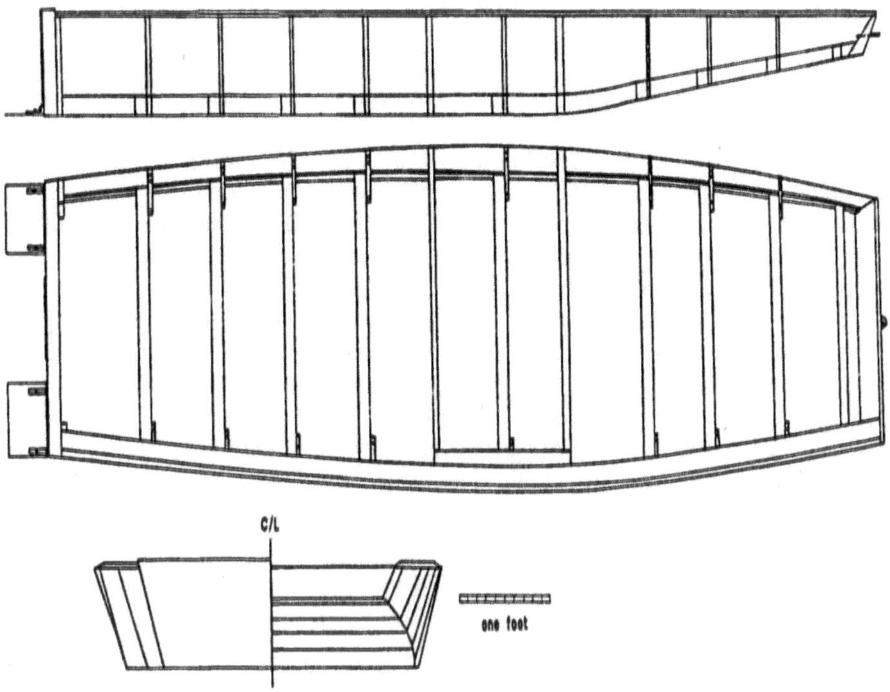

Fig. 7.4 Lines drawing of modern *bateau* (Pearson and Saltus 1991:98)

(3.6–24.4 m) though usually constructed between 20 and 40 ft. (6.1–12.2 m) long (Birchett et al. 2001:52). Saltus described this form as round-chined, distinctive from the square-chined *chalands*, scows, *radeaux*, flatboats, and barges, and the angular-chined planked pirogues, skiffs, and yawls (Saltus 1988:44). The early form of *bateau* was larger than a canoe and had a greater carrying capacity than the pirogue (Pearson et al. 1989:80). It was rowed, poled, or sailed on rivers and lakes and likely evolved from the early flatboat appearing in the eighteenth century (Pearson et al. 1989:249; Birchett et al. 2001:115).

The term *bateau* is used in the modern context to describe a vessel measuring 15 ft. (4.6 m) or more in length and at least 5 ft. (1.5 m) wide. Modern *bateau* are flat-bottomed vessels with a blunt bow and stern and forward sheer (Fig. 7.4). Also referred to as john boat, joe boat, launch, put-put, or gas boat, these vessels are partially decked fore, aft, and along the sides, creating an open space in the middle (Pearson et al. 1989:249; Edwards and Pecquet du Bellay de Verton 2004:222–223). Larger forms may have a cabin and can be virtually identical to "flatboats." They are propelled by inboard motors and constructed of aluminum or fiberglass (Birchett et al. 2001:115). Due to their inability to plane at speed, caused by their slender but heavy form, *bateaux* began to fade in popularity after the middle of the twentieth century (Comeaux 1985:170). These vessels have all but disappeared from the modern fleets of vernacular watercraft though the term remains in use (Brassieur 2003e; Edwards and Pecquet du Bellay de Verton 2004:222–223).

Chaland

Chaland, French for "barge," is a flat-bottomed, square-chined, rectangular craft with sharp, angular, upward-slanting ends, and no sheer. These vessels generally measured 10–14 ft. (3.04–4.26 m) in length and resembled the barge form (Pearson and Saltus 1991:26). Typically operated as ferries and propelled by paddling, *chalands* transported people and goods short distances, such as across a river (Pearson et al. 1989:248). A variation of the *chaland* is the "plank boat," typically used for logging. This vessel had a narrower beam than the *chaland*, at less than 2 ft. (61 cm) wide (Pearson et al. 1989:248). The *chaland* form is considered a primitive type of flatboat and likely evolved from the early form of barge in French Louisiana. English speakers refer to this same vessel form as a punt or scow (Brassieur 1989). Another form of *chaland*, the *chaland a boeufs*, appeared as a much larger flatboat with a cabin and was used for transporting cattle (Birchett et al. 2001:115). The heyday of the *chaland* occurred in the late eighteenth and early nineteenth centuries, after which its popularity diminished due to changes in technology and desired watercraft forms (Birchett et al. 2001:52).

Radeau

The *radeau*, a French term meaning "raft," was a square-chined vessel used by eighteenth-century explorers on the Mississippi River and its tributaries. This craft, resembling the modern flatboat, was used primarily for transporting bulk cargo and freight (Saltus 1988:48). One early eighteenth-century example measured 40 ft. (12.19 m) long, 9 ft. (2.74 m) wide, and 4 ft. (1.21 m) deep (Rowland and Sanders 1927:348). Other terms applied to this form of craft included flatboat and *bateau plat* (Edwards and Pecquet du Bellay de Verton 2004:223). Nancy Surrey, however, noted that the early eighteenth-century vessel referred to as *bateau plat* was "not a 'flatboat' of the type which became common on the Mississippi River in later years" (Surrey 1916:60). It "had a sharp bow and stern and was of light draft and narrow beam. It was made of several pieces of timber with a broad flat bottom, was larger than a canoe and of greater capacity than the large pirogue" (Surrey 1916:61).

Scow

Scow is a term generally describing a boxy hull form and has been applied as a descriptor to watercraft such as scow barges, scow boats, and even scow schooners. The term "scow" was used in exchange with "flatboat" to describe a category of vessels with flat bottoms and square ends such as *chalands*, *radeaux*, barges, and flats (Krause et al. 2004:165; Edwards and Pecquet du Bellay de Verton 2004:223). Saltus (1988:49), quoting Chapelle (1951), described a scow as "a rectangular box with narrow ends sloping outward from the bottom, usually with shoal upright transoms or end timbers finishing off these sloping ends on top." Chapelle (1951:45) further expounded, "the scow was of elementary form… rough, simple, and undistinguished in character and appearance." Used to ferry passengers and cargo across and along waterways, the scow was propelled with poles, paddles, oars, or by following the currents. Some barges are merely a variation of the scow with their sides extending upward (Saltus 1988:49). Saltus recorded a scow in Springfield, Louisiana (site 16LV66), which measured 10.3 ft. (3.13 m) in length by 3.3 ft. (1 m) in beam by 1.5 ft. (45 cm) in depth (Saltus 1985, 1988). Other archaeological examples of Louisiana scows include the Morgan City Floodwall Boat (Goodwin and Selby 1984), the Bayou Colyell Ferry (16LV75) (Saltus 1986), and the Rushing boat in the Amite River (Fig. 7.2) (Saltus 1986).

Flatboat

Although the term can be loosely applied to any vessel with a flat bottom, the traditionally defined flatboat has an oblong or rectangular shape, blunt and raked bow and stern, square chine, and vertical or slightly flared sides (Comeaux 1985:168). Early flatboats were propelled by paddle or oar and were renowned for their stability and maneuverability. Flatboats drifted downriver with the current to their destination where they were often sold off and dismantled (Birchett et al. 2001:54). These vessels probably evolved from the early barge form, since barges were commonly referred to as "flatboats," though were usually smaller. Historic flatboats and barges shared construction techniques. They were built upside down with bottom planking oriented transversely rather than longitudinally (Comeaux 1985:168).

Flatboats were known by other names such as ark, barge, broadhorn, *bateau de cent*, *bateau plat*, ferry, wharf boat, quarterboat, and New Orleans boat (Saltus 1988:49; Edwards and Pecquet du Bellay de Verton 2004:223). The term *chaland* has also been applied to the flatboat, further blurring the lines between watercraft types. As flatboats were often constructed with the intent to sell the hull with the cargo or dismantle upon arrival downriver, they were built using oak or a cheaper wood such as pine (Saltus 1988:54). To further minimize costs, boatbuilders utilized wooden pins or treenails rather than iron fasteners (Saltus 1988:54). Early flatboats generally measured 12–14 ft. (3.6–4.3 m) in length with a 3-ft. (91 cm) beam. Later flatboats were constructed up to 100 ft. (30.5 m) long and 20 ft. (6.1 m) wide (Fig. 7.5) (Birchett et al. 2001:52). They had flaring sides, a raked bow and stern, contained horizontal and elbow braces for hull strengthening, and had neither cabins nor decks (Pearson et al. 1989:249; Birchett et al. 2001:115). Baldwin (1941:48) further described their construction:

> The flatboat was built on sills or gunwales of heavy timbers about six inches thick and was strengthened by sleepers. The gunwales were a foot or two high, and on top of them were mortised studs three inches thick and four to six inches wide. At the top of these studs were fastened the rafters that were to bear the roof. The planks of the floor were about two inches thick, but the siding boards were of ordinary thickness. The bow was raked forward so that it would offer less resistance to the water.

Fig. 7.5 Flatboat on the Mississippi river circa 1870 (Courtesy of the collections of the Louisiana State Museum)

After the introduction of the small internal combustion engine at the dawn of the twentieth century, the design of flatboats changed. The overall length of the vessel generally increased though its beam was not altered significantly. Forward sheer dramatically increased, which caused the bow to rise high out of the water. A rudder was introduced along with a transom-mounted motor for propulsion allowing the boat to plane at speed (Comeaux 1985:170). Modern flatboats are constructed of marine plywood, average 16 ft. (4.9 m) in length, have a wide bottom, raked bow, and a broad but unraked stern (Fig. 7.6) (Comeaux 1985:170; Pearson et al. 1989:249; Birchett et al. 2001:115). The ability to plane above the water rather than plow through it gave an immediate advantage to the flatboat over the historic *bateau*, ensuring the *bateau*'s disappearance from modern fleets (Comeaux 1985:170; Pearson et al. 1989:249; Brassieur 2003i).

Saltus reported the discovery of a flatboat with a cabin at site 16ST135 on the Tchefuncte River (Fig. 7.2) (1988). This vessel was originally interpreted as the deckhouse to a steamboat, but was later identified as a flatboat, measuring 44.2 ft. (13.47 m) in length, 14.2 ft. (4.32 m) in beam, with a 2-ft. (61 cm) depth of hold (Saltus 1988:149). He reported that the sides were mortised with tongue and groove planking. Another flatboat, constructed of plywood and coated in fiberglass, was recorded by archaeologists at the Adams Place site (Watercraft 2, 16SMY55) on Bayou Shaffer. This flatboat would have measured 15.7 ft. (4.78 m) in length, 6.1 ft. (1.85 m) in beam, and 1.5 ft. (45 cm) deep if complete. Evidence for outfitting with a motor was reported by the archaeologists, thereby dating this vessel to the early to mid-twentieth century (Pearson and Saltus 1991:94–97).

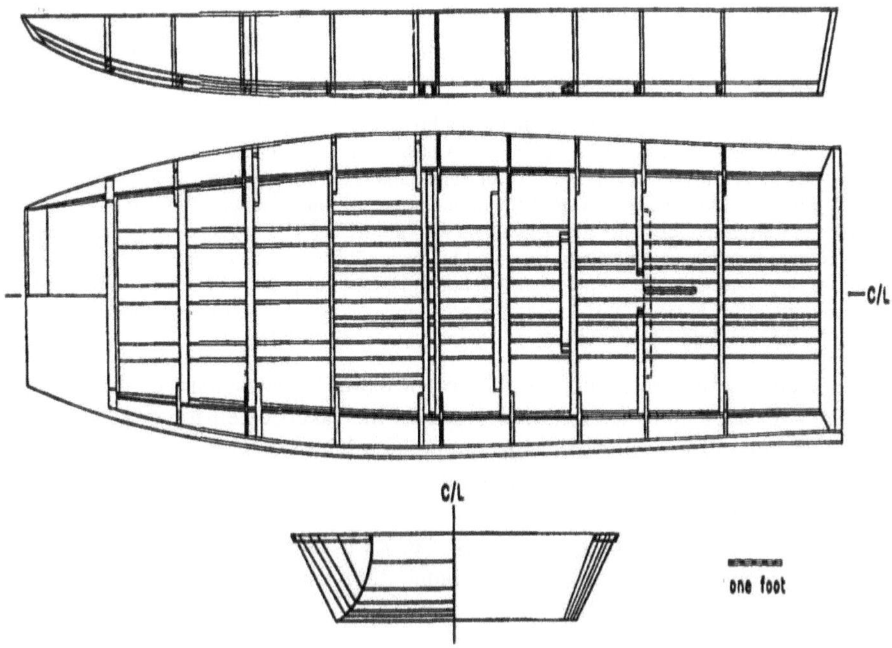

Fig. 7.6 Line drawing of a modern motorized flatboat (Pearson and Saltus 1991:97)

Barge

The barge is another type of flat-bottomed vessel and relative of the flatboat. Usually exhibiting a boxy appearance, built-up sides, and square chine, barges were often towed to transport large quantities of goods (Saltus 1988:54). Constructed with or without a deck, the flat-bottomed variety appears in stark contrast to the keeled barge and had a very shallow draft. Though different in form, keeled barges were constructed for the same purpose as flat-bottomed barges, to transport cargo. The keeled barge was steered by a rudder and often had a cabin on the rear deck. The terms "flatboat" and "barge" have often been used interchangeably (Edwards and Pecquet du Bellay de Verton 2004:223). The flat-bottomed type can appear as small as tens of feet in length to as large as hundreds of feet. Barges have been and continue to be one of the most important forms of watercraft in Louisiana's history. These vessels have served as the proverbial workhorses, carrying goods and merchandise between inland farms and plantations, the Port of New Orleans, and beyond Louisiana's borders.

Barges have been recorded in the archaeological record throughout the Maurepas basin and Lake Pontchartrain's associated waterways (Fig. 7.2) (Saltus 1987, 1988). One such site is a deckless barge recorded by Saltus during his 1984–1985 investigations in the Blood River in Livingston Parish (Fig. 7.2) (Saltus 1985). He described this partially exposed wooden vessel as "a rectangular box 49.3 ft. long, 18.3 ft. wide, with a 3.4 ft. depth of hold. Raked extensions, 5.5 ft. in length, were added to each end, giving the barge an overall length of 60.3 ft." (Saltus 1985:128). He continued:

> The sides of the barge, unlike those in the flat boat tradition, were 3 in. wide planks placed on upright frames spaced on 28 in. centers. A datto secured the upright frames at the chine to a 5 by 10 in. timber. In between the upright frames, the floor timbers also notched into this 5 by 10 in. timber. They were not contiguous as in normal boat construction where the upright frame, futtocks, and floors make up a single frame. The upper ends of the frames were held together by an interior band and by diagonal braces which were secured to this upper band and to the floor timbers at 4 ft. intervals. Exterior planking also helped hold the frames together. Five rods 6 ft. long and overlapping planks from the sides secured the ends across the top (Saltus 1985:128).

Vernacular watercraft, such as this, often exhibit nonstandard or unique construction features and modifications (such as the raked extensions). Evidence of repairs to maximize a craft's use-life can be identified. These craft may also retain evidence of construction innovations employed during times of economic stress and raw material depletion.

Damour and Biddiscombe returned to the site in 2005 and recorded the barge's condition after 20 years of repetitive inundation and exposure. The barge was situated at the edge of the Blood River near the bank and was almost entirely contained within riverine sediments. One end was completely buried and inaccessible; the exposed length measured 52.8 ft. (16.1 m) with a 17.9-ft. (5.45 m) beam (Damour and Biddiscombe 2005:114). While its upper hull has undergone considerable erosion during the past 20 years, the lower hull surrounded by sediments appeared to be intact and well preserved. The barge was double-hulled, constructed of vertical planks 2.8 in. (7.1 cm) thick, creating sides that were 10.2 in. (25.9 cm) deep. Upright wooden frames 5.9 in. (15 cm) thick extended between the inner and outer hull planks, with a room and spacing of approximately 23.6 in. (59.9 cm) and 28 in. (71 cm), respectively (Damour and Biddiscombe 2005:114). The thick iron plate diagonal braces, originally recorded by Saltus, remained affixed to the sides of the barge though much of the stern extension has eroded away. Three existing iron rods, one of which is capped with a hexagonal head 1.5 in. (3.8 cm) in diameter, extend aft of the stern and are all that remain of the stern extension (Damour and Biddiscombe 2005:115). The barge may have been abandoned sometime in the early to mid-twentieth century.

Skiff

Skiffs, or *esquiff*, are identifiable by their small size, sharp and pointed bow, angular chine, and squared or blunt stern. Resembling the traditional rowboat form, skiffs often range in length from 14 to 25 ft. (4.3–7.6 m). Rising in popularity in the late eighteenth and early nineteenth centuries, skiffs were typically employed to move goods and supplies (Pearson and Saltus 1991:90). This term has often been applied

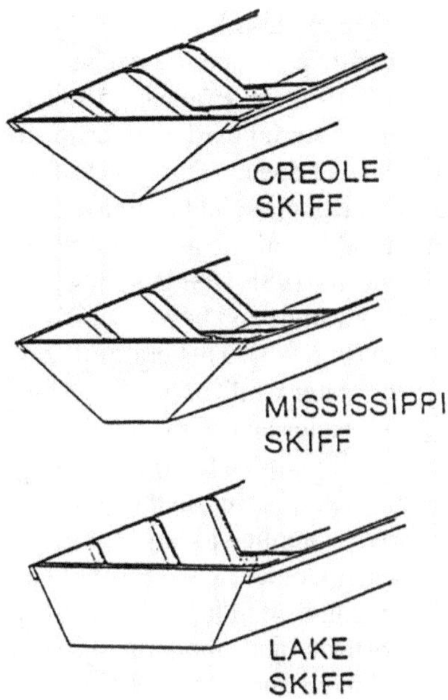

Fig. 7.7 Typical skiff stern morphologies (Comeaux 1985:163)

to any small vessel with a pointed bow and has been generally applied to vessel types also identified as *peniche*, *chaloupe*, and *galere* (Comeaux 1985:166; Edwards and Pecquet du Bellay de Verton 2004:223). True historic skiffs can be divided into three varieties based on the morphology of the stern: Creole skiffs, Mississippi skiffs, and lake skiffs (Fig. 7.7) (Comeaux 1985:163).

Creole skiffs are small vessels with a very narrow beam, a V-shaped transom, and have the greatest sheer and rake at the stern than the other varieties. The Mississippi skiff is slightly larger with a wider beam, less sheer and rake at the stern, and a less pronounced V-shaped transom. The lake skiff is longer and beamier than the other varieties with a wide and nearly rectangular transom. Comeaux explained that each variety evolved to operate in particular environments (Comeaux 1985:166). The Creole skiff is operated in calm, inland waters as it is the least stable of the three but is the easiest to row (Comeaux 1985:166). The Mississippi skiff operates in inland waters, rougher coastal lakes and bays, or larger rivers and offers greater stability than the Creole skiff (Comeaux 1985:166). Lastly, the lake skiff provides the most stability, is seaworthy, and has the heaviest cargo carrying capacity. It is operated specifically in coastal waters or larger rivers (Comeaux 1985:166).

Skiffs are rowed through the use of a *joug*, or yoke, allowing the rower to stand as the tholepins and straps are elevated and extend out past the gunwales (Knipmeyer 1976:140–142; Comeaux 1985:168; Brassieur 2003g). The *joug* is often used with the Creole skiff but can appear on the other varieties. Skiffs could also be propelled by sail or by cordelling, a method of pulling the boat upriver by using two ropes stationed on shore (Tunis 1961:66).

More modern versions of skiffs, designated by their geographical origin, include the Lafitte skiff and the Atchafalaya skiff as well as other regional variations of the basic skiff form (Comeaux 1985:168; Brassieur 1989). Lafitte skiffs are designed to operate with inboard engines in shallow waters along the Gulf coast allowing them to plane at speed. The typical form exhibits substantial flare and sheer at the bow and a distinctive fantail transom (Brassieur 1989, 2003j). These vessels are widely used for the shrimp and seafood industry and are popular for recreational use as well (Comeaux 1985:168; Brassieur 2003p). The Atchafalaya skiff developed to operate in the swamps and bayous of the southern coastal region, originally the Atchafalaya Basin (Fig. 7.3). This craft utilizes an outboard motor and is also designed to plane on the water while underway. This form has a pointed bow, little or no rake between the gunwales and stern, and a broad bottom extending aftward from the center of the vessel, allowing it to cruise over thick aquatic vegetation (Comeaux 1985:168). Due to the lack of sheer on the gunwales, this type of skiff is easy to construct. The term *canotte* is applied to larger skiffs with an inboard engine and often with a cabin and decking (Comeaux 1985:168; Pearson et al. 1989:249). *Canottes* are very similar in form to the small lugger, which will be described later.

Historic and modern skiffs have been recorded by archaeologists in the Tchefuncte and Blood Rivers (Fig. 7.2) (Saltus 1988). Skiffs in the Atchafalaya Basin recorded by archaeologists include a small cypress skiff (16SMY61) and a motorized Lafitte skiff at the Adams Place site (Watercraft 1 at 16SMY55) on Bayou Shaffer (Fig. 7.3). The flat-bottomed cypress skiff was located onshore and measured 18.8 ft. (5.73 m) long with a 4.6-ft. (1.4 m) beam and 11.5-in. (29.2 cm) depth of hold (Pearson and Saltus 1991:88). The Lafitte skiff was partially exposed alongside dock pilings and measured 28.6 ft. (8.71 m) long with a 9.5-ft. (2.9 m) beam and 3.4-ft. (1.03 m) depth of hold (Pearson and Saltus 1991:92). The cypress skiff exemplified a variety popular between 1910 and 1940, while the Lafitte skiff, with the use of cypress and plywood in its construction, likely dated to the early to mid-twentieth century (Pearson and Saltus 1991:91–94). Their historic prevalence and continued popularity to this day throughout southern Louisiana emphasizes their importance to Louisiana's maritime cultural heritage.

Yawl

Yawls are small boats that served as service boats for large, ocean-going sailing ships, though the term has been applied to skiffs as well. The riverine version of the yawl is flat-bottomed with an angular chine and constructed of oak or yellow pine (Saltus 1987: 83; Saltus 1988:56). Often used as a lifeboat or service boat for steamboats, the yawl had a pointed bow, wider between the gunwales than the bottom, and had a large, squared stern to carry line ashore or to steamboats (Saltus 1988:56). Not to be confused with the keeled type of yawl that serviced larger ships, the river yawl was propelled by oars but could also utilize a sail (Saltus 1987:83).

Keeled Boats

Keelboat

Keelboat is a generic term used to describe vessels constructed with a bottom keel and designed for travel up as well as downriver. Baldwin defined two types of keelboats: keelboats proper and barges (Baldwin 1941 in Saltus 1988:68). The true definition refers to a specific type of vessel with a keeled and rounded bottom. Also referred to as *bercha*, berge, barge, or *berchita*, the keelboat was double-ended with a pointed bow and stern, shallow keel, and a 12–18 in. (30.5–45.7 cm) cleated footway constructed around the gunwales (Birchett et al. 2001:53; Edwards and Pecquet du Bellay de Verton 2004:223). They typically measured 40–80 ft. (12.2–24.4 m) long, 7–10 ft. (2.1–3 m) wide, with a 3–4 ft. (91.4 cm to 1.2 m) depth of hold, and a draft of 2 ft. (61 cm) when fully loaded (Saltus 1987:49; Pearson et al. 1989:98). Pearson and Saltus (1991:29) reported that many of the keelboats used in Louisiana's waterways were built up north; those built locally tended to be smaller, about 60 ft. (18.3 m) in length.

The average keelboat could transport between 15 and 50 tons of cargo and could have a cabin or cargo box in the middle (Pearson et al. 1989:98; Birchett et al. 2001:54). Some had seats for rowers while others were fitted with sailing rigs (Fig. 7.8). Sweeps, or oars, located on either side provided a means of propulsion

Fig. 7.8 Sketch depicting keelboats in the background (Baldwin 1941)

while a sweep at the stern served as a rudder (Birchett et al. 2001:54). Keelboats were also propelled by poling or cordelling upriver and then drifting downstream with the current. This watercraft form rose to prominence in the late eighteenth century and was subsequently replaced by the steamboat in the mid-nineteenth century. They continued to be used, though, by those operating in the narrow and shallow bayous and inland rivers where many steamboats could not navigate (Birchett et al. 2001:54).

Barge

The keeled barge, different in hull form than the flat-bottomed barge, was commonly used throughout the eighteenth and nineteenth centuries for transporting cargo and supplies. Also known as *bercha*, *lanchon*, or wherry, the keeled barge appeared similar to the double-ended keelboat though was wider, longer, and heavier (Pearson et al. 1989:98; Edwards and Pecquet du Bellay de Verton 2004:223). They were constructed in lengths up to 170 ft. (51.8 m) and drew roughly 3 ft. (91 cm), though the average barge measured 32–57 ft. (9.8–17.4 m) in length in the eighteenth century and 46–125 ft. (14–38 m) in the nineteenth century (Saltus 1988:65; Pearson et al. 1989:98). Barges were fitted with a mast and square sails for propulsion and rudder for maneuverability. Poling or cordelling provided additional methods for upstream travel. Often, a small cabin was constructed on the rear deck. The cargo capacity of the larger barges was typically between 50 and 150 tons (Saltus 1988:65). Much of the confusion arising when describing the keeled form versus the flat-bottomed, rectangular form was summed up by Saltus (1988:65):

> The barge is one of the most interesting and most misunderstood of riverine craft. Part of the confusion stems from the pervasive notion of the modern square scow river barge. … The barge was similar in construction to a keelboat, but was intended for use on the larger main trunk routes of the river.

The term "barge" is often applied to other vessels of both keeled and flat-bottomed varieties, thereby causing additional confusion.

Longboat

Other keeled boats used throughout Louisiana include the longboat, launch, *chaloup*, pinnace, yawl, cutter, and *felouque* (Saltus 1988). Few, if any, have been found and documented by archaeologists. As such, a broad description of each vessel will be given. Saltus (1987:58) described the longboat as "deep, with broad bows and a wide belly; it was double-banked, occasionally decked, and fitted with mast and sails." Steered with a rudder and typically fitted with a davit for lifting anchors, the longboat usually served as a ship's boat during the early colonial period until it was gradually replaced by the launch toward the end of the eighteenth century (Saltus 1988:73). The longboat was typically constructed in lengths between 19 and 36 ft. (5.8–10.97 m) but required a wide beam to ensure stability while maneuvering in coastal or bay waters to service much larger sailing vessels (Saltus 1988:77). This form of watercraft has also been known as *bateau plat* or *chaloup* (Saltus 1988:77; Edwards and Pecquet du Bellay de Verton 2004:223), terms used to indicate other forms of vessels as well, lending to the versatility of vernacular nomenclature.

Launch

The launch, measuring 19–26 ft. (5.8–7.9 m) in length, also served as a ship's boat and increasingly usurped the longboat's popularity toward the end of the eighteenth century. Referred to as *chaloup* in French and *lancha* in Spanish, the launch was smaller than the *lanchon*, or keeled form of barge, and wider with a relatively flat bottom (Saltus 1988:73). This craft was typically used to transport cargo or small parties of crew to and from an anchored ship. Its rise in prevalence was due to its proportionally larger size yet shallower draft compared to the longboat. Though it was apparently not as adept a small sailing vessel as the longboat, the launch was propelled by rowing (Saltus 1988:73). Several launches, though of more modern forms, have been recorded by Saltus in the Maurepas basin (Saltus 1985). The rise in popularity of this vessel over its predecessor, the longboat, into the nineteenth century led to the longboat's declining numbers among service boats.

Chaloup

The term *chaloup*, meaning "launch" in French, has been used to describe watercraft as small as those operating as ship's boats to larger sailing vessels of up to 60 tons burden (Pearson et al. 1989:93). The term, when applied to the larger form, indicates a vessel of a length generally between 27 and 29 ft. (8.2–8.8 m) in the eighteenth century, or 27–34 ft. (8.2–10.4 m) in the nineteenth century, and closer in hull appearance to the pinnace or yawl (Saltus 1988:80). First used in Louisiana by early eighteenth-century explorers and settlers, the round and deep hull of this craft was better suited for the deeper coastal bays than the rivers and bayous of the interior due to its tendency to drag through the water and create a greater resistance in a current (Pearson et al. 1989:93). The *chaloup*, known to English speakers as "shallop," carried its foremast near or at the stem and could also be rowed (Saltus 1987:75).

Pinnace

The pinnace was very similar to the contemporaneous *chaloup* (shallop) of the eighteenth century though it was longer, approximately 40–60 ft. (12.2–18.3 m) in length, and included a deck (Saltus 1988:83). The pinnace was a small sailing craft that also contained oars for rowing. This vessel type served a much larger ship and steadily increased in size over time (Saltus 1988:83). Little additional information is available as the pinnace is not well represented in the archaeological record.

Yawl

The small sailing yawl, not to be confused with the term applied to small, flat-bottomed riverine craft also called skiffs, appeared similar in hull form to the pinnace yet was narrower and carried a smaller complement of oars (Saltus 1988:83). The yawl, often used as a pilot boat, appeared as early as the eighteenth century, measuring 11–26 ft. (3.35–7.9 m) in length. By the early twentieth century, the sailing yawl was constructed in lengths between 26 and 60 ft. (7.9 and 18.3 m) (Saltus 1988:83).

Cutter

The cutter, appearing in Louisiana as early as the eighteenth century, was similar in size to the yawl; 11–26 ft. (3.35–7.9 m) in length. After the eighteenth century and into the twentieth century, cutters could be constructed as long as 60 ft. (18.3 m) (Saltus 1988:83). It was designed with a straight keel, nearly vertical stem, and a transom shaped similar to a yawl that hung over the stern. This vessel, along with the lugger, was preferred by smugglers partly because of its stability, due to its deep draft and relative short length. It sailed at an easy tack and could be produced cheaply (Saltus 1988:83).

Felouque

The *felouque* (also *felucca* or *falua*) was a keeled boat with a flat bottom and shallow draft (Saltus 1987:41). This vessel type was usually employed as a ship's boat, like the longboat and launch, but had a different form. Wider than a barge, the *felouque* was rowed but could be sailed as well (Saltus 1987:43). When fitted with its lateen sails and oars, it resembled the *chaloup* but was designed to allow placement of the helm at either the bow or stern (Saltus 1987:43). Appearing in Louisiana as early as the eighteenth century, the *felouque* endured until the early nineteenth century though never attaining the popularity of other similarly sized watercraft (Saltus 1988:85). A nineteenth-century example, registered in New Orleans, measured 33 ft. (10 m) long, with a 9.6-ft. (2.9 m) beam, and 3-ft. (91 cm) depth of hold, two masts, one deck, and a square stern (Saltus 1988:85).

Sailing and Motorized Vessels

Sailing vessels, for the purpose of this discussion, are characterized as those larger watercraft designed for use in coastal and deeper waters and primarily dependent upon sails for propulsion. Exhibiting similar hull construction techniques and overall form, sailing vessels were often classified by size but could be designated by variations in hull form and sail configuration. Large sailing ships generally appear as sturdy vessels with heavy construction, a large and durable keel, use of floor timbers and futtocks to form the hull, keelson atop the floors and keel, longitudinal stringers for hull strengthening, deep draft and depth of hold, single or multiple decks, greater angle of deadrise than flat-bottomed vessels for stability at sea, tumblehome to increase stability by reducing the weight of the vessel and lowering the center of gravity above the waterline, and a complex stern and transom built to articulate with a large rudder system (Harland and Myers 1984:44; Steffy 1994). With more extensive masting, rigging, and sail configuration, and deeper drafts than riverine craft, these vessels were larger than the small sailing craft utilized in the shallow coastal waters, narrow rivers, and bayous of southern Louisiana.

Sailing vessels were used along the coast of Louisiana for transatlantic crossing by French and Spanish explorers, for cargo transport and military use through the era of the steamship, for the early fishing and seafood industries, and into modern times as pleasure craft. Versatile sailing craft, such as the schooner and sloop, appeared in diverse sizes and could sail the coastal waters of southern Louisiana or travel upriver toward the inland ports. Sailing vessels were often serviced by ship's boats of various types, described above, to transport cargo or personnel from the larger ship to shore.

Sailing vessel types are discussed according to the forms that are known to have been utilized within southern Louisiana and its coastal waterways. Other vessel types such as the galley, tartan, bark, barque longue (double shallop or barcassa), brig, ketch, smack, and frigate have been documented in the historical record, however, little is known about these vessel forms and their association with Louisiana's waterways may have been brief (Saltus 1987, 1988). Steamboats, ubiquitous to Louisiana, have been well documented in the historical record as well as by archaeologists; their description is beyond the focus of this discussion.

Schooner

The schooner, known as *goelette* in French or *goleta* in Spanish, was a versatile craft that operated in the open ocean, shallow bay waters, rivers, and inland lakes of Louisiana. Nineteenth-century schooners throughout coastal Louisiana typically measured 28–87 ft. (8.5–26.5 m) in length, while twentieth-century versions usually measured 46–74 ft. (14–22.5 m) (Saltus 1988:89). Schooners can be categorized according to type of rigging, function, hull form, or region of use. When defined by function, schooner types included pilot schooners, trading schooners, fishing schooners, and packet schooners. Those defined by hull form included scow schooners, barge schooners, pungy schooners, file bottom schooners, and ram schooners (Saltus 1988:90). Centerboard schooners contained a centerboard that was placed either through or alongside the keel, which dropped through the bottom of the hull and allowed the vessel to sail faster, steer easier, and tack closer to the wind (Barkhausen 1990:34). Saltus argued that schooners used in Louisiana's rivers and inland waterways, especially the Maurepas basin, could be called river trading schooners based on their function. He continued, "The diagnostic attribute is the vessel's shallow draft and wide beam, dictated by the environment, depth, and functional need" (Saltus 1988:90).

Schooners have been documented by archaeologists in Louisiana's inland waterways. During his 1987 archaeological survey and investigation of watercraft in the waterways of Lake Pontchartrain's North Shore, Saltus recorded several schooners (Fig. 7.2). In the Tchefuncte River, Saltus (1988:149–154) recorded two schooners with offset centerboards and square sterns, measuring 61.5 × 20.7 × 3.5 ft. (18.7 × 6.3 × 1 m) and 49.5 × 19.5 × 3 ft. (15 × 5.9 m × 91 cm). He recorded two other schooners, one measuring 59.65 × 20.9 × 4.68 ft. (18.2 × 6.4 × 1.4 m) and representing a riverine schooner with multiple deck hatches, and another measuring 75.2 × 22.6 × 6 ft. (22.9 × 6.9 × 1.8 m) (Saltus 1988:156–157). Saltus also recorded the Civil War-era schooner *James Stockton* (site 16LV63) in the Blood River, Livingston Parish. According to historical sources, the vessel was built in 1851 in Springfield, Louisiana, measured 55 ft. 4 in. (16.8 m) in length, 18 ft. 6 in. (5.6 m) in beam, and a 4-ft. 6-in. (1.4 m) depth of hold (Saltus 1985:119).

Sloop

Sloops were capable of sailing in a range of environments including narrow inland rivers and open ocean. A versatile sailing craft like the schooner, sloops generally were single masted and designated by sail configuration. Sloops can be defined as "a vessel with one mast like a cutter; but having a jib stay, which a cutter has not" (Brande 1856 quoted in Saltus 1987:71). Other varieties of the sloop include the sloop-of-war, ship-sloop, brig-sloop, and corvette (Blackburn 1978; Saltus 1988:92). Historical records show that sloops built for Louisiana's inland waterways were nearly as common and reliable as the schooner. Their array of size included general ranges of 30–77 ft. (9.1–23.5 m) in length, 11–20 ft. (3.4–6.1 m) in beam, with a 3–6-ft. (91 cm to 1.8 m) depth of hold (Saltus 1988:92). An archaeological example of a historic sloop or schooner was recorded in Bayou Shaffer, St. Mary Parish (site 16SMY61), and was investigated by Coastal Environments, Inc. (Pearson and Saltus 1991:87).

Lugger

The lugger is a widely used sailing or motorized vessel, popularized in the nineteenth century in coastal Louisiana (Brassieur 2003n). The early lugger, whose name and form is derived from the rig of Mediterranean sailing boats and keel yawlboats, had rounded hulls and used centerboards (Chapelle 1951; Comeaux 1985:172; Pearson et al. 1989:198). Chapelle (1951:284) added:

> The construction of the boats was conventional: sawn frames, carvel planking, and the usual plank keel of the centerboarder. The timbering and plank were often local longleaf pine and cypress. The boats usually had a long and well-formed run and trimmed by the stern, which reduced the bluntness of the rather full bow. These luggers sailed very fast, were powerful boats, and were reputed very close-winded. … The luggers ranged in size from about 18 to 45 feet in length, and it is claimed that it was the practice of many builders to construct all of their boats on one model, varying the scale to suit the owner's requirements and pocketbook. … The curve of the stem below the water line was very slight, which made the lower forebody very fine; the hollow in the forefoot was often very marked. This was supposed to help the boats to hold on close-hauled in shallow water where the board could not be lowered very much.

Luggers were also descended from, and are still referred to as, canots (or *canottes*), a general term used to describe a ship's auxiliary boat utilized by French colonists (Edwards and Pecquet du Bellay de Verton 2004:223). According to Brassieur (1989), canots had rounded hulls, a shallow keel, hourglass-shaped transom, and employed a lug sail or oars. Often employed as work boats for oystering and shrimping activities, luggers operated in the shallow coastal lakes, bayous, and marshes as well as the deeper bays (Pearson et al. 1989:198). Small sailing luggers frequented the Mississippi River and the Port of New Orleans carrying out a variety of tasks (Krause et al. 2004:149). Davis (1984:51) argued, "They have to stem the swift current of the Mississippi and twist and turn up the many inlets and bayous… This requires a minimum of draft to pass over the shoal places and to operate on the oyster reefs and yet they have to go off in deep water and negotiate all kinds of waves." In the early twentieth century, the sailing lugger was modified by adding an engine and screw propeller (Saltus 1988:89). With the advent of the motorized lugger, older sailing luggers were surpassed in quantity and popularity (Brassieur 2003h). Motorized luggers, omitting the centerboard, allowed for rapid transport of seafood to the market unlike the slower sailing luggers (Comeaux 1985:172). This variety of lugger appears as small, flat-bottomed craft, generally

20–30 ft. (6.1–9.1 m) long, and included a cabin to house the engine and operating controls. More seaworthy luggers of 40–50 ft. (12.9–15.2 m) were later introduced to access offshore oyster beds and fishing resources for the burgeoning seafood industry (Comeaux 1985:172). Since the mid-twentieth century, motorized luggers have declined in popularity (Brassieur 1989).

Trawler

In the early twentieth century, the exploitation of shrimp as part of the seafood industry brought the motorized shrimp trawler to the fleets of vessels traveling to deeper waters in the Gulf of Mexico. Initially introduced by outsiders in the 1930s, the South Atlantic trawler was modified by Louisiana boatbuilders to become the shrimp trawler (Brassieur 2003a, h). The inshore version was smaller, less than 50 ft. (15.2 m) in length, and had a flat bottom designed to trawl the bay and near-shore waters (Comeaux 1985:172, Brassieur 1989). Larger versions designed for deeper waters are known as Florida-type trawlers, or *floridiane*, and average 50–70 ft. (15.2–21.3 m) in length (Brassieur 2003a). Trawlers exhibit substantial forward sheer; high, flaring bows with a nearly vertical stem; semi-V hulls; and include a forward cabin at or near the bow (Brassieur 2003b, d).

Trawlers are constructed of wood (cypress or Spanish cedar) or steel and have been readily adopted and adapted by Louisianans to suit the needs of the seafood industry and the demands of the coastal environment (Brassieur 2003c, k). The South Atlantic trawler, steadily declining in numbers, has given way to the locally designed, flat-bottomed South Lafourche trawler (Brassieur 2003b, h). The predominant form of shrimp trawler in Louisiana's modern fleet is the wooden-hulled, flat-bottomed skiff (Brassieur 2003o). Exhibiting a flared bow and broad hull, the trawling skiff retains the forward cabin similar to the South Atlantic trawler but trades a deep, soft-chined hull for a shallower, hard-chined hull (Brassieur 1989, 2003o).

Immigrant populations often bring their cultural traditions with them; this includes boatbuilding methods and design. When traditional watercraft forms that evolved to operate within the native environment are brought into a new environment, the traditional form can be adapted to operate more effectively in the new environment. Brassieur (1989) cited a more recent, mid-twentieth-century example of Vietnamese immigrants and their boatbuilding tradition. Vietnamese boatbuilders in Louisiana initially designed and built their fishing boats in the same manner as their tradition dictated. These forms appeared quite different from Louisiana's fishing luggers and trawlers. He stated, "as they became more familiar with the demands of the region's waterways, their boats became increasingly similar to those which had been used for generations in Louisiana" (Brassieur 1989).

While there are a plethora of forms of watercraft and vernacular terms to describe them, the versatility of watercraft nomenclatures must be kept in mind. The frequent use of vernacular terms for more than one form of watercraft can undermine the ability of the archaeologist to develop watercraft typologies. It is important to illuminate the differences as well as the similarities between many of Louisiana's vernacular constructed watercraft to recognize the continuing evolution of boatbuilding whereby new techniques and styles modify or replace older ones. The pirogue, introduced by Native Americans and adapted by French and Spanish explorers, continued in use for centuries. Vessel types of French influence, such as the *bateau*, *radeau*, and *chaloupe* of the eighteenth century, were slowly replaced in the nineteenth century by the flatboats and keelboats introduced by Anglo-American immigrants (Pearson et al. 1989:93). This synthesis of watercraft types, by no means all-inclusive, creates a baseline from which to facilitate identification of Louisiana's submerged cultural resources, a maritime tradition as rich and complex as any other in North America.

Acknowledgments I would like to acknowledge Allen R. Saltus, Jr., Malcolm Comeaux, C. Ray Brassieur, and Charles Pearson for their decades of archaeological investigations and historical research. These individuals have been at the forefront of recording the oral histories, boatbuilding traditions, and archaeological remains of Louisiana's watercraft. Without their research and dedication, this article would not have been possible. I would also like to thank Allen Saltus and Malcolm Comeaux for generously granting permission for use of their figures.

References

Baldwin, L. D. (1941). *The keelboat age on the western waters*. Pittsburgh, PA: University of Pittsburgh.

Barkhausen, H. N. (1990). *Focusing on the centerboard*. Manitowoc, WI: Wisconsin Maritime Museum.

Birchett, T. C. C., Pearson, C. E., & Castile, G. J. (2001). *Historic navigation and shipwreck study, lower Atchafalaya basin reevaluation study, South Central Louisiana*. Report to the US Army Corps of Engineers, New Orleans District, Baton Rouge, LA. Baton Rouge, LA: Coastal Environments.

Blackburn, G. (1978). *The illustrated encyclopedia of ships, boats, vessels, and other waterborne craft*. Woodstock, VT: The Overlook Press (Reprinted 1999).

Brassieur, C. R. (1989). Louisiana boatbuilding: An unfathomed fortune. In *Louisiana folklife festival booklet*. Baton Rouge, LA: Louisiana Division of the Arts, Department of Culture, Recreation, and Tourism. Retrieved from http://www.louisianafolklife.org/LT/Articles_Essays/creole_art_boatbuild_unfat.html.

Brassieur, C. R. (2003a). *Cap'n Bozo, South Atlantic Trawler. Creole State Exhibit Artifacts*. Retrieved from http://www.louisianafolklife.org/FOLKLIFEimagebase/showlisting.asp?ImageID=2.

Brassieur, C. R. (2003b). *Master Flint, Lafourche Shrimp Trawler. Creole State Exhibit Artifacts*. Retrieved from http://www.louisianafolklife.org/FOLKLIFEimagebase/showlisting.asp?ImageID=4.

Brassieur, C. R. (2003c). *Pappy's Gold, Lafourche Shrimp Trawler. Creole State Exhibit Artifacts*. Retrieved from http://www.louisianafolklife.org/FOLKLIFEimagebase/showlisting.asp?ImageID=5.

Brassieur, C. R. (2003d). *Southern Star, Terrebonne Shrimp Trawler. Creole State Exhibit Artifacts*. Retrieved from http://www.louisianafolklife.org/FOLKLIFEimagebase/showlisting.asp?ImageID=6.

Brassieur, C. R. (2003e). *Albert Latiolais in his Bateau or Putt-Putt. Creole State Exhibit Artifacts*. Retrieved from http://www.louisianafolklife.org/FOLKLIFEimagebase/showlisting.asp?ImageID=10.

Brassieur, C. R. (2003f). *Boating in the Atchafalaya Basin. Creole State Exhibit Artifacts*. Retrieved from http://www.louisianafolklife.org/FOLKLIFEimagebase/showlisting.asp?ImageID=90.
Brassieur, C. R. (2003g). *Creole skiff on Little Grand Bayou, ca. 1969. Creole State Exhibit Artifacts*. Retrieved from http://www.louisianafolklife.org/FOLKLIFEimagebase/showlisting.asp?ImageID=246.
Brassieur, C. R. (2003h). *Docked shrimp trawlers in Terrebonne Parish, circa 1960. Creole State Exhibit Artifacts*. Retrieved from http://www.louisianafolklife.org/FOLKLIFEimagebase/showlisting.asp?ImageID=426.
Brassieur, C. R. (2003i). *Flat boat at Pierre Part, 1969. Creole State Exhibit Artifacts*. Retrieved from http://www.louisianafolklife.org/FOLKLIFEimagebase/showlisting.asp?ImageID=498.
Brassieur, C. R. (2003j). *Lafitte skiff. Creole State Exhibit Artifacts*. Retrieved from http://www.louisianafolklife.org/FOLKLIFEimagebase/showlisting.asp?ImageID=666.
Brassieur, C. R. (2003k). *Making a wooden boat. Creole State Exhibit Artifacts*. Retrieved from http://www.louisianafolklife.org/FOLKLIFEimagebase/showlisting.asp?ImageID=701.
Brassieur, C. R. (2003l). *Malcolm Comeaux examines a Pierre Part Dugout, 1969. Creole State Exhibit Artifacts*. Retrieved from http://www.louisianafolklife.org/FOLKLIFEimagebase/showlisting.asp?ImageID=709.
Brassieur, C. R. (2003m). *Mr. Wiltz's plywood pirogue, ca. 1969. Creole State Exhibit Artifacts*. Retrieved from http://www.louisianafolklife.org/FOLKLIFEimagebase/showlisting.asp?ImageID=751.
Brassieur, C. R. (2003n). *Sailing lugger at Vermilionville. Creole State Exhibit Artifacts*. Retrieved from http://www.louisianafolklife.org/FOLKLIFEimagebase/showlisting.asp?ImageID=877.
Brassieur, C. R. (2003o). *Shrimp trawler rigged with wing nets. Creole State Exhibit Artifacts*. Retrieved from http://www.louisianafolklife.org/FOLKLIFEimagebase/showlisting.asp?ImageID=895.
Brassieur, C. R. (2003p). *Trawling for shrimp. Creole State Exhibit Artifacts*. Retrieved from http://www.louisianafolklife.org/FOLKLIFEimagebase/showlisting.asp?ImageID=998.
Bremer, C. (1907). *The Chata Indians of Pearl River*. New Orleans, LA: Picayune Job Print.
Chapelle, H. (1951). *American small sailing craft*. New York: W.W. Horton.
Comeaux, M. L. (1985). Folk boats of Louisiana. In N. R. Spitzer (Ed.), *Louisiana folklife: A guide to the state* (pp. 160–178). Baton Rouge, LA: Louisiana Folklife Program, Division of the Arts and the Center for Gulf South History and Culture.
Damour, M., & Biddiscombe, J. (2005). *Cultural resources investigations at Warsaw Landing (16LV60), Livingston Parish, Louisiana*. Report to D&S Environmental Services, Inc., French Settlement, LA. New Orleans, LA: Earth Search.
Damour, M., Bissett, T., Maygarden, B. D., & Lintott, K. B. (2005). *Phase I marine remote sensing survey of the lower Atchafalaya River and portions of Bayous Black, Boeuf, and Chene, Assumption, St. Mary, and Terrebonne Parishes, Louisiana*. Report to the US Army Corps of Engineers, New Orleans District, New Orleans, LA. New Orleans, LA: Earth Search.
Davis, E. A. (1971). *Louisiana: A narrative history*. Baton Rouge, LA: Claitor's (Reprinted 1976).
Davis, C. G. (1984). *American sailing ships: Their plans and history*. Mineola, NY: Dover.
Edwards, J. D., & Pecquet du Bellay de Verton, N. K. (2004). *A Creole lexicon: Architecture, landscape, people*. Baton Rouge: Louisiana State University Press.
Goodwin, R. C., & Selby, G. W. (1984). *Historical archeology of the Morgan City floodwall boat*. Report to the US Army Corps of Engineers, New Orleans District, New Orleans, LA. New Orleans, LA: RC Goodwin and Associates.
Harland, J. H., & Myers, M. (1984). *Seamanship in the age of sail*. London: Conway Maritime Press (Reprinted 2006).
Knipmeyer, W. B. (1976). Folk boats of eastern French Louisiana. In D. Yoder (Ed.), *American folklife* (pp. 105–149). Austin: University of Texas Press.
Krause, K., Coyle, K., Turner, S., Pelletier, J. B., Pincoske, J., George, D., et al. (2004). *Phase I cultural resources survey and archaeological inventory of the proposed Carrollton revetment project, Orleans Parish, Louisiana*. Report to the US Army Corps of Engineers, New Orleans District, New Orleans, LA. New Orleans, LA: RC Goodwin and Associates.

Louisiana State Museum (1998). *Louisiana and the mighty Mississippi River online exhibit.* Various photos and images. Retrieved from http://www.crt.state.la.us/louisiana-state-museum/online-exhibits/louisiana-and-the-mighty-mississippi/index.

McWilliams, R. G. (Trans. and Ed.). (1953). *Fleur de Lys and Calumet: Being the Penicaut narrative of French adventure in Louisiana.* Baton Rouge: Louisiana State University Press. (Reprinted 1988 by The University of Alabama Press, Tuscaloosa, AL).

McWilliams, R. G. (Trans. and Ed.). (1981). *Iberville's Gulf journals.* Tuscaloosa: The University of Alabama Press.

Pearson, C. E., Castille, G. J., Davis, D., Redard, T. E., & Saltus, A. R. (1989). *A history of waterborne commerce and transportation within the US Army Corps of Engineers New Orleans District and an Inventory of Known Underwater Cultural Resources.* Report to the US Army Corps of Engineers, New Orleans District, New Orleans, LA. Baton Rouge, LA: Coastal Environments.

Pearson, C. E., & Saltus, A. R. Jr. (1991). *Remote sensing survey of the Atchafalaya Basin main channel, Atchafalaya channel training project, Sts. Martin and Mary Parishes, Louisiana.* Report to the US Army Corps of Engineers, New Orleans District, New Orleans, LA. Baton Rouge, LA: Coastal Environments.

Rowland, D., & Sanders, A. G. (1927). *Mississippi provincial archives 1729–1740, French dominion 1 and 2.* Jackson, MS: Mississippi State Department of Archives and History.

Saltus, A. R. Jr. (1985). *Submerged cultural resources investigation of the western portion of the Maurepas Basin with intensive surveys at Warsaw Landing, Blood River, and Springfield area, Natalbany River, Louisiana.* Report to US Dept. of the Interior and Louisiana Division of Archaeology, Department of Culture, Recreation, and Tourism, Baton Rouge, LA.

Saltus, A. R. Jr. (1986). *Submerged cultural resources investigation of the western portion of the Maurepas Basin with intensive underwater surveys at Hoo Shoo Too Landing, 16EBR60, Colyell Bay, Catfish Landing, and at the Mouth of Bayou Chene Blanc.* Submitted to the Louisiana Division of Archaeology, Department of Culture, Recreation, and Tourism, Baton Rouge, LA.

Saltus, A. R. Jr. (1987). *Submerged cultural resources investigation of the western portion of the Maurepas Basin.* Submitted to the Louisiana Division of Archaeology, Department of Culture, Recreation, and Tourism, Baton Rouge, LA.

Saltus, A. R. Jr. (1988). *Submerged cultural resources investigations of various waterways of Lake Pontchartrain's north shore.* Submitted to the Louisiana Division of Archaeology, Department of Culture, Recreation, and Tourism, Baton Rouge, LA.

Steffy, J. R. (1994). *Wooden ship building and the interpretation of shipwrecks.* College Station: Texas A&M University Press.

Surrey, N. M. (1916). *The commerce of Louisiana during the French regime, 1699–1763. Columbia University Studies in History, Economics, and Public Law, 61(1).* New York: Columbia University Press.

Tunis, E. (1961). *Frontier living: An illustrated guide to pioneer life in America, including log cabins, furniture, tools, clothing, and more.* New York: Thomas Y. Crowell (Reprinted in 2000 by The Lyons Press, Guilford, CT).

Chapter 8
The Watercraft of Castle Island, Washington, North Carolina

Bradley A. Rodgers, Nathan Richards, Franklin H. Price, Brian Clayton, Andrew Pietruszka, and Heather White

Introduction

From the late 1990s to the turn of the decade, three East Carolina University archaeological summer field schools (1998, 1999, and 2000) documented submerged watercraft near Castle Island in Washington, North Carolina. Over these 3 years Dr. Bradley A. Rodgers, the Principal Investigator for the Castle Island study, taught archaeology and site documentation in the murky waters of the Tar River. Professor William Still first suggested that submerged cultural resources may be located near Castle Island in 1985, while he was Director of the Program in Maritime History and Underwater Archaeology (PMHUA). Dr. Still conducted a

B.A. Rodgers, Ph.D. (✉) • N. Richards, Ph.D.
Program in Maritime Studies, East Carolina University,
Admiral Ernest M. Eller House, Greenville, NC 27858-4353, USA
e-mail: rodgersb@ecu.edu; richardsn@ecu.edu

F.H. Price
Bureau of Archaeological Research, Florida Department of State Tallahassee, Florida 32301, USA

B. Clayton
Deep Sea Acoustic Surveyor, Fugro Chance Inc., Houston, TX
e-mail: claytonbt@icloud.com

A. Pietruszka
Underwater Archaeologist, School of Marine Science and Policy
College of Earth, Ocean, and Environment, University of Delaware,
700 Pilottown Road, Lewes, DE 19958, USA
e-mail: andrewpi@udel.edu

H. White
Assistant Director for Assessment & Engagement, Joyner Library,
East Carolina University, Greenville, North Carolina 27858, USA
e-mail: whiteh@ecu.edu

A.M. Evans (ed.), *The Archaeology of Vernacular Watercraft*,
When the Land Meets the Sea, DOI 10.1007/978-1-4939-3563-5_8

walking survey of the island in the early 1980s and identified the remains of at least two near-shore vessels. He also noted that he could see the remains of two center-board schooners at low water, an estimated 200 yards (182.88 m) west of a point between the island and an unidentified south shore pier ruin. These vessel remnants were assigned North Carolina State site numbers 0014PMR and 0015PMR and are not included in the 1998–2000 survey. Later investigation by Rodgers determined that these vessels were but two among many abandoned wooden vessels that lay mostly submerged in the area surrounding Castle Island and the south shore of the Pamlico River near Washington.

During the 1998 and 1999 field seasons, ECU documented ten vessels. The damaging effects caused by flooding after Hurricane Floyd in 1999 led to a third field season in 2000. The 2000 investigation of the dynamic underwater environment sought to determine what changes, if any, had occurred to the archaeological sites during hurricane related flooding. After preliminary reconnaissance and diving, the primary investigator determined that divers could not conduct data recovery at the ten sites located near or adjacent to Castle Island. This was for two reasons: divers could only relocate two of the sites, and the water depth had changed dramatically from the gently sloping sand beach of the island in June 1998–1999, to 25 ft. (7.62 m) of depth with a new hard pan sand bottom, no silt, and very little loose sand. It seemed the raging currents created by Hurricane Floyd had undercut the upstream or north side of the island. Since the island appears to be composed of sand, major cave in and sloughing of the face of the island made for an ever-present danger to any divers working in the area. In all, the island had become a very unwelcoming place to dive. Of the ten vessels located from the previous year, only one could be immediately relocated underwater by divers, and another above water. Researchers felt Vessel Seven's entire submerged 60 ft. bulk rocking back and forth in the current on the bottom, teetering on an unseen fulcrum, created a dangerous, and unnerving situation in the zero visibility water. Notwithstanding their possible loss, archaeologists assigned individual North Carolina State site numbers to the vessels located in the 2 previous years for the completion of the research.

The overall goal of the 3 year survey was to complete a Phase II pre-disturbance reconnaissance of the vessels in the Castle Island Ships' Graveyard. The objective of a Phase II survey is a detailed site map, photographic imaging and interpretation of the site, as well as an examination of individual artifacts for diagnostic purposes. In all very few artifacts remained on these vessels, indicating quite clearly that they represented, for the most part, abandoned watercraft rather than shipwrecks. In addition, with few exceptions, river conditions had fully exposed the vessels on the bottom, and they contained very little silt and sand. In most cases, construction details became obvious with a small amount of hand fanning.

Investigation of each vessel was an extension of previous work, and generated documentation sufficient to confirm the construction and architectural record associated with the surviving hull structures. Personnel conducted additional historical research in museums, archives, and libraries to identify primary and secondary source material associated with the 11 vessels, and the maritime history of Washington. Researchers also undertook an environmental study to investigate the various geological processes underway within the vicinity of Castle Island. The greatest lesson learned through this study concerned the effects of Hurricane Floyd, which impacted North Carolina. The combination of archaeological and historical documentation preserves construction and design data at each site that may now be lost, and has somewhat mitigated the disastrous effects of Hurricane Floyd on the Castle Island area.

The sites at Castle Island are a diverse and significant cross section of the rivercraft, coastal vessels, and working boats of North Carolina. Such vessels played a significant role in the development of commerce in both the region, and the state. They range from flats such as oyster shell barges for the island's lime kilns, to sailing vessels of the coastal fisheries, and even include a stern wheel steam vessel, probably employed in upriver commerce and trade. The beauty of the Castle Island Ships' Graveyard, as the project area is known, is that it represented a plethora of vessel types all in one spot, where the archaeologist could record and document the activities of the cultural landscape through the remains left behind.

Historians believe that Europeans first settled the Washington area by the early eighteenth century (Hodges 1976:9–11; Paschal 1976:1–2; Hill 1984:2–3). It was a prominent center for supply and privateering during the Revolutionary War, and an important center for shipbuilding up to the Civil War (Attmore 1922:28–29; Paschal 1976:2–3; Worthy 1976a:6, 1976b:8–9; Hill 1984:4). The town was also significant because of its central place in the timber milling, naval stores industries, and various fisheries (Myers 1937:35; Litchfield 1976:230; May 1976; Hill 1984:4–6). Today's Washington economy has dedicated use of the Tar-Pamlico waterway for recreational purposes, promoting fishing, boating, and water sports. In 1974, the State of North Carolina contributed to this effort by purchasing land and creating the nearby Goose Creek Park area (Morgan and Abeyounis 1976:512).

Throughout its early history, Washington was an important trading center and played an important role in the economic development of eastern North Carolina. Washington's location as a port at the confluence of the Tar and Pamlico Rivers made it a prosperous center of trade and the major port of the Tar/Pamlico River system from the nineteenth century onward. When the railroads advanced through the state, however, and the agricultural commerce on the river declined, the importance of water borne commercial traffic disappeared. In addition, as commercial vessels grew larger and ship building materials changed to the use of iron and steel toward the end of the nineteenth century, shipwrights abandoned Washington, for all practical purposes, as a center for their business activities. It could not accommodate the ever larger deep draft vessels or the complexities and cost of steel ship construction. Shipbuilding, therefore, moved to both Wilmington and Morehead City, ports endowed with deeper harbors. Yet Washington continues to thrive into the twenty-first century. Its once impressive late nineteenth-century business district continues to change, metamorphosing from large business, banking, and commercial concerns to restaurants and businesses that cater to visitors and tourists enamored of the recreational opportunities afforded by the waterfront community and the beauty and ambiance of the small port town. Many of the watercraft abandoned at Castle Island represent the ability of the local population to adapt to changing conditions and convert or adaptively reuse past technologies and equipment to work in new economic settings and circumstances (Bridgers 1974; McCabe 2007:123, 154, 156).

The diversity of the vessels located at Castle Island gives archaeological testament to the community's broad economic base, that ranged from the transportation of agricultural commodities and people on the Tar River to the use of local fisheries, and even the broader picture of ocean borne trade and commerce via what may be, large, locally built sailing schooners. Abandonment surveys are part of a growing international trend in underwater archaeology to invest time and effort in the documentation of local working craft, the vessels that invariably made or broke the economic fortunes of an individual community. Researchers see these craft for the important historic characters that they were, and recognize their potential to fill gaps in knowledge concerning working ships and their relationship to an area's cultural landscape (Richards 1997, 1998, 1999, 2002).

The changes visited on the Castle Island Ships' Graveyard during the flood of 1999 seems significant, and although the final project report was originally intended as a survey and listing of archaeological resources available for further in-depth study, it may now be the only archaeological work that can be conducted on some of the wrecks previously located there. Research suggests that Hurricane Floyd may have moved, scrambled, or buried these sites. Fortunately, it seems that the other abandoned vessels and wreck sites located on the south shore of the Tar River mouth may have escaped the destruction of Floyd, and await documentation in further projects in the years to come.

Project Location and Environment

Castle Island is located in the Tar-Pamlico River Basin across from the waterfront of the town of Washington, North Carolina (Fig. 8.1). The Tar-Pamlico River Basin originates in the Piedmont of North Carolina near the city of Roxboro. The Pamlico River section of the basin starts in Washington. The Tar-Pamlico River Basin is 167 miles in length (268 km) and covers a relief of 590 ft. (180 m). Two miles upstream from Greenville, North Carolina, the gradient of the basin and its water velocities decrease significantly. The total area of drainage is 3050 mile^2 (7900 km^2) (Fournet 1990:1).

The identified cluster of 11 vessels was located slightly north and northwest of Castle Island across from the Washington waterfront (Fig. 8.2). During the 1998 investigation, the sites were located in an average of approximately 5 ft. (1.52 m) of water. Following Hurricane Floyd, divers discovered that Vessels One, Two, and Five from the cluster were missing or moved and the water was now 20 ft. (6.09 m) deep. Wreckage was located at the sites of Vessels Three, Four, Six, Seven, Eight, and Nine but whether this wreckage represents the same craft documented in 1998 and 1999 is unknown. Vessel Eight, the only other vessel near the island that survived the flood, is located on shore northwest of the project's primary datum point in the same location it was before the hurricane. Vessel Eleven also survived and is located underwater northwest of Castle Island, and appears unharmed, possibly due to its shallow location away from the island and main channel.

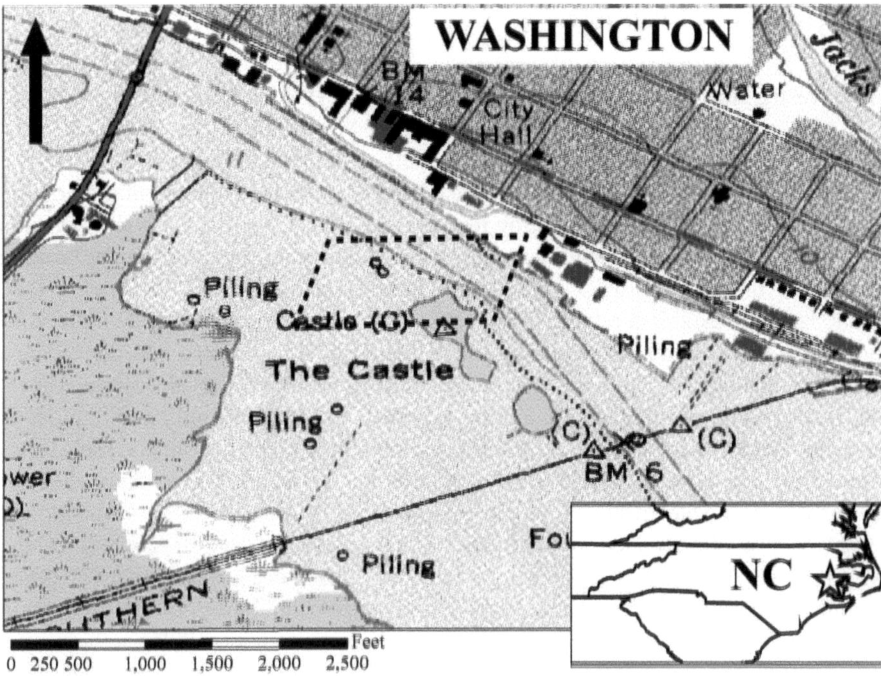

Fig. 8.1 Map of Washington, North Carolina, showing the Washington waterfront, the location of Castle Island, and the area of study (Plan by Maritime Studies Program, East Carolina University, adapted from USGS topographic map, 2005)

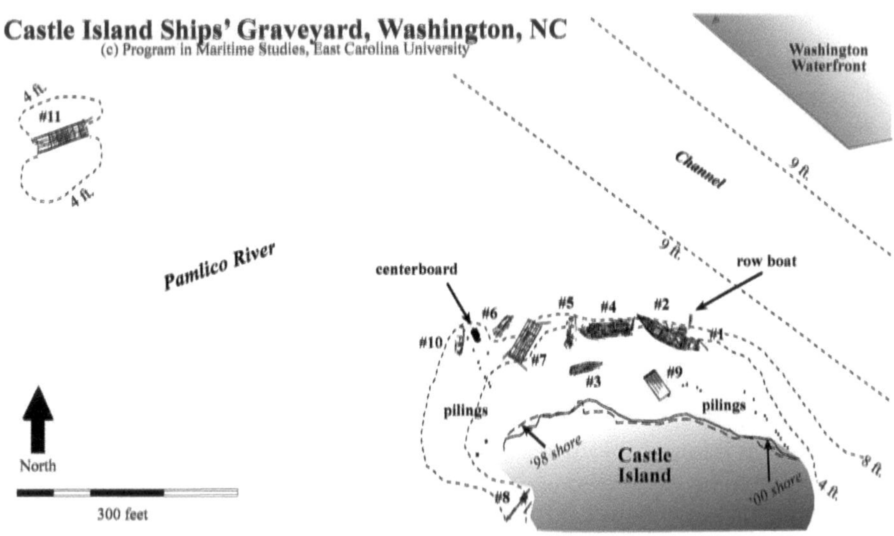

Fig. 8.2 Master plan of the watercraft studies during the ECU Maritime Studies field schools, 1998–2000 (Plan by Maritime Studies Program, East Carolina University, 2005)

Description of Findings

The investigation of the Castle Island Ships' Graveyard commenced in 1998 as a part of East Carolina University's Program in Maritime Studies Summer Field School. Since researchers could see the vessels in the area at low water, the project called for a diver survey followed by Phase II pre-disturbance documentation of any located sites. Phase II documentation in maritime archaeological methodology calls for a plan view map of the area with sites and anomalies plotted within the broader context of the archaeological site and nearby land formations. Following this, researchers can establish relationships between the sites that are located on the foreshore of the island and the waterfront of Washington. Phase II is a non-disturbance archaeological study which calls for no permanent artifact retrieval. Instead, divers opt to document artifacts in situ, and no excavation occurs beyond hand fanning or sweeping debris from the site. At the time, researchers identified nine vessels (later found to be ten) close to Castle Island, between the island and the northern shore of the Washington waterfront. As per the methodology, the wreck sites, or abandoned vessels, were documented in plan view and plotted within the context of the Washington maritime district or waterfront. Since this was a pre-disturbance assessment, the sites would remain intact for closer inspection, and possible excavation for Phase III work would only occur if research questions warranted it in the future.

The 1998 field season documented Vessels Three, Five, Six, Seven, Nine, and Ten. Investigations continued during the 1999 field season with an examination of Vessels One and Two. The 2000 field season reexamined previously documented watercraft to determine what disturbances Hurricane Floyd may have caused in the fall of 1999. Project investigators hypothesized that significant flooding may have relocated some, if not all of the vessels. An additional site (Vessel Eleven) was located southwest of the island during the 2000 investigation and drawn, using Phase II procedures, in plan view. The description and interpretation of these vessels follows (Table 8.1).

Vessels One and Two

The site originally designated Vessel One was later determined to be two separate watercraft, a coasting type schooner (Vessel Two) with a flat or barge (Vessel One) lying on top of the aft section of the schooner (Fig. 8.3). Examination of the site confirmed that only the port side of the schooner remained intact. The rest of the hull, including its centerline (keel, keelson), appears to have been destroyed or pulled apart. It is impossible to determine the draft or depth of hold of the vessel due to its condition. Its size, however, is consistent with a coasting schooner configuration that would likely have contained a centerboard for stability and added sailing capability on a shallow draft.

Table 8.1 List of individual submerged sites near Castle Island

ECU No.	State designation	Type	Salient features	Approx. dimensions	Tentative date of construction
1.	None	Flat	Chine log construction	W: 15′	Late nineteenth century
2.	0014PMR	Schooner	Midships wrecked no centerboard	L: 95′	Early-mid nineteenth century
3.	0054PMR	Sharpie schooner/ terrapin smack	Chine log construction	L: 35′ B: 8′	Last quarter nineteenth century
4.	0055PMR	River steamboat *Alma*?	Western river design	L: 43′ B: 15′ D: 3′	Turn of twentieth century
5.	0056PMR	Oyster sloop	No extant centerboard	L: 30′ B: 10′	Last quarter nineteenth century
6.	0057PMR	Oyster sloop	Centerboard offset to port	L: 30′	Last quarter nineteenth century
7.	0058PMR	Flat	No chine log	L: 50′ B: 15′	Last quarter nineteenth century
8.	0059PMR	Motor boat Sophie wood?	Engine mount saddle	L: 53′ (keelson)	Turn of twentieth century
9.	0060PMR	Flat	No chine log	L: 38′ B: 17′	Last quarter nineteenth century
10.	0015PMR	Sailing log canoe	Centerboard and cabin combing	L: 40′ B: 12′	Third quarter of nineteenth century
11.	0061PMR	Flat	Chine log construction	L: 75′ B: 30′	Turn of twentieth century

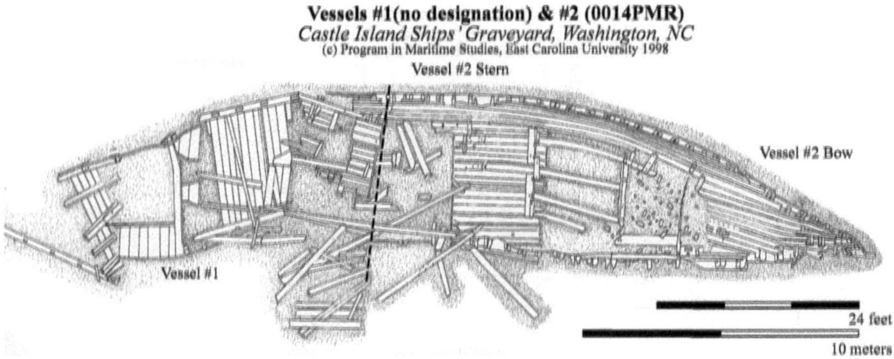

Fig. 8.3 Plan view site map of Vessels One and Two (Plan by Maritime Studies Program, East Carolina University, 2005)

The vessel is approximately 95 ft. (28.96 m) in length with an approximate beam of 30 ft. (9.1 m). The starboard side of the vessel is completely gone, potentially separated during channel dredging or breaking operations. The vessel was large enough to have participated in ocean-going trade at 300–350 t. and contained brass fasteners, indicating that a copper alloy had once sheathed it below the waterline as an anti-fouling measure, a sure indication of ocean trade capacity. Rosehead fasteners used throughout the site date it to within the first half of the nineteenth century. The vessel had double framing and butt scarfs, suggesting it may be a post-colonial or nineteenth-century vessel (Steffy 1994:293). If a centerboard was originally present on the craft, salvagers may have removed it around the time they destroyed the starboard portion of the vessel. Interestingly, divers mapped a disarticulated 15 ft. (4.6 m) centerboard between Vessels Six and Ten. The size appeared to be consistent with that of the centerboard that may have originally been present in Vessel Two.

Confusion regarding the orientation of planking in the stern of the ship, originally thought to be from the ship's transom, was eventually determined to be part of another vessel, a flat (Vessel One), lying on top of the schooner. Low visibility conditions never allowed researchers to see the superimposed vessels; they only discovered them in the drafting stage. Tell-tale signs for the archaeological team included the fact that planking from Vessel One ran counter to the planking on the schooner.

Standard cut nails and extruded iron fasteners suggest that the construction of the flat, Vessel One, dates from the later part of the nineteenth century when these industrial techniques were common (Sutton and Arkush 2002:164). Two chine logs were present on either side of the vessel at the 90° turn of the bilge, indicating standard scow-type construction, rectangular with a flat bottom and slab sides. An inclined apron on Vessel One is a further indicator of this utilitarian vessel type.

Sources suggest that schooners like Vessel Two typically replaced earlier eighteenth-century vessel types such as sloops (Olsberg 1973:189–299; Swanson

1991:57–58; Southerly 2003:53). These coastal schooners normally had two mast steps compared to the one found on sea going sloops. Smaller schooner sails were more manageable since they were divided between two masts, a characteristic that made it possible to build nineteenth-century schooner rigged vessels of a much larger tonnage than the eighteenth-century sloops (Alford 1990:33). Though larger in tonnage, ship-builders often constructed these vessels with a relatively shallow draft. Patented in 1811, the centerboard achieved widespread popularity in the 1830s and 1840s. The centerboard allowed operators to raise or lower a retractable board depending on water depth, helping the vessel sail on and off the wind without much leeway (sideslip), even with a shallow draft (Barkhausen 1990:9).

Archaeological analysis suggests that someone converted the large coasting schooner (Vessel Two) to a workboat or barge late in its working life. This adaptive reuse was a common practice for vessels that had reached the end of a useful life in oceanic trade in lieu of total abandonment (Richards 2008:118, 122–136). It seems likely the ship could have undergone conversion to a barge for harbor use, but only after powered craft such as tugboats could manage and maneuver the hulk toward the end of the nineteenth century. Industrial groups on Castle Island would put such barges to use transporting commodities such as lime or lumber to shore for loading on railroad cars or transshipment to ocean going ships. Ultimately, the ship's owners abandoned the vessel near the shoreline of Castle Island. It remains unclear if it burned accidentally, or if salvagers torched the vessel intentionally in order to facilitate better recovery of valuable fasteners.

Unfortunately, the site formation process for Vessels One and Two may now be undetectable as divers could not relocate either watercraft in 2000. During diver inspection of the site it became obvious from bottom scouring, the replacement of silt with hard pan sand, and the vastly deeper water depth, that both vessels had been moved by the river current and perhaps redeposited. The presence of hard pan sand rules out the concept that environmental processes reburied the vessels in place at a deeper depth.

It is clear, however, from preliminary research that the schooner is older than the flat that came to rest on it, and that the schooner shows signs of sea going use in trade based on its ocean type brass fasteners. It was built possibly as early as the Federal period (if it did not originally have a centerboard), or mid-century (if it did have a centerboard). Though introduced in the early part of the nineteenth century, centerboards were not common until the late 1830s and 1840s. The presence of a large centerboard between Vessels Six and Ten perhaps tips the scales in favor of the later mid-century date for Vessel Two. As mentioned, the hull of the vessel shows obvious signs of burning. Burning was a typical way to salvage fasteners from a hull undergoing scrapping in the nineteenth century (Matthews 1987a:160, 231, 1987b:3, 151, 186, 229, 301, 332), but was also an all too common end for vessels, particularly barges, carrying lime (Labadie and Herdendorf 2004:10–11). The chemical reaction to the introduction of water from a leaky hull infiltrating large amounts of lime created tremendous heat, and the danger of flash fire. Without further study, requiring the relocation of Vessels One and Two, there will be no definitive answers to the questions of age, use, and site formation processes.

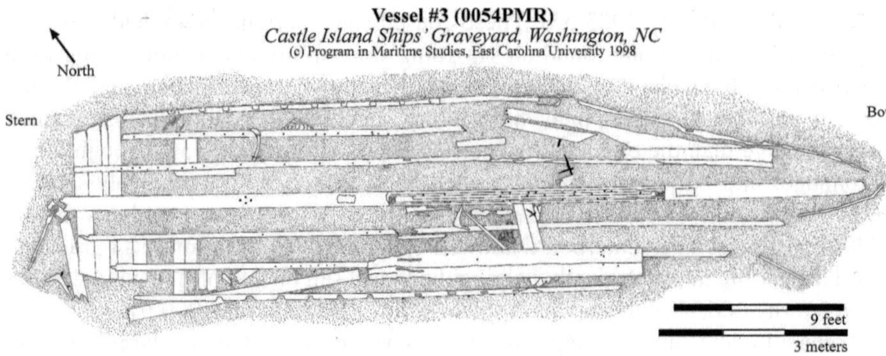

Fig. 8.4 Plan view site map of Vessel Three (Plan by Maritime Studies Program, East Carolina University, 2005)

Vessel Three

Vessel Three was determined to be the remains of a "bottom built" vessel akin to a sharpie, also known as a North Carolina sharpie schooner or a terrapin smack (Fleetwood 1995:148) (Fig. 8.4). "Bottom built" or "bottom based" refers to the fact that this type of hull is based entirely on a flat bottom with the sides built up on a hard chine, or nearly 90° turn of the bilge, while the side supports are patterned differently than the bottom (Hocker 2005:66). The hull is remarkably similar to a scow schooner except outwardly at the bow and stern, where its appearance is more conventional with a pointed bow and a raked transom stern. Internally the vessel resembles a scow, complete with athwartship bottom planking, longitudinal stringers, and chine logs. It also contains an internal keel (central longitudinal stringers) on which the two masts are stepped. Unlike a scow, however, this vessel has no flat ramp or apron at each end. The chine logs, or stringers, have notches for king posts that rise from the turn of the bilge to support the sides of the vessel, much as frames do on a conventional hull. Given that the hull remains were intact only to the chine, researchers estimated the overall dimensions, which are approximately 35 ft. (10.67 m) in length with a beam measurement of 8 ft. (2.44 m). It is impossible to determine the depth of hold of the hull due to the lack of structural remains of the sides, but it would probably have had a shallow draft, not exceeding 3 ft. (1 m).

Vessel Three could be a type of centerboard workboat, with the centerboard clearly visible through the keel, and the rudder still attached at the stern. The keel (there is no keelson) runs the length of the vessel with two mast steps, and is accompanied by two bilge stringers per side for added longitudinal support and to act as nailers for the athwartship bottom planks. The straight athwartship planking style is slightly different from the usual herringbone pattern of a Chesapeake sharpie style hull, perhaps indicating a local variant. This type of work boat was used during the last quarter of the nineteenth century and well into the twentieth century in the shell fishing industry when, as time passed, many of these vessels were motorized. While the vestiges of the bottom of the ship are identifiable, the structural remains of the sides disappeared before field documentation commenced. Vessel Three represents a local variation on the theme of a Chesapeake Bay "sharpie schooner" that likely flourished toward the end of the nineteenth century in this area. Typically sharpies were used as oyster dredges along the inland waterways and sounds because of their steadiness and wide beam (Chapelle 1951:110, 117, 121). It seems probable that someone abandoned this craft on the shoreline of Castle Island after its useful life as a working boat in the shellfishery was over. After abandonment, the vessel remained relatively undisturbed until the arrival of Hurricane Floyd in 1999. The depth of the water near this site has changed dramatically since that time. Before the hurricane's arrival, divers found the vessel at a depth of 5 ft. (1.52 m). After hurricane flooding, the water in this area is approximately 20 ft. (6.09 m) deep, and the site can no longer be located.

Vessel Four

Vessel Four represents the remains of a small stern paddle wheeler that measured approximately 43 ft. (13.1 m) in hull length with a beam of approximately 15 ft. (4.57 m), and oriented approximately southeast to northwest (Fig. 8.5). The draft and depth of hold were impossible to determine from the remains as the interior of the boat was detritus-laden. Nonetheless, the slope and depth of the cylinder timbers indicate the riverboat had a three foot depth of hold and a draft of about 2 ft. (0.66 m). The characteristic traits of this vessel include its shallow draft hull based on a system of internal bracing not seen on any of the other vessels located near Castle Island. This vessel, although of a smaller size, is internally similar to construction

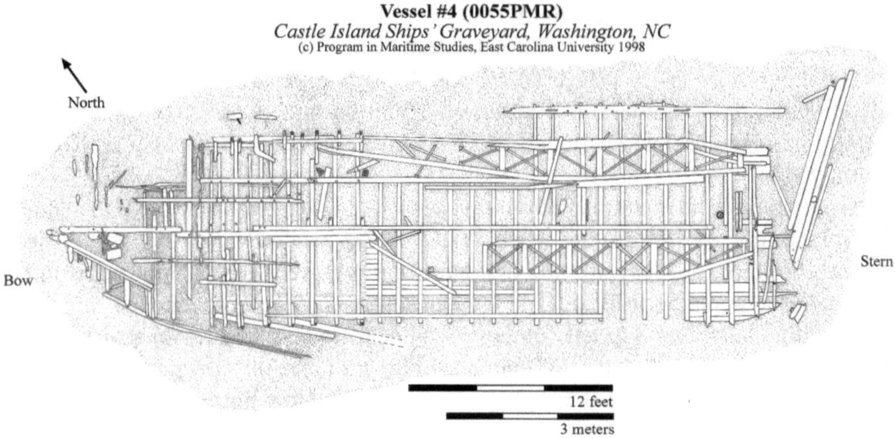

Fig. 8.5 Plan view site map of Vessel Four (Plan by Maritime Studies Program, East Carolina University, 2005)

Fig. 8.6 Unidentified stern wheel paddle steamer on the Tar River (date unknown) (Courtesy of the North Carolina Department of Cultural Resources)

demonstrated on "western river" or "mountain" steamers (Corbin and Rodgers 2008:74). People used these western river steamers on the great rivers of the American West, such as the Missouri, and Mississippi. A small version of this type of river boat would have been well suited to the shallow waters of the upper Tar River, and could have carried significant deck cargoes of tobacco, cotton, and other agricultural commodities (Fig. 8.6). It seems likely, on careful comparison, that Vessel Four is what is classified as an "Up River Boat," a phrase coined by Washington, N.C. builders to describe their flat-bottomed vessels exhibiting outwardly flaring sides and shallow draft (Rodgers et al. 2008:78; George and Brown Library 2010).

At the time of its documentation, this vessel contained no remaining machinery within its hull, and the paddle wheel could not be located within the survey area. Machinery items are typically the first hardware removed from a ship that is in the process of scrapping or salvage, and it is not surprising that these elements are missing. Additionally, two internal composite cylinder timbers that lie on their sides on either side of the stern section of the vessel hint at the vessel's original purpose and construction. In their original upright positions, the cylinder timbers would have supported horizontal steam cylinders, one per side. The after part of the cylinder timbers would have extended aft of the stern of the hull and supported the pillow blocks for the stern paddle wheel. The slope of the vessel's stern apron appears to be mirrored in the incline of the aft part of the cylinder timbers. The only difference between this ship and its larger western cousins is that in the west the cylinder timbers were massive constructions of solid timber. This vessel, therefore, may show a later construction date because of the use of more modern composite (wood and iron) cylinder timbers. These composite timbers would also have been lighter but likely stronger than solid timbers because of the diagonal bracing within their structure, seen as a series of "X-shaped" structures within the cylinder timbers in plan view.

The hull of Vessel Four demonstrates some classic western river construction with floors that extend across the beam of the boat. It was not possible, because of the amount of debris and the damage done to the sides of the craft, to determine if the turn of the internal bilge demonstrated the classic chine clamp and cocked hat construction arrangement of a western steamer, but it seems more likely that the outwardly flaring chine was formed from obtuse angled knees attached to the floors as demonstrated in the Old Sparta Vessel, another of the "Up River Boats" documented in 2008 (Rodgers et al. 2008:55–57, 67). Three stringers per side rested on and stabilized the floors, while acting to stiffen the hull longitudinally. This ship was lightly constructed, even by western steamer standards, and would have been able to float in very shallow water, likely less than 2 ft.

There is no evidence of decking except for a piece of the starboard sponson located just aft of the main section of the vessel. Sponsons, or guards, are narrow side platforms that would have extended down each side of the deck from the bow area to past the paddle wheel in the stern. The guard not only added a few feet to the beam of the deck (and consequently also to the ship's cargo capacity), but it allowed the ship's engineers to walk out and lubricate the pillow blocks supporting each end of the paddle wheel shaft, as well as the pitman arm and crank bearings connecting the engines to the paddle wheel. There are no signs of boiler supports in the internal structure of this steamer. This is expected of a vessel based on the western river model because the design of these vessels normally called for the mounting of the boiler or boilers on deck.

Vessel Four is an oddity among the watercraft located near Castle Island in that it appears to have been steam powered as evidenced by the composite wood/iron cylinder timbers lying flat within the hull. It is likely that this steamer dates to the late nineteenth century or even twentieth century because of the sophisticated structure of the composite cylinder timbers only recently corroborated archaeologically (McKay and Pollack 2009), but similar to the "trussed wood keelson" described by Charles Desmond in his classic *Wooden Ship-Building* (Desmond 1998:96). In this case they are likely cylinder timbers rather than a keelson, do not extend the length of the vessel, and there are two of them rather than a single keelson down the centerline. No doubt, its owners used this vessel to transport goods, commodities, and people up and down the Tar-Pamlico River system, though it could also have been used in other nearby rivers. Additionally, a number of scotch boilers are located in the river on the south side of Castle Island. While beyond the scope of this survey, these boilers could easily be of the correct vintage to have performed in this vessel. Unfortunately, Vessel Four may have joined the list of missing or scrambled sites since the visitation of Hurricane Floyd.

Though it would be much too early to pronounce a match, one steamer that fits the description of this abandoned steam vessel is *Alma*. For its entire short career,

1897–1899, *Alma* operated out of Washington and carried agricultural goods under command of Captain George R. Jones. Research indicates that *Alma* had an extremely shallow draft of 2 ft. (0.66 m), and displaced 16 t. *Alma's* hull dimensions of 41 by 13 ft. fit well within the parameters of the approximate dimensions of this archaeological site. *Alma*'s period of operation also coincides with that of the archaeological record estimated for Vessel Four. After too brief a career, one historical source lists *Alma* as being "torn up and abandoned, 1899" (Bridgers 1978:211). No other historical information currently pinpoints the place of this vessel's breaking.

If subsequent historical research can prove that Vessel Four is *Alma*, it lends credence to the notion that vessels surrounding Castle Island represent the remnants of a marine scrapping operation. Certainly, most of the vessels surveyed, including Vessels One, Two, Three, Four, Five, Six, Eight, and perhaps Vessel Ten appeared in 1999 to be in various stages of being broken up. Whether this is a result of natural site formation process or breakers tools cannot immediately be deduced or proven.

Vessel Five

The remains of Vessel Five represent a small, locally built North Carolina type sailing vessel, possibly an oyster sloop (Fig. 8.7). The length of the site measures 30 ft. (9.14 m) with a 10 ft. (3.04 m) beam. Some difficulties were associated with recording the site due to the amount of debris covering the vessel. Though originally built as a single-masted shallow draft sailing vessel, it is odd that it contained no discernible centerboard, generally a requirement in such a shoal draft vessel. Although it is possible that someone removed the centerboard and trunk, or that it broke away from the vessel, a slot in the bottom of the boat was not visible because of sand and debris.

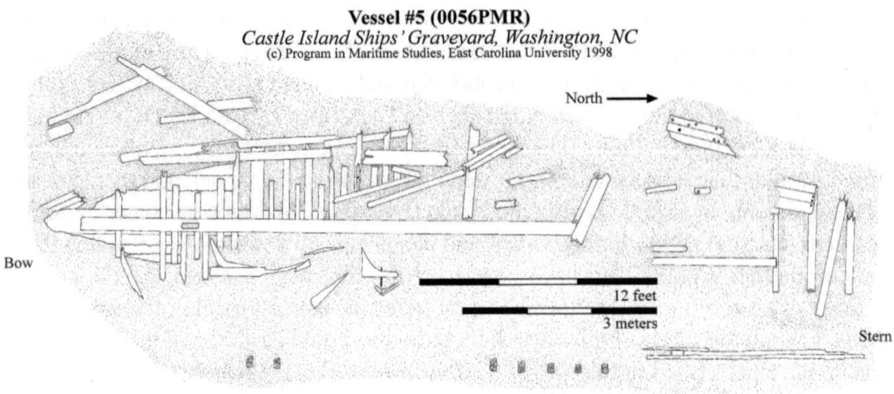

Fig. 8.7 Plan view site map of Vessel Five (Plan by Maritime Studies Program, East Carolina University, 2005)

A single mast step, indicating its sloop rig, was present 7 or 8 ft. (2.1–2.4 m) from the bow of the vessel. The uniform construction, mill cut lumber of standard size, and hardware fasteners including screw and wire nail fastenings loosely date this vessel to the end of the nineteenth and the beginning of the twentieth century (Fleetwood 1995:305–310). The vessel is flat-bottomed with what must have been a hard chine and slab sides. Framing is standard with floors running under the keelson, and planking running the axial length of the craft. The sides and stern of the vessel are not extant in the archaeological record, but the shape of the bottom indicates a sharp bow with a gentle side curve expanding to the midships bend and tucking in aft to a flat transom.

Though small, Vessel Five appears to represent a riparian, or (on calm days) a Pamlico Sound working vessel. Oyster sloops were shallow draft craft used primarily as day sailing working craft. Construction of these boats was an art, and generally did not involve plans or calculations. Invariably, ship builders constructed these vessels with a mind to maximizing cargo space. As Fleetwood (1995:307) notes: "It can be said that the construction of these boats was as simple and uncomplicated as it was possible to achieve." In the Castle Island area, the sloop may have worked in the shellfish industry or as a lighter to transfer shells for lime processing. The placement of the craft within the concentrated cluster of other vessels supports the idea that it had served out its useful lifetime and was abandoned, or was in the process of being salvaged.

Vessel Six

Vessel Six was very similar to Vessel Five except that this craft did have a centerboard (Fig. 8.8). The length of the working boat measured 30 ft. (9.14 m). It was impossible to determine the beam, but it is likely that it was 10–12 ft. (3.0–3.7 m). The starboard side of the vessel is oriented toward the island. An offset centerboard was present on the site, as was a single mast step located far forward on the keelson. Divers identified a few frames, some floors, and planking. The vessel has a standard build with floors sandwiched between keel and keelson and planking and ceiling planking running fore and aft on the axial plane. The bottom has no deadrise, so it was likely hard-chined and slab-sided like Vessel Five.

Vessel Six is likely another small oyster sloop of the late nineteenth century, very similar if not identical to that of Vessel Five with the same overall characteristics (Fleetwood 1995:305–310). The offset centerboard is a peculiar construction detail that generally disappeared on larger vessels by the middle of the nineteenth century (Barkhausen 1990:14). Theoretically, it is possible that the keel/keelson structure is too small in its molded and sided dimensions on this craft to withstand the weakening imposed by cutting the centerboard through these structures, hence the offset. The offset centerboard may also reflect local building practices or even indicate that the sloop is older by some 30 or 40 years than its overall structure indicates.

As with the other watercraft sites surrounding Castle Island, its placement may indicate that it represents an abandonment event. The vessel's observed rough condition, with missing sides and stern, could also indicate salvage or high energy site formation process.

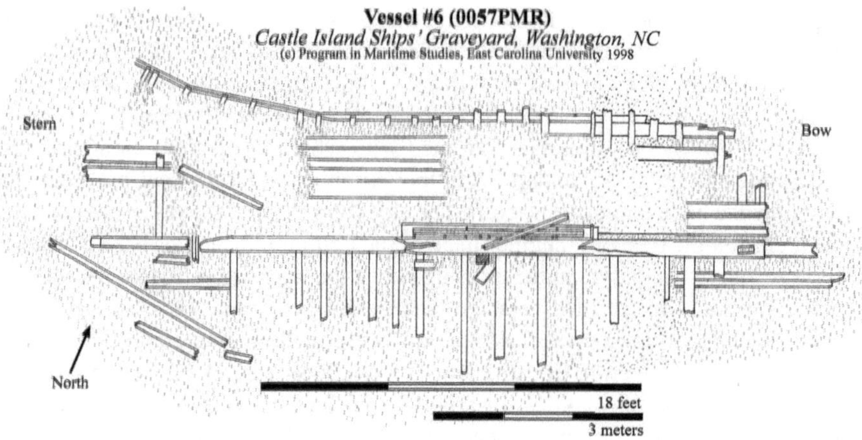

Fig. 8.8 Plan view site map of Vessel Six (Plan by Maritime Studies Program, East Carolina University, 2005)

Vessel Seven

Vessel Seven was located during the preliminary investigations conducted on Castle Island by Maritime Program Director William Still prior to the 1998 field season (Fig. 8.9). Historical interpretation suggests that this is an oyster barge, flat, or less likely, a ferryboat (Watts and Hall 1986:26–34). It is 50 ft. (15.24 m) long and 15 ft. (4.57 m) in beam, with a typical rectangular shape. Five longitudinal stringers give the vessel longitudinal support. There are no indications of chine logs or stringers located at the 90° turn of each bilge, and nothing to show that king posts (a flat's version of frames) supported its flat sides. The craft is largely intact with side and bottom planking in place but an absence of the characteristic ceilings or railings of a ferry.

The rectangular shape, overall dimensions, and construction suggest that the vessel may have once operated as a towed oyster barge, scow, flat, or plantation flat. A flat, unlike most other ship names, is a literal description. It has a flat-bottomed, barge-like hull with a very shallow draft, and ends that ramp up called aprons. Archaeologists have found similar vessels throughout both North and South Carolina (Watts and Hall 1986; Fleetwood 1995:102). Their simplistic build and utility make them a universal working boat as lighters and shallow water freight haulers (Watts and Hall 1986). People constructed flats to carry maximum loads down current with large sweeps or poles providing steering. While in port, steam or motor type vessels would be engaged to tow them.

There is no indication of exactly how anyone used Vessel Seven, and no indication from the archaeological data that it was ever sailed. There are, for instance, no mast steps in the middle longitudinal stringer and no indication of standing or running rigging. Nonetheless, in North Carolina, the flat or barge could be schooner-rigged and, if narrow enough in beam, could sail reasonably well (Fleetwood 1995:51).

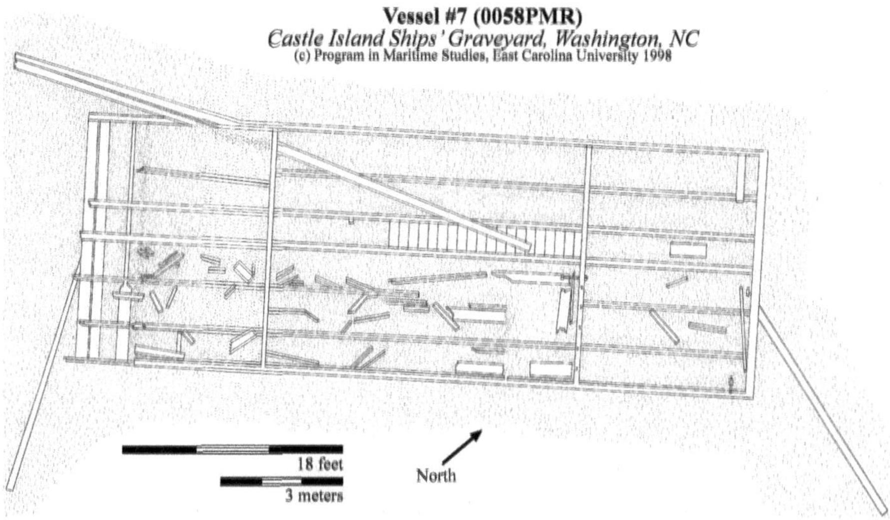

Fig. 8.9 Plan view site map of Vessel Seven (Plan by Maritime Studies Program, East Carolina University, 2005)

These vessels were well suited for riverine commerce and not intended for offshore use. Like the other Castle Island vessels, this craft probably represents an abandoned watercraft. Inexplicably, divers were not able to relocate Vessel Seven after Hurricane Floyd.

Vessel Eight

Vessel Eight was located during the preliminary investigation of the 1998 field season (Fig. 8.10). The wooden remains were limited to a rectangular wooden engine support saddle (4 by 6 ft. or 1.22 by 1.82 m) connected to a keelson of approximately 53 ft. (16.2 m) in length. The keelson is supported by two sister keelsons running nearly the entire length of the keelson. Without excavation, it was impossible to tell if floors or much outer hull planking exists under the sand. In some places, iron keel straps were evident. There is no sign of a boiler or boiler saddle or supports, making it likely that an internal combustion engine powered the vessel. Most notable on this site is the engine mount saddle with shaft hole located in the deadwood near the stern. The stern is the closest feature to Castle Island. This is an unusual type of beaching as the stern hog (deeper draft) usually mandates that the bow beach first. Other features include iron pins, hex head nuts with washers, scarf joints, and minimal remaining hull planking.

Vessel Eight most likely represents a motorized propeller driven fishing or transport vessel of the early twentieth century. This site could not be relocated in 2000 but has since reemerged from the sand, and is visible at the time of this report.

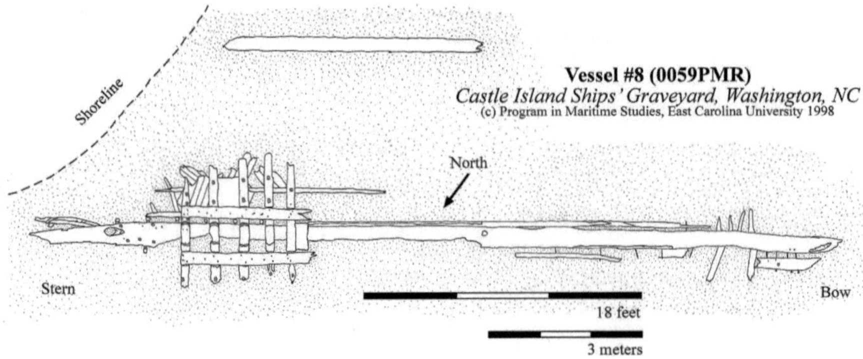

Fig. 8.10 Plan view site map of Vessel Eight (Plan by Maritime Studies Program, East Carolina University, 2005)

Oddly enough, archival sources list a vessel of near these dimensions named *Sophie Wood*, as having worked for the Pamlico Transportation Company out of Washington, having begun its career out of Edenton in 1891, owned then by John G. and F. Wood. The vessel, built in East Lake, worked an unspecified number of years out of Edenton before its transfer to the Pamlico Transportation Company. It worked under several masters including Howard L. Brooks, Robert B. Jackson, Macon H. Bonner, R.S. Griffin, and John A. Roberts. In 1905, W.H. Whitley of Durham Creek gained title, with James H. Harris as the vessel's master. *Sophie Wood*'s final owners beached the vessel on "the Castle" that same year when in a sinking condition. The vessel was a propeller, measuring 63 ft. in length by 12 by 4 ft. (19.2 by 3.7 by 1.2 m). Its final enrollment of September 28, 1914, lists the vessel as having been out of commission since 1905 and destroyed by the great Hurricane of 1913 (Bridgers 1978:209–210).

The condition of Vessel Eight precludes gaining its exact dimensions without excavation; therefore, naming the remains as *Sophie Wood* is premature. Nonetheless, speculation indicates that the wreckage may represent a motorized propeller vessel of standard build for the late nineteenth and early twentieth century (Fleetwood 1995:250–255). As the keel dimensions do not include features like a fan tail (that tend to extend the overall vessel length), the wreckage could indeed represent a vessel of 63 ft. in overall length. Finally, this seems to be the only vessel haphazardly beached on the island stern first, separated physically from the other vessels, perhaps indicating that its owners beached it in haste. Richards (2008:9) calls this a "consequential abandonment," indicating that the circumstances of this vessel's destruction are perhaps much different from those located nearby. Regardless of Vessel Eight's actual name, there can be no doubt that the hurricane of 1913 all but destroyed this hapless abandoned craft while wreaking havoc on all of the abandoned watercraft off Castle Island.

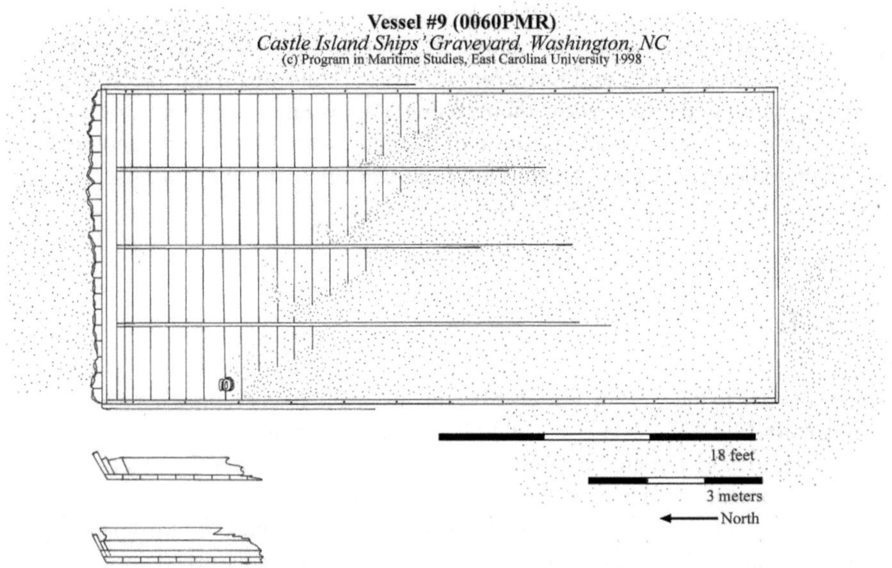

Fig. 8.11 Plan view site map of Vessel Nine (Plan by Maritime Studies Program, East Carolina University, 2005)

Vessel Nine

Vessel Nine was located during preliminary investigations prior to the 1998 field season (Fig. 8.11). The flat lies embedded in the Castle Island north shore, near Vessel One with only a few feet of one end exposed. The remains represent yet another working flat estimated at some 38 ft. (11.7 m) long and 17 ft. (5.2 m) in beam. Except for the north end, Vessel Nine lies embedded in the island, and filled with miscellaneous debris such as sand, brick, oyster shell, glass, and a ship's deadeye. This flat resembles Vessels One and Seven in its rectangular shape and construction detail except that it has fewer (only three) longitudinal stringers. Like Vessel Seven, it has no chine logs or stringers at the turn of the bilge. The outer-planking on the sides of the craft were edge fastened to one another with through bolts and are 2 in. (5.08 cm) thick by 10 in. (25.4 cm). Since the ends of both craft face upriver, one can theorize that Vessels Seven and Nine may represent intentional abandonment events, designed to extend the deep face of the landing wharf at Castle Island. Vessel Seven now lies in deeper water and is debris-laden; therefore, it is not possible to prove the circumstances of abandonment. It is unclear what affects Hurricane Floyd brought to Vessel Nine; at the time of recording, sand and debris covered two-thirds of the vessel. Divers could not relocate this vessel after the hurricane of 1999.

Flats of this type (Vessels One, Seven, Nine, and Eleven) all resemble the North Carolina flat barge tradition, which can be traced to European flats and barges, and used all over North America (Watts and Hall 1986:23, 31; Fleetwood 1995:102;

Rodgers and Corbin 2003:210–211). The shallow draft, beamy and wedge-ended designs are extremely useful in shallow North Carolina waterways for lightering shell, agricultural goods, and fertilizer. The use of flats was multi-faceted; people could use them under tow, allow them to float downstream in currents, or even sail them. The good condition and position of the flats near Castle Island lend credence to the hypothesis that they possibly had outlived their original purpose, and had become extensions to the island's wharves.

Vessel Ten

Vessel Ten lies to the west of Vessels Six and Seven and likely represents a deep water fishing craft (Fig. 8.12). It is just over 40 ft. (12.19 m) long and 12 ft. (3.66 m) in beam. This vessel may be a centerboard sloop or schooner, perhaps of a log canoe variety of bugeye or brogan design most common in the Chesapeake region. Though it was mostly buried, divers discovered a centerboard trunk surrounded by the coaming of a deck house, and ballast rock piled aft at the site. Researchers discerned very few other details during the preliminary survey of the vessel due to the depth of sand overburden. A small number of frame ends protruded from the bottom, and no mast steps were visible. The keel of this sailing canoe is likely a single carved U-shaped log, a feature consistent with the construction techniques for sailing log canoes of the time. The frames and hull appear built up from the keel in a traditional manner, with floors sandwiched by the keelson. The only actual dugout log vestige of these craft are their keels, otherwise they look like a traditional sailing vessel. Unfortunately, divers could not examine the keel of Vessel Ten because of the degree to which sand had buried it.

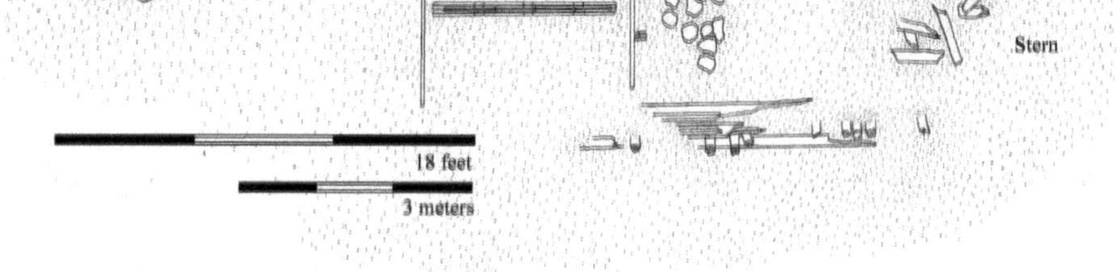

Fig. 8.12 Plan view site map of Vessel Ten (Plan by Maritime Studies Program, East Carolina University, 2005)

While deep sea fishing may have been the original intent for this craft, how people employed it in Washington remains a mystery. Nonetheless, the vessel would have been a fast and seaworthy sailor that could have easily traversed the sound to fish off the coast. It is possible that in its later days its owners transferred it to the local oyster industry, or alternatively that they converted it into a barge. The vessel dates from mid to late nineteenth century. Historical records point to numerous vessels of this type present in Washington, but few if any examples remain. This vessel is unique in the archaeological collection near Washington and would be worth further investigation to gain information concerning this particular variation of a Chesapeake Bay craft. Divers did not relocate its remains after Hurricane Floyd.

Vessel Eleven

In both 1998 and 1999 a public outreach segment of the Washington field school included setting up a sheltered booth in the town's waterfront area for the purpose of educating the public about the survey and soliciting information from the community. Residents, happening by the public outreach booth mentioned that a ferry barge used by the public to cross the river was lost during the storm of 1913. The location of the loss, toward the south shore across from the Washington waterfront, is where Vessel Eleven was located during a survey in 2000. Vessel Eleven, easily detected on sonar, is in such shallow water that people can see it from shore when a low tide combines with a west wind to lower the water in the sound (Fig. 8.13). The vessel is a flat measuring 75 ft. (22.8 m) long, and 30 ft. (9.14 m) wide and is constructed of yellow pine (Fig. 8.14). With the exception of a few missing bow and

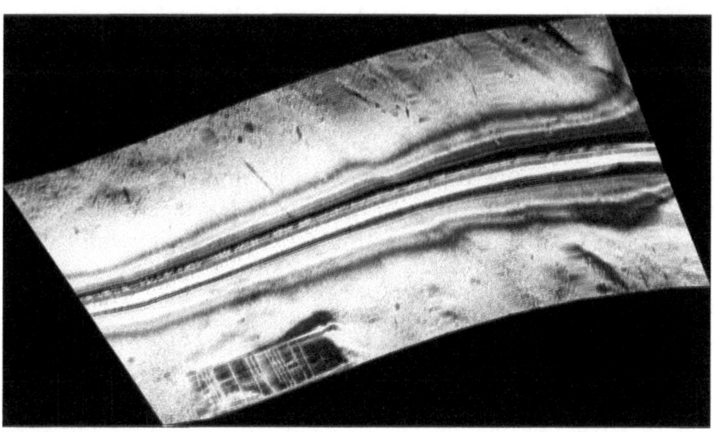

Fig. 8.13 Side scan sonar image of Vessel Eleven (0061PMR) (Image by Maritime Studies Program, East Carolina University, 2005)

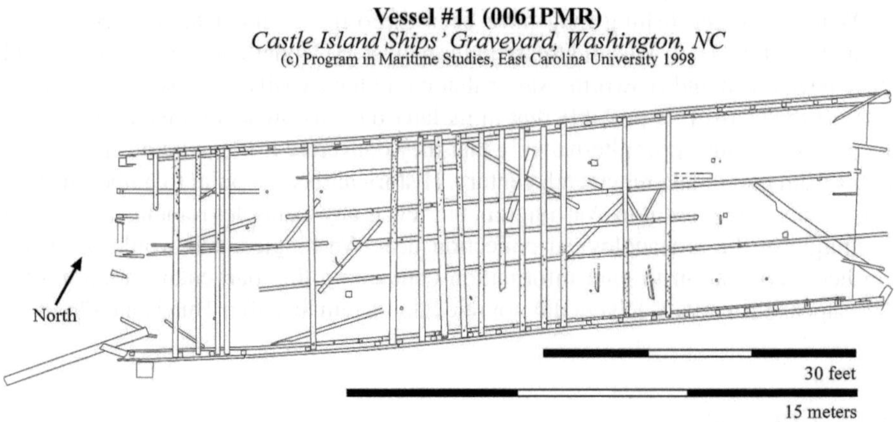

Fig. 8.14 Plan view site map of Vessel Eleven (Plan by Maritime Studies Program, East Carolina University, 2005)

stern apron planks, the vessel is intact and survived Hurricane Floyd. Most fasteners were made of iron, though investigators also discovered a number of treenails. Through-hull fasteners were round wire nails, as were the nails found on deck beams. This is a fastener not prevalent until the latter stages of the nineteenth century, and continues to be the common nail today (McCarthy 2005:90–91). One may also interpret a large number of in situ nail heads as a sign of salvage activity, suggesting that at some time prior to abandonment salvagers retrieved the entire internal deck by simply prying the wood free of the nails while leaving the nails attached.

Vessel Eleven's construction does deviate somewhat from some of the other flats located near Castle Island. Vessels Seven and Nine do not contain chine logs and are somewhat smaller than Vessel Eleven. The chine log or stringer is located at the turn of the bilge and supports the vessel longitudinally while it contains pockets in which king posts are set to serve as frames and supports for the side planking of the vessel. Originally some of these king posts may have continued past the shear to serve as a deck railing and guide for the ferry cable leads. Like Vessel Eleven, Vessel One also contains chine logs, though little else of the burned craft was evident. In addition to the chine logs on Vessel Eleven, the ship's builder laid cross keelsons on the top of the five internal longitudinal stringers, as well as the two chine logs to serve as a base for the traffic deck. This deck, with its longitudinal planks, took the weight of passengers, animals, and carts. The traffic deck is missing from the site and presumed salvaged.

The storm of 1913 must have rivaled Hurricane Floyd in intensity and devastated parts of Washington. Destruction included damage to all of the bridges in the area, requiring the temporary use of ferry barges that may have included Vessel Eleven. It is unclear, therefore, whether the Hurricane of 1913 sunk Vessel Eleven, if someone used it to mitigate the effects of the hurricane by letting traffic continue across

Fig. 8.15 Ferry running between north and south shores of the Tar River, Washington, NC (ca.1913). Ferry may correspond to remains of Vessel Eleven (Photograph courtesy of Mr. Hugh Sterling)

the river while the bridges underwent repair, or if Vessel Eleven is unrelated to this historic event. A photograph provided by Hugh Sterling of Washington shows this ferry, or one very much like it in operation, complete with passengers (Fig. 8.15). It is unclear how the ferry is operating. The horse seems attached to the cart in the ferry, and is not providing power for the cable. Additionally, the ferry operator does not appear to be walking the cable across the deck. By 1913, a powered winch system may have been set up on shore to pull the craft across the river.

Vessel Eleven lies just on the other side of the river from where the photographer took his or her image. Analysis of the three flat examples located near Castle Island may determine that chine log construction on a flat indicates it is a ferry, whereas the absence of a chine log indicates a working barge or plantation barge. Other archaeological examples do not contribute to this conclusion. The two ferries studied in the Northeast Cape Fear River do not have chine logs, and even though both were from an earlier era, they do not show traits contrary to those discovered on Vessels Seven and Nine, making conclusions concerning the use of chine logs difficult (Watts and Hall 1986:23, 33).

Divers recorded Vessel Eleven in 2000. The wreckage does not in any way appear to be associated with the cluster of vessels located near Castle Island; therefore, its deposition in the archaeological record is likely not associated with the island watercraft. The archaeological evidence provides no clue as to whether it is an abandoned or wrecked vessel, other than the fact that someone had removed the traffic deck. Furthermore, divers discovered no artifacts associated with passengers, or ferrymen adjacent to the site.

Conclusion

The wrecks or abandoned watercraft that surround Castle Island in Washington, North Carolina, offer an invaluable glimpse into the life and times of this bustling nineteenth-century port town. Of the eleven vessels documented in this survey, all represent various aspects of the area's economic base, and relate in a material way concerning how area ship and boat builders designed craft to reflect the work that they were to accomplish. These vessels vary from swift, ocean fishing watercraft to plodding, sound shellfishing vessels. They also reflect modest agriculture transports for the river, both steam-powered and non-powered, as well as large coastal and international trading ships. The watercraft of Castle Island also show how the historic community extended road transportation across rivers and carried on lime production through the extended use of utilitarian bottom built flats. In fact, the only major nineteenth-century industries not represented in the remains of these vessels that contemporary Washington entrepreneurs were engaged in are the timber, lumber, and naval stores industry. These activities are of course represented archaeologically in other areas in or near Washington, but are not reflected in the ruined craft on the north and northwest side of the island, except in the predominantly softwood timber that makes up their hulls.

The Castle Island wreck and abandonment sites run the gamut of named North Carolina and import vessel types including flats (barges and ferries), coasting schooners, North Carolina sharpie schooners, oyster sloops, a sailing log canoe, and an Up-River Steamer. In some instances, the Castle Island study reveals the internal construction of these small vessels for the first time making it an invaluable archaeological resource for anyone wishing to study these vernacular working craft. Research suggests that people brought many of these vessel designs from other areas, probably working in local variations; therefore, the Castle Island examples may have shown just how the North Carolina versions differed from their namesakes (Fleetwood 1995:iv). The shallow draft Up-River Steamer is also atypical of the eastern seaboard type of steamer, and shows a distinct relation to mountain and western river designs, again pushing the knowledge envelope for types of local steamboats.

In all, therefore, Castle Island represents a smorgasbord of vessel types widening the definition of "typical" work boats of the area. This survey also demonstrates that the ship and boat building prowess of the local builders was far more complex than had been anticipated, not in their engineering plants, but rather in the design and construction of shallow up-river hulls. This, in turn, demonstrates the burgeoning economic and cultural importance of the city of Washington near the turn of the twentieth century. It truly was a hub of commerce, agriculture, industry, and technical acumen—particularly in vessel design and use.

Unfortunately, the work carried out at Castle Island from 1998 to 2000 was only a pre-disturbance archaeological survey. There was no excavation or in-depth archaeological study. The authors intended this survey to describe the resources available for further study in the survey area and do not consider it the final word concerning each of these sites.

Ironically, one of the intentions behind this 3 year Phase II survey was mapping and documentation that could demonstrate site formation process and the impact of both human and natural forces on the sites at Castle Island. Nature, however, proved that it is the final arbiter of archaeological site formation process, fully capable in this instance of scrambling a site or possibly wiping the slate clean of large amounts of cultural material. Humbling as it may seem, even embedded shipwrecks may disappear through natural and energetic site formation, and natural disaster scenarios even after a century of preservation and seeming protection.

The full extent of the damage done to the sites at Castle Island is still unknown. Future survey work in the Tar-Pamlico River and drainage area is necessary, and East Carolina University researchers may carry this out in the future. Remote sensing (sonar and magnetometer) surveys of Castle Island undertaken in May 2010 have revealed that a majority of the submerged vessels remain at the Castle Island location, but have either moved or are now mostly buried, providing a better idea of site formation in this periodic high energy environment. Future underwater archaeological work near the island is not feasible until the north face of the island sloughs and stabilizes. Even a return to a stable environment will affect the cultural resources adjacent to the island. The slumping and movement of the island will swallow and bury ruined structures and artifacts near the north face just as it buries structures and artifacts on the south and east face. Rivers are indeed dynamic, ever changing environments, and so are the islands that inhabit them.

Finally, the 1998–2000 survey data may assist future archaeologists and historians to better plan and conduct meaningful work, and serve as a springboard to more sophisticated studies. The vessel remains located in and near Washington, North Carolina, represent a valuable, nonrenewable, and as documented following Hurricane Floyd, an unfortunately all too easily impacted resource. Nevertheless, there are other remains of historical working craft still located on or near the bottom lands of Washington that were not impacted by the flood. This work increases awareness of these archaeological sites, their value, and most importantly, points researchers toward asking proper questions into the future. Castle Island provided a case study in which to formulate these questions, and may help archaeologists find the answers.

Acknowledgements We would like to thank a number of people for their assistance with the archaeological survey, the preparation of the original technical report (Rodgers et al. 2005), and this chapter. In particular, thanks to Director Blount Rumley and staff of the Washington Estuarium, for their donation of resources and facilities for the project as well as dock space for our research vessels. In addition, many Master's students in the Program in Maritime Studies at East Carolina University worked long hard hours under some difficult conditions to insure the accuracy of the documentation and fieldwork at Castle Island between 1998 and 2000. In alphabetical order, the 1998 field crew was composed of William C. Aycock, Cathy Fach, Doug Jones, Jason Lowris, Rod Linder, Sarah A. Milstead, Suzanna Pavelle, Larkin Post, Greg Purdy, John Rossi, Chris Southerly, Kathy Southerly, and Kim Williams. The 1999 field crew included Eric Bruning, Garry Byrd (Diving Safety Office), Tane Casserley, Jen Dorton, Kim Eslinger, Russ Green, Mike Hughes, Doug Jones, Matt Lawrence, Dede Marx, Sarah Milstead, Matt Muldorf, Mike Plakos, Larkin Post, and Steve Sellers (Diving Safety Officer). Finally, the 2000 field crew included, Sam Belcher, Heather Cain, Suzanne Finney, Russ Green, Stephen A. Hammack, M.J. Harris, Mike Hughes, Matthew Lawrence, David Miller, Marc Porter, Giovanni Wagemans, and Scott M. Whitesides.

Public outreach for this project included setting up an information tent along the waterfront with project drawings and diagrams manned by graduate students who explained to interested townsfolk how the survey was proceeding and what researchers had discovered to that point. The outreach segment was a great success that brought in more historical information from the town's longtime citizens, folks who knew the waterfront intimately. Special thanks, therefore, goes to the townspeople of Washington for their hospitality and interest in the survey. In this light, long time residents of Washington, Mr. and Mrs. Hugh Sterling were especially helpful with photographs and information concerning the waterfront. Other helpful informants include Mr. Whiting Toler and Ms. Bee Morton.

Thanks also to the staff and faculty of the Maritime Studies Program for their support and good advice, including Dr. Larry Babits, Dr. Tim Runyan, Dr. Annalies Corbin and especially Frank Cantelas for his logistical and technical help on the project and his bravery in operating the research vessels. Thanks also go to Chris McCabe, dedicated graduate student and the researcher who acquired information on vessels *Alma* and *Sophie Wood*. In addition, this work could not have been completed without research and editing by Theresa Hicks and Steve Williams. All work conducted during the archaeological survey documented in this report was done under North Carolina State Permit Number 00PMR608.

References

Alford, M. B. (1990). *Traditional work boats of North Carolina*. Harkers Island, NC: The North Carolina Maritime Museum, Hancock Publishing.

Attmore, W. (1922). Journal of a tour to North Carolina by William Attmore, 1787. *James Sprunt Historical Publications, 17*(2), 1–46.

Barkhausen, H. N. (1990). *Focusing on the centerboard*. Manitowoc, MI: Manitowoc Maritime Museum.

Bridgers, H. C., Jr. (1974). *Extracts from certificates of enrollment*. Manteo, NC: Outer Banks History Center.

Bridgers, H. C. Jr. (1978). *Steamboats on the Tar*. Unpublished manuscript. Manteo, NC: Outer Banks History Center.

Chapelle, H. I. (1951). *American small sailing craft*. New York: W.W. Norton and Company.

Corbin, A., & Rodgers, B. A. (2008). *The steamboat Montana and the opening of the west. Leviathan of the plains: History, excavation, and architecture of the steamboat Montana*. Greenville, NC: East Carolina University.

Desmond, C. (1998). *Wooden ship-building*. Vestal, NY: Vestal Press. (Original work published 1919).

Fleetwood, W. C., Jr. (1995). *Tidecraft: The boats of South Carolina, Georgia and Northeastern Florida—1550–1950*. Tybee Island, GA: WBG Marine Press.

Fournet, S. C. (1990). *Pleistocene-Holocene sediment aggradational history of the lower Tar river and the upper Pamlico river Estuary, North Carolina*. Master's thesis, Department of Geology, East Carolina University, Greenville, NC.

George, H., & Brown Library, L. E. (2010). Abstract of articles in *Washington Daily News*, Lucretia Hughes Files (Vol. 5). Washington, NC.

Hill, M. R. (1984). *Historical research report: The waterfront area of Washington, North Carolina*. Microfilm, Historical Research Reports: Series 2, Number 34, North Carolina Division of Archives and History, Raleigh, NC.

Hocker, F. M. (2005). Bottom-based shipbuilding in Northwestern Europe. In F. M. Hocker & C. A. Ward (Eds.), *The philosophy of shipbuilding: Conceptual approaches to the study of wooden ships* (pp. 65–93). College Station: Texas A&M Press.

Hodges, N. D. (1976). *Historical highlights of Washington and Beaufort County North Carolina bicentennial edition 1776–1976*. Major Reading Blount Chapter, National Society, Daughters of the American Revolution, Washington, NC.

Labadie, C. P., & Herdendorf, C. E. (2004). *Wreck of the steam barge adventure: An archaeological investigation in Lake Erie at Kelleys Island, Ohio*. Vermillion, OH: Great Lakes Historical Society.

Litchfield, Y. D. (1976). Shipping. In U. F. Loy & P. M. Worthy (Eds.), *Washington and the Pamilco* (pp. 225–247). Raleigh, NC: Washington-Beaufort County Bicentennial Commission.

Matthews, F. C. (1987a). *American merchant ships 1850–1900: Series 1*. New York: Dover Publications.

Matthews, F. C. (1987b). *American merchant ships 1850–1900: Series 2*. New York: Dover Publications.

May, L. G. (1976). Lumber, Part 1: The story of Beaufort County's lumber industry. In U. F. Loy & P. M. Worthy (Eds.), *Washington and the Pamilco* (pp. 329–352). Raleigh, NC: Washington-Beaufort County Bicentennial Commission.

McCabe, C. P. (2007). *The development and decline of Tar-Pamlico river maritime commerce and its impact upon regional settlement patterns*. Master's thesis, Department of History, East Carolina University, Greenville, NC.

McCarthy, M. (2005). *Ships fastenings: From sewn boat to steamship*. College Station: Texas A&M Press.

McKay, J. W., & Pollack, J. C. (2009). *Interpretive reconstruction of the hull of Seattle no. 3.*. College Station, TX: Institute of Nautical Archaeology.

Morgan, J. I., & Abeyounis, B. (1976). Since World War II. In U. F. Loy & P. M. Worthy (Eds.), *Washington and the Pamilco* (pp. 510–513). Raleigh, NC: Washington-Beaufort County Bicentennial Commission.

Myers, L.W.W. (1937). Myers Family Papers (#431), East Carolina Manuscript Collection, J. Y. Joyner Library, East Carolina University, Greenville, North Carolina, USA.

Olsberg, R. N. (1973). Ship registers in the South Carolina archives, 1734–1780. *South Carolina Historical Magazine, 74*, 189–299.

Paschal, H. (1976). In the beginning. In U. F. Loy & P. M. Worthy (Eds.), *Washington and the Pamilco* (pp. 1–6). Raleigh, NC: Washington-Beaufort County Bicentennial Commission.

Richards, N. T. (1997). *The history and archaeology of the Garden island ships' graveyard*. Honors dissertation, Department of Archaeology, Flinders University, Adelaide, Australia.

Richards, N. T. (1998). Inferences from the study of iron and steamship abandonment: A case study from the Garden island ships' graveyard, South Australia. *Bulletin of the Australian Institute for Maritime Archaeology, 22*(1), 75–80.

Richards, N. T. (1999). The Garden island ships' graveyard: Results and findings of archaeological fieldwork 1996–1998. *Proceedings of the National Archaeology Students' Conference 1998* (pp. 11–18). Australian National University, Canberra, Australia.

Richards, N. T. (2002). *Deep structures: An examination of deliberate watercraft abandonment in Australia*. Doctoral dissertation, Department of Archaeology, Flinders University, Adelaide, Australia.

Richards, N. T. (2008). *Ships graveyards: Abandoned watercraft and the archaeological site formation process*. Gainesville: University Press of Florida.

Rodgers, B., & Corbin, A. (2003). Mud box filled with stone: The wreck of the scow schooner Dan Hayes. *International Journal of Nautical Archaeology, 32*(2), 210–224.

Rodgers, B. A., Hicks, T. R., & Wyllie, E. (2008). *The enigma of old Sparta*. Greenville, NC: Program in Maritime Studies, East Carolina University.

Rodgers, B. A., Richards, N., Price, F. H., Clayton, B., Pietruska, D., White, H., et al. (2005). *The Castle Island ships' graveyard: The history and archaeology of eleven wrecked and abandoned watercraft—the 1998–2000 Castle Island field seasons*. East Carolina University Program in Maritime Studies Research Report #14, Greenville, NC.

Southerly, C. W. (2003). *Cedar on the reef: Archaeological and historical assessment of eighteenth-century Bermuda sloops exemplified by the wreck of the* Hunter Galley. Master's thesis, Department of History, East Carolina University, Greenville, NC.

Steffy, J. R. (1994). *Wooden ship building and the interpretation of shipwrecks*. College Station: Texas A&M University Press.

Sutton, M. Q., & Arkush, B. S. (2002). *Archaeological laboratory methods: An introduction* (3rd ed.). Dubuque, IA: Kendall/Hunt Publishing.

Swanson, C. E. (1991). *Predators and prizes: American privateering and imperial warfare, 1739–1748*. Columbia: University of South Carolina Press.
Watts, G., & Hall, W. (1986). An investigation of blossom's ferry on the Northeast Cape Fear river. East Carolina University Program in Maritime History Research Report #1, Greenville, NC.
Worthy, P. (1976a). They fought for freedom. In U. F. Loy & P. M. Worthy (Eds.), *Washington and the Pamilco* (pp. 6–7). Raleigh, NC: Washington-Beaufort County Bicentennial Commission.
Worthy, P. (1976b). The town develops. In U. F. Loy & P. M. Worthy (Eds.), *Washington and the Pamilco* (pp. 8–14). Raleigh, NC: Washington-Beaufort County Bicentennial Commission.

Chapter 9
The Shallop of Hart's Cove

David C. Switzer[*]

Introduction

In New Hampshire's Piscataqua River off New Castle Island lie the remains of a small colonial-era vessel. Hart's Cove is near the mouth of the river and has been a safe haven and cargo offloading location for lighters and small vessels since the 1600s (Fig. 9.1). Discovered in 1980 during a remote sensing survey of the Piscataqua Basin by Kittery Historical and Naval Museum under the auspices of the University of New Hampshire's Sea Grant program, the wreck provided archaeologists and historians with a rare look at the ubiquitous genre of colonial vessel, known as *chalupa* or *shallop*, that migrated from Europe with Basque whalers of the sixteenth century and over the intervening centuries became the workhorse of New World coastal trade (Grenier et al. 2007:1–52; also Loewen this volume).

The Hart's Cove remains suggest a vessel of about 35–45 ft (10.66–13.7 m). The associated artifacts, primarily ceramic, included North Devon Sgraffito ware, redware, stoneware, and ubiquitous kaolin pipe fragments, which suggested a sinking date in the late seventeenth or early eighteenth century. The medallion on a fragment of Westerwald mug bearing the name "Wilhelmus Rex" referred to King William III, who came to the throne in 1688 as part of the "Glorious Revolution" (Fig. 9.2). The pipe stem bores cluster around 6/64, and following methods developed by Binford (1978), provide an estimated sinking date in the late 1690s (Switzer 1980). Although evidence points to a probable sinking date, the launch date of the shallop could have easily been years earlier, perhaps as much as a decade, as illustrated by evidence of hull repair.

---

[*]Author was deceased at the time of publication

D.C. Switzer (✉)
Department of History, Philosophy, and Social Studies Education, Plymouth State University, 17 High Street, MSC #30, Plymouth, NH 03264, USA

A.M. Evans (ed.), *The Archaeology of Vernacular Watercraft*, When the Land Meets the Sea, DOI 10.1007/978-1-4939-3563-5_9

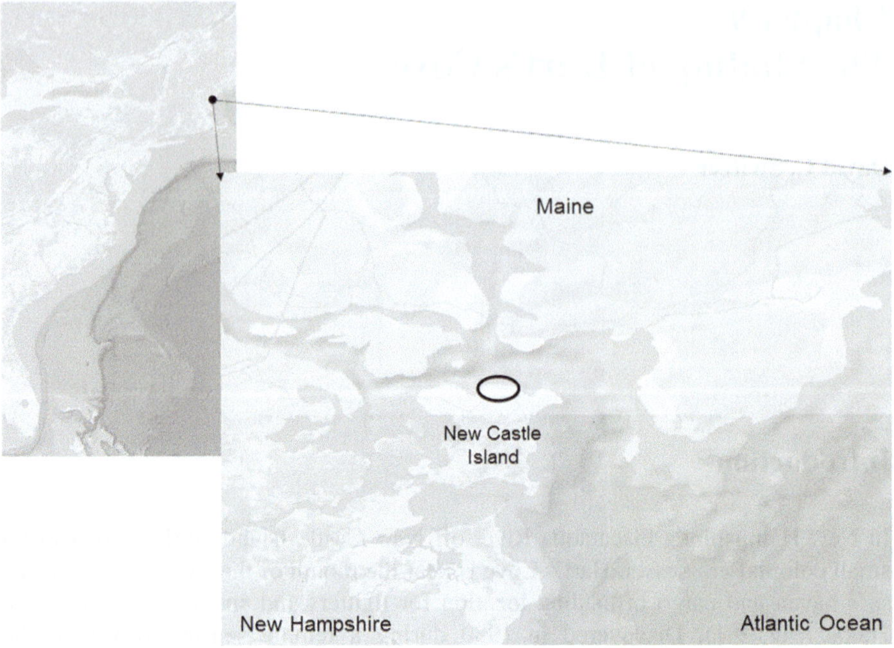

Fig. 9.1 Hart's Cove is located on New Castle Island, along the Piscataqua River. Image modified from Google Earth

Fig. 9.2 Westerwald mug fragments

Far from being a barrier between early American colonies, the coastal waters and rivers were the highways down which communication and trade flowed. Early journal references tell how Native Americans lent their piloting skills to the Basque and, later, the English colonists. They also tell that after a very short period, Native mariners were regularly seen moving up and down the Atlantic seaboard in shallops. Small- or medium-sized watercraft like the shallops were extremely important in the colonies, but very few vessels dating from this period have been examined in the United States. Vessels like the Hart's Cove shallop were used for fishing in the inshore waters, trading voyages throughout the colonies, defense, and general transportation. The wreck in Hart's Cove offers a tantalizing suggestion of the,range of such boats. A lump of coral, for example, was excavated from between the floors. The coral's location illustrates that, at least on one voyage, the shallop was using coral as ballast. The more important question is where did the captain find the coral? Perhaps the coral was brought to New Hampshire in a larger vessel, dumped onto a ballast pile, and reused by the shallop captain, but it is possible that the shallop had traveled as far south as Bermuda and picked up the coral at the source. Shallops were the threads that stitched widely spaced colonies together, and the study of their remains allows some insight into the lives of seventeenth-century settlers, the resources available to them, and how they adopted a style of vernacular watercraft to suit their needs.

Repairs and modifications made to the hull of the Hart's Cove shallop present information on the economy of the time. While timber was abundant in the Piscataqua region, labor was not. Unskilled labor was necessary to turn timber into lumber with axes, pitsaws, and adzes; semiskilled or skilled labor was required to fashion raw timber into ships' construction components. Rather than construct a new vessel to replace the worn-out Hart's Cove shallop, seventeenth century shipwrights strengthened the hull with crudely finished frames. Instead of repairing the hull by replacing all or most of the rotted or broken plank, the vessel's owner trimmed it back the barest minimum. A patch was tacked into place without overlapping it to the midpoint of the frame that secured it (Fig. 9.3). Surely this saved money, but it might have

Fig. 9.3 Photograph of hull patch on the Hart's Cove vessel

compromised safety, too. Was the shallop at the end of a long life when these repairs were made? Was it serving restricted duty in the river, lightering goods from larger vessels instead of engaging in long-distance trade itself? Or, were these types of repairs standard practice in seventeenth century colonial boatyards?

An examination of the Hart's Cove shallop's construction and repairs provides a glimpse of both the economy and the shipwrighting skills of the region. Shipbuilding was a major component of the economy in coastal colonies. The towns of Piscataqua and Durham, located upriver from New Castle, had access to the hinterlands and timber, enjoyed a robust shipbuilding industry, and were closely connected to the pottery industry of North Devon (Switzer 1991, 1985; Grant 1983). The remains of what is believed to be the 1626 shallop *Sparrowhawk*, put on display at Boston Common in 1865, reveal a vessel with a curved bow and a duck-tail stern that tucks into a sternpost (Holly 1969:11) (Fig. 9.4). Although *Sparrowhawk* predates the Hart's Cove wreck and was more intact, there are a number of shared features, particularly the scantlings.

The carvel-built hull of the Hart's Cove wreck was fashioned from pine hull strakes and ceiling planking; the frames were made of oak (Fig. 9.5). The remains measure 31 ft (9.3 m) along the keel, which would have rendered an overall vessel length of approximately 35–45 ft (10.66–13.7 m). A short 8.2 ft (2.5 m) oak keelson contains a mortise close to the vessel's midpoint; the mortise is believed to be a mast step (Fig. 9.6). Many similar vessels carried a small foremast as well as the central main mast, but

Fig. 9.4 Remains of the 1626 shallop *Sparrowhawk*, on public display

Fig. 9.5 Site plan drawing of the Hart's Cove vessel, as of 1987

Fig. 9.6 Photograph of the mortise believed to be a mast step

conclusive evidence for a second mast was not found on the Hart's Cove site. The vessel has a rounded hull with a projected overall beam from the offsets of 11–12 ft (3.4–3.7 m). Scantling comparisons of similar type vessels that had an estimated length-to-breadth ratio of 4–4.5:1 between perpendiculars suggest that the Hart's Cove shallop would have measured around 48 ft (14.6 m) in length, which is very close to the projected length of the vessel from the remnant keel. The broad, round beam of shallops made this type of vernacular boat very stable in the choppy coastal waters and able to carry large heavy cargos without worry of swamping in heavy seas.

The framing system of the Hart's Cove shallop alternated between floors and half-frames that crossed the keel (Fig. 9.5). The frames and half-frames are independently placed and vary in spacing at centers between 21 and 23 in. (Fig. 9.7). The floor timbers extend out to the third outer hull strake where they "toe" into an

Fig. 9.7 Photograph of the hull illustrates the frame and half-frames

adjoining futtock, which continues up, around the turn of the bilge. Although adjoining floors and futtocks are not attached to one another, they appear to have been placed upon the keel before being trunneled to the outer planking. Frames 13 and 14 make up the midship frame, where the futtock placement reverses, indicating the change in hull shape.

The sided and moulded dimensions of the frames recorded at the centerline range between 3 and 5 in. in sided dimension. The moulded dimension is consistently 4.5 in. The most common sided dimension is 4 in. Building in a base of 4–4 in., 4 ft, or multiples thereof—was a common colonial characteristic (Baker 1966; Goldenberg 1976). All of the frames are oak and were shaped from naturally curved "compass pieces". Each frame has an axe-cut limber hole located next to the keel to allow accumulated water in the bilge to seek the lowest point in the hull and be pumped out. The short keelson rides over the frames at the center of the hull and is notched on the underside to accommodate the floor timbers. Four large spikes at each end, along with trunnels along the top face of the keelson, secure the keelson to the keel.

The notches on the bottom of the keelson and the trunnel pattern reveal that the frames were laid first, and then the keelson, to sandwich them in place. Onto these defining frames, the garboard and next three or four outside hull strakes were attached. Once the shipwright defined the bottom hull curvature, the half-frames were hooked under the keelson. Two large ceiling planks were then laid atop the floors and half-frames and trunneled through the hull, clamping the lower structure together for strength. After rounding the turn of the bilge, the shipwright alternated between attaching futtocks, then outer planking, and then sandwiching with ceiling planks. It is not clear how far up the inside of the hull the ceiling planking ran. The leap-frog construction, however, continued until the sides of the hull reached the maximum height.

The outside of the hull planks reveal evidence of horsehair and pitch. Thin planks of pine sheathing or "firring" would have been nailed over the pitch/hair concoction as a sacrificial skin. The presence of this feature, used as a form of retardant against *Teredo* worm damage, suggests that the Hart's Cove shallop sailed in both fresh and saltwater environments where *Teredo* worms thrive. Firring as a form of outer sheathing was a common form of hull protection in the seventeenth century (Goldenberg 1976).

Four large spikes and a series of small 1/4-in. square nail holes are located on the first two ceiling planks and are the only evidence of metal fasteners found on the shallop. The presence of small iron nails in the ceiling suggests these planks were regularly lifted to clean the bilge of accumulated debris. The lack of metal in shipbuilding reflects both the cost of the building material and its accessibility in the colonies. Numerous varieties of wood such as oak, hickory, and pine were readily available and affordable, and therefore were used more often in the colonies than in Europe, where iron smelting made iron and bronze available at affordable prices. The wooden trunnels had a wedge driven into the head after they had been sheared off smooth. The use of wedges, and later diamond tees, driven into trunnels to tighten the fit at the end of the wooden nail are often associated with American shipbuilding.

The Hart's Cove shallop more than likely had a long career. The patch in the hull planks indicates that the craft was repaired at least once. In addition, the fastening patterns suggest that several frames were added or used to replace damaged frames during the vessel's career. These added frames were more roughly fashioned than the original timbers, and their limber notches were triangular instead of rectangular, like those of the original, more prevalent style of floors.

The portable artifactual remains found with the hull reflect a long-lived workhorse that probably had a mixed career sailing up and down the New England coast as well as lightering cargo to and from larger ships that could not make it up to the river ports of the Piscataqua River. The hull patching and evidence of replacement frames reflect a colonial economy rich in resources but with limited available labor, necessitating a preference for repair over replacement. Together the shape of the hull and its repairs reflect a style of vernacular watercraft well suited to the needs of the time and people, creating a use for the type of sailing craft even if it was not in tip-top shape. The shape and size of the shallop were well suited to the trade needs of colonial North America and thus flourished as a genre of watercraft in New England, evolving over time as the style kept pace with the needs of the regional trade.

References

Baker, W. A. (1966). *Sloops & shallops*. Barre, MA: Barre Publishing.

Binford, L. (1978). A new method of calculating dates from kaolin pipe stem samples. In R. L. Schuyler (Ed.), *Historical archaeology: A guide to substantive and theoretical contributions* (pp. 66–67). Farmingdale, NY: Baywood Publishing.

Goldenberg, J. A. (1976). *Shipbuilding in Colonial America*. Charlottesville, VA: University Press of Virginia & Mariners' Museum.

Grant, A. (1983). *North Devon pottery: The seventeenth century*. Manuscript, University of Exeter, Devon, England.

Grenier, R., Bernier, M.-A., & Stevens, W. (2007). *The underwater archaeology of Red Bay: Basque shipbuilding and whaling in the 16th century*. Ottawa, ON: Parks Canada.

Holly, H. H. (1969). *Sparrow-Hawk: A seventeenth century vessel in twentieth century America*. Boston: Nimrod Press/Pilgrim Society.

Switzer, D. C. (1980). Piscataqua basin survey: Phase I. *INA Newsletter, 7*(4), 5–7.

Switzer, D. C. (1985). Archaeology under New Hampshire waters: The present and the future. *Historical New Hampshire, 40*(1–2), 34–46.

Switzer, D. C. (1991). An example of a Carvel constructed Colonial American shallop-type vessel. In R. Reinders & P. Kees (Eds.), *Carvel construction technique: Fifth International Symposium on boat and ship archaeology, Amsterdam 1988*. Oxford, England: Oxbow Books.

Chapter 10
Cultural Transmissions of the "Biscayne Shallop" in the Gulf of St. Lawrence, 1560–1750

Brad Loewen

Introduction

Since the 1980s, the archaeology of small craft in the St. Lawrence basin has been dominated by two spectacular finds, one at Red Bay on the southern Labrador coast, and the other at Québec City, at the head of the St. Lawrence estuary (Fig. 10.1). At Red Bay, three Basque whaleboats named *chalupas*, dating from about 1565, came to light during Parks Canada excavations at the site of a large Basque whaling station and multiple wrecks. Soon after, seven craft from about 1751 were unearthed from a shoreline site at Québec City, including three vessels corresponding to historical descriptions of *chaloupes* used for fluvial and estuarial transport during the French colonial period. While the Québec City boats have been studied in considerable depth (LaRoche 1985, 1986, 1987, 2009; Dagneau 2004; Larochelle 2009), only recently has publication of the Red Bay findings been possible (Grenier 1985; Grenier et al. 2007; Harris and Loewen 2007). The most complete examples from each site are now exhibited respectively at Red Bay and Québec City.

Discovered on opposite sides of the St. Lawrence basin and separated by nearly two centuries, the two shallops bracket a space-time of great originality in North American navigation. Since their discovery, they have invited comparison, as archaeologists were struck by the similar form and name of the sixteenth century Basque *chalupa* and the eighteenth century Canadian *chaloupe*. These similarities soon raised questions about the two shallops' cultural parentage and link. In the St. Lawrence basin, Basque seasonal fishermen and French permanent colonizers formed distinct cultural groups whose interaction, for economic and geopolitical reasons, was often limited and at times adversarial (Bélanger 1971; Nadon 2004).

B. Loewen (✉)
Département d'anthropologie, Université de Montréal,
6092, Avenue Somerled, Montréal, QC, Canada, H3X 2A7
e-mail: Brad.Loewen@umontreal.ca

A.M. Evans (ed.), *The Archaeology of Vernacular Watercraft*,
When the Land Meets the Sea, DOI 10.1007/978-1-4939-3563-5_10

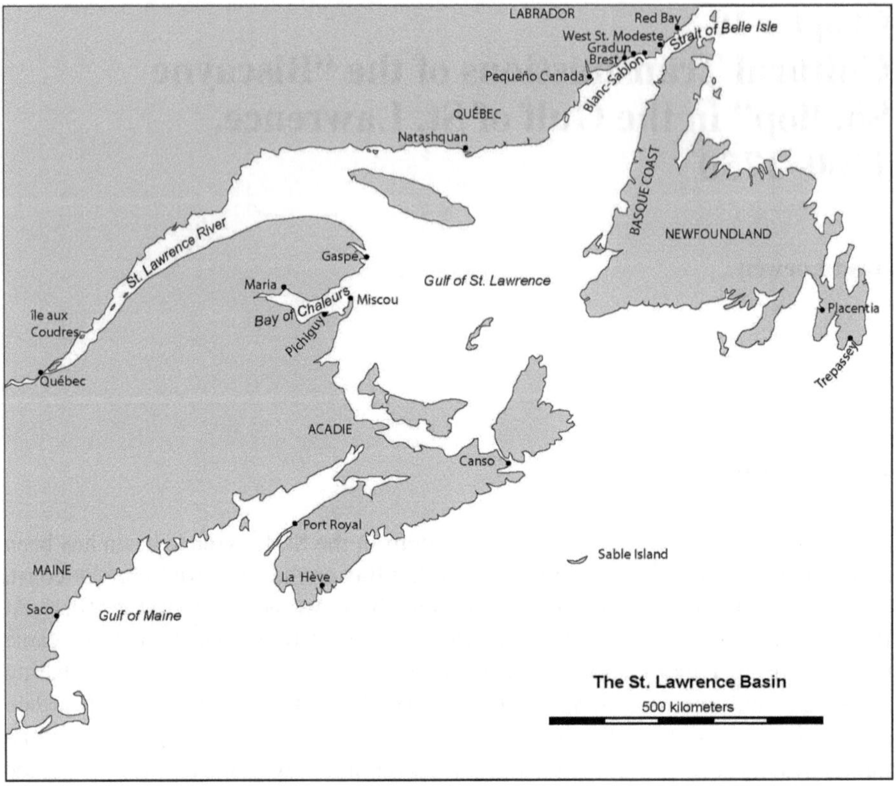

Fig. 10.1 Map showing the places mentioned in the text

Canadian historical thought tends to see the relation between sixteenth century seasonal fishermen and seventeenth century permanent colonists as a fracture line and not a smooth continuity (Turgeon 1995). Perhaps not surprisingly, efforts to find evidence of the shallop's direct transmission from Basque to Canadian boat builders, whether in Europe or North America, have been unsuccessful (LaRoche 2009; Larochelle 2009). An exploration of the two boats' broader context, however, reveals an indirect path of transmission that also uncovers the shallop's place at the core of early Canadian material culture, within a common cultural context that included both the fisheries and the early colonies.

The beginning and the end of the shallop's path of transmission are well known from archival sources. In the 1540s and 1550s, the Basque *chalupa* was only one of several boats used to hunt whales in Labrador, before it emerged as the principal whale-hunting craft in the 1560s (Barkham 1988; Proulx 1995). The shallop was made of European wood and was shipped across the Atlantic. Cod fishermen also adopted it and, after Labrador whaling collapsed about 1580 (Loewen 1999:117–118), it remained as a workhorse of the transatlantic fishery. As for the three boats from Québec City, they were discovered in a 1751 context and archival research indicated that such craft were called *chaloupes* (LaRoche 1985, 1986, 1987, 2009). These boats were slightly larger than their sixteenth century Basque homonyms and were built with local materials, but similarities between the two archaeological examples were nevertheless apparent.

While more examples are needed to fully understand the evolution of the shallop between 1565 and 1751, the Red Bay and Québec City finds are signposts in a cultural transmission whose exact path has long eluded researchers. The Basque Country is the only European region where this boat type is documented prior to 1700, but none of the approximately 20 known boat builders in colonial Québec City was Basque. One was from Bayonne, a mixed Basque and French port, while the others came from La Rochelle, Normandy or were born in Canada. None of them is a likely candidate to have introduced the *chaloupe* to Québec City, yet most were able to build such craft. Given the lack of an apparent European connection between the *chalupa* and the *chaloupe*, the New World appears as the only possible context for the shallop's transmission from Basque fishermen to French colonists. In the Gulf of St. Lawrence, the shallop transcended the European fishery, spilling into a larger cultural context and gaining popularity among all cultural and professional groups in the Gulf of St. Lawrence. Native societies, especially the Micmac in Nova Scotia and the western Gulf, adopted the shallop and transformed it into an icon of a new cultural identity. It was this reinvented shallop, along with its cultural cargo, that appears to have been transmitted to French colonists in the first half of the seventeenth century.

The Two Shallops

The Red Bay whaling station is the most extensive and best-preserved sixteenth century archaeological site in Canada. Known as Buitres—the name of a migratory bird species—to Basque whalers who occupied the harbor on a seasonal basis from about 1543 to about 1600, Red Bay lies in Labrador on the Strait of Belle Isle. Archaeologists from Memorial University of Newfoundland, led by James Tuck, excavated a shore station on Saddle Island, while Parks Canada archaeologists led by Robert Grenier investigated the wreck of the *San Juan*, a Basque whaler from Pasajes that sank off Saddle Island in 1565 (Grenier et al. 2007). Three small craft identified as *chalupas* lay underneath the wreckage of the *San Juan*. The most complete example was restored and is exhibited at Red Bay. It measures 8.02 m in length, 2.01 m in breadth, and 0.72 m high at midship (Fig. 10.2). Its shape is characterized by a rounded bottom and by a stem post and sternpost that curve smoothly from the keel into a vertical line. Nineteen frames spaced at 36 cm intervals each include a floor timber and two futtocks, mostly of oak but also other European hardwoods. The strakes fall into two groups, a lower group of carvelled strakes and an upper group of two clinkered strakes. A fixed mast step at midship indicates the main sail, while a mobile step in the bow, as well as holes in the upper strake for passing lines, indicates a lugsail rig. Well-worn thole pins and boards show that the

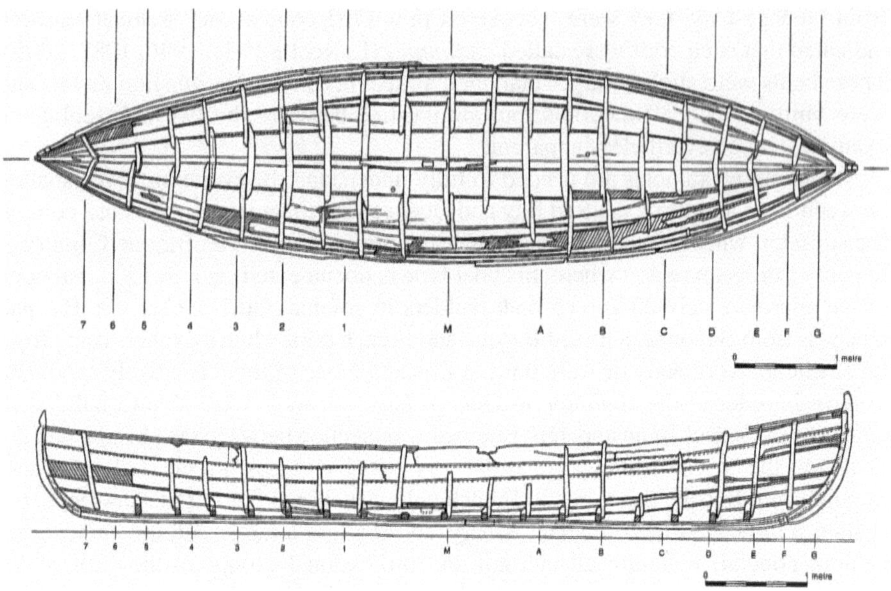

Fig. 10.2 The Basque chalupa found at Red Bay, Labrador. Parks Canada

chalupa was frequently rowed. The stern area shows evidence of both a rudder and a stern oar (Harris and Loewen 2007). The steersman may have used a stern oar during whale-hunting maneuvers and a rudder when the boat was under sail (Figs. 10.3 and 10.4). Hull analysis shows that the boat was designed using the typical geometric methods of the Renaissance, as widely found in naval treatises and wrecks from the fourteenth to the seventeenth centuries (Rieth 1988; de Castro 2005; Loewen 2007; Barker et al. 2009).

Used to hunt whales offshore and tow them to the whaling station for processing into oil, the *chalupa* was part of the extensive sixteenth century Basque transatlantic fishery. Basques were not the first to fish off Newfoundland but by 1550 they were perhaps the most numerous and best equipped group. Most Basque crews and ships came from Guipuzcoa in Spain, with the rest supplied by the neighboring provinces of Vizcaya in Spain, and Labourd in France. Basque whaling sprang into prominence about 1543, reached its peak in the 1570s and declined abruptly about 1579, although it continued sporadically until the 1730s. As whaling faded, cod fishing expanded, especially in the Gaspé Peninsula and in western Newfoundland, which came to be known as the Basque Coast. Basque cod fishing is documented from 1517 to 1767, with a peak around 1550–1580 and a higher one in 1650–1713, cut short by the Treaty of Utrecht that ceded Newfoundland to England and excluded "Spanish" ships and crews (de Gandía 1942; Bélanger 1971; Egaña Goya 1995; Turgeon 1995; Loewen 1999:117–118; Loewen and Delmas 2012). The Red Bay *chalupas* thus belong to an early phase of the Basque transatlantic fishery, namely the whale hunt concentrated in the Strait of Belle Isle.

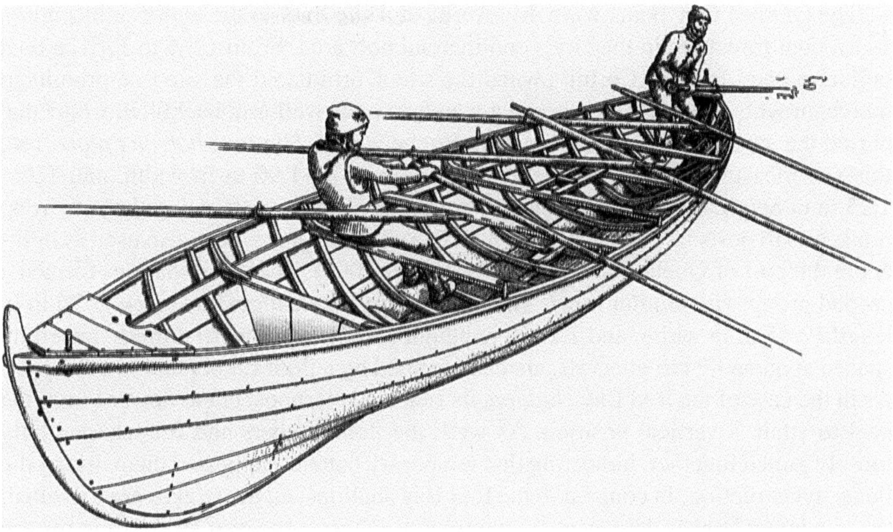

Fig. 10.3 Rowing scheme of the Basque chalupa, with a stern oar for steering. Parks Canada

Fig. 10.4 Lugsail rig with rudder, reconstructed on the Basque chalupa. Parks Canada

The Québec City boats were discovered at a site that, in the eighteenth century, was a tidal foreshore in the city's commercial port area. From 1736 to 1751, a boat builder named Étienne Corbin owned the site. Corbin sold the site to a prominent merchant who built a residence and a warehouse, as well as a backfilled wharf that buried the seven small boats (Daniel LaRoche 2009). The four *bateaux plats* from this site measure from 9.5 to 10.5 m in length, 1.70–1.90 m in width, and 1.22–1.25 m in height. The army purchased hundreds of these craft, using them as troop transports to posts far inland (Dagneau 2004). *Bateaux plats* were also used as lighters in the port of Québec (Loewen and Cloutier 2003). The three shallops formed a second group with similar traits. The most complete example measures 12.76 m in length, 2.45 m in width, and 1.20 m in height (Fig. 10.5). The 27 frames, generally spaced at about 47 cm intervals, are each formed by a floor timber and two futtocks. As in the case of the Red Bay *chalupa*, its stem and sternpost curve upward from the keel to attain a vertical position. As well, the floor timbers and futtocks are only loosely joined together, indicating that temporary battens supported them during the boats' construction. In contrast to the Red Bay shallops, all the strakes are carvelled; the two upper strakes, however, are continuous whereas the next strake from the top is cut down diagonally at the ends (Fig. 10.6). Two mast steps, at midship and in the bow, indicate a lugsail rig. No evidence of oared propulsion can be found, but the stern area includes fastening evidence to suggest the use of a rudder (LaRoche 2009). The hull's geometry reveals the same ancient design methods found in the Red Bay *chalupa* (Larochelle 2009). Québec City's notary records, when they begin in 1663, reveal a flourishing boat building industry already in place, with *chaloupe* construction in full bloom. About 1715 however, the boat-name *chaloupe* disappeared and, in its place, a new name appeared, the *bateau*, designating a craft identical to the former *chaloupe* (Brisson 1983). This name change corresponds to the arrival of new notaries in Québec City, as well as Acadian and French merchants and sailors who were expelled from Newfoundland in 1713. These influential newcomers did not follow traditional Québec City nautical terminology. The change in notarial vocabulary that ensued leaves some uncertainty as to the boats' common name at the time of their abandonment in 1751.

The shallop's function in colonial St. Lawrence navigation, between the Saguenay and Montréal, is well documented (Larochelle 2009). It provided loyal service in the redistribution and collection of packaged merchandise around the port of Québec City. It also brought hay, wood, and fresh meat into Québec City from surrounding rural areas. Its owner was typically a professional sailor who hired one or two crewmen for longer voyages. The master might take time off during the fall harvest and haul the boat ashore for the winter season. Little is known of the *chaloupe*'s history in Québec City before 1663. The city was founded in 1608 and grew slowly until about 1640; it was even abandoned in 1628–1630 following an English raid. Its military, religious, administrative, and commercial roles blossomed in the 1660s, bringing rapid population growth. These roles intensified after the Treaty of Utrecht in 1713 and even more so in the 1740s and 1750s when Franco-English war became endemic and French immigration peaked. The city fell to an English fleet in 1759 and the colony capitulated the following year. The Québec City *chaloupes* thus fit into the final phase of the French colonial period.

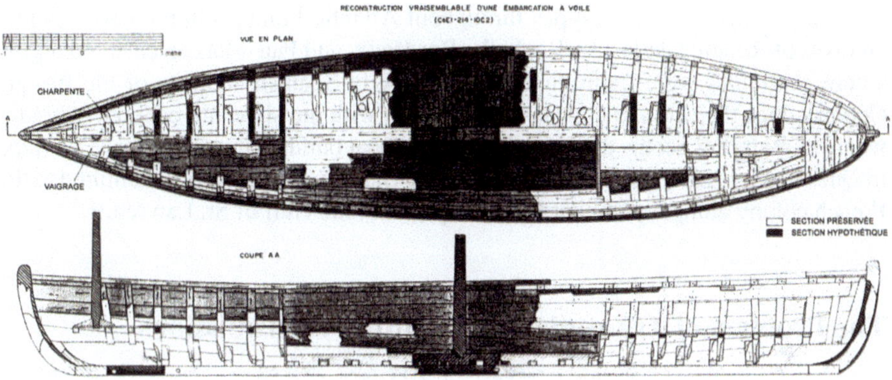

Fig. 10.5 The Québec City chaloupe. Daniel Laroche

Fig. 10.6 Display of the Québec City *chaloupe*, showing the joint of the upper and lower strakes. Musée de la Civilisation.

The Red Bay *chalupa* and the Québec City *chaloupe*, despite their similar form and name, are each part of very distinct cultural and historical contexts. Similarities between the two craft are not in themselves astonishing, considering that maritime material culture globalized at the expense of regional traditions from the sixteenth to the eighteenth centuries (L'Hour and Veyrat 2003; Dagneau 2009). Problems of understanding the boats' similarity arise when one seeks their precise cultural link.

A census of historical boat types throughout Atlantic France—in the departmental archives of Rouen, Nantes, La Rochelle, Bordeaux, and Pau—has revealed no region where *chaloupes* were traditionally built. Only in Spanish archives of the Basque Country does the *chalupa* appear in connection with the sixteenth century Labrador whale hunt. It thus seems doubtful that French boat builders introduced the *chaloupe* to Québec City, leaving the possibility that the Basque *chalupa* was transmitted to the French colony along a path that lay wholly within the Gulf of St. Lawrence.

The Problem of Transatlantic Maritime Transmissions

Throughout the colonial period, maritime traditions were transplanted from Europe to North America, took root and regenerated as new forms. Many authors have sought to trace the European ethno-cultural origins of colonial boats and small ships, and have come to grips with the complexity of maritime cultural transmissions. Where Howard Chapelle (1980) broadly depicted the New World as a crucible of originality and innovation, Charles Moore (1993) focused on function, local environment, and ethno-cultural tradition to explain the existence of various boat types in a given context. In another approach, Michel L'Hour and Élisabeth Veyrat (2003) observed a growing internationalization of maritime technology from the sixteenth to the eighteenth century. Pursuing this idea, Charles Dagneau (2009) studied the regional, national, and international provenances of artifacts from eighteenth century shipwrecks, using a world-economy approach. These studies show that, even as the Atlantic was a highway for the diffusion of maritime material culture, it also acted as a filter that allowed some European traditions to spread to the New World, but blocked others. They also reveal the New World's ability to conserve, transform and create maritime forms in a dynamic cultural context.

Where all these approaches to transatlantic maritime transmissions converge is in their Euro-centricity. It is true that assembled wooden boats were European in origin, but their New World implantation involved Native builders and navigators whose cultural references were rooted in a local context that remained primary for them, and which Europeans perceived only partially. Native societies formed an intermediary context that also had the ability to filter European material culture, adapt it to new functions and transform its meaning to produce new cultural identities (White 1991; Turgeon 1996; Moussette 2002, 2003). Such cultural processes operated ceaselessly beyond the colonial pale, producing a "contact culture" that had a major enabling and structuring effect on the colonization that followed it. In the St. Lawrence basin, the shallop was a focal point of this cultural dynamic. By dividing its history into pre-contact (Basque), contact (Native), and post-contact (Canadian) phases, it is possible to trace its cultural transmission over two centuries and, in so doing, come to a better understanding of the earliest periods of historical archaeology in North America.

Seen from this angle, available data take on a new significance and a new set of data come to light, in which the shallop appears as one of the most significant elements of European material culture to be transferred to North America in the sixteenth century.

Efficient, versatile, and affordable, it was valued equally by European and Native sailors. Natives' enthusiasm for the shallop, and for the mobility and other advantages that it offered, lay at the heart of their many formal and informal exchanges with European fishermen. In Native hands, the shallop grafted the Native economy onto the transatlantic fishery, an economy that included Native boat building for European fishing crews and off-season care of fishing facilities and equipment. The shallop also transformed Native social structures, bringing influence to chiefs who could consolidate a power base of nautical skills. On a larger geographic scale, the shallop enabled commercially minded Natives to weave trade networks between Europeans on the coast and other Natives living in regions far inland. As Atlantic groups extended their activities westward, they turned the tables on St. Lawrence Valley groups that had travelled eastward to meet the Europeans in the mid-sixteenth century.

In the longer term, the Biscayan shallop was instrumental in creating a vast "middle ground" of mixed European and Native influences prior to actual colonization. The term "middle ground" belongs to Richard White (1991) who theorized it as a space of cultural contact where an original, blended culture emerged and flowered. White based his ideas on a study of the Great Lakes region in the seventeenth and eighteenth centuries, including many examples of recontextualized French and Native objects that formed the expression of a new bicultural identity. Anglo-American colonists, as they sought to move westward from their coastal colonies, encountered this bicultural identity and recognized it as a significant, singular Other. Their expressions such as "French and Indian Territory" and "French and Indian Wars" encapsulate the blended culture that thrived in the Mississippi Valley and around the Great Lakes (Vaugeois 2002; Havard 2003). Material culture studies have brought out the intense originality of middle ground phenomena. Laurier Turgeon (1996; Turgeon et al. 1992) has followed the cultural itinerary of copper trade pots from the Basque Country to Native societies that incorporated these common trade objects into their mortuary practices. Marcel Moussette (2002, 2003) has linked decorative designs found on trade musket hardware to the Baroque style in Europe, and has shown how these vegetal, symmetrical motifs drew their inspiration from the New World. White viewed the middle ground as a transitory phase in colonial history, leading to various cultural outcomes. One of its outcomes in the nineteenth century was the emergence of a politically conscious plains Métis culture. A long-term outcome was the re-separation of European and Native societies. As European colonists gained strength and numbers, they were able to isolate and marginalize Native lifeways and, at the same time, they closed down the shared cultural space of the middle ground. Cross-cultural Middle Grounders were assimilated either as Europeans or as Natives.

The middle ground can thus be seen as a transitory, westward-moving space that preceded formal European colonization (White 1991). Its existence in a given territory may have been a necessary precondition for successful colonization. Early attempts at colonization in the sixteenth century may have failed because they lacked the supporting social and economic "infrastructure" of a middle ground that seventeenth century colonists could count on. The successful colonies nested in a middle ground and relied on its habits of sharing, adaptation, and inventiveness for survival during their first years. Colonists arriving from Europe learned about their new environment from resident Middle Grounders who were familiar with European technology and had adapted it to the New World context. Arriving colonists, lacking nuance, often took Middle Grounders for friendly Natives. These cultural processes are well understood in central and western North America (Brown 1980; Van Kirk 1980; Vaugeois 2005), but little research has been carried out on the easternmost, sixteenth century middle ground that preceded all successful European colonization. Indeed, the idea of a sixteenth century middle ground runs counter to the deeply entrenched theme in colonial history of unprepared, idealistic colonists arriving in pristine territory and surviving only by the grace of the helping but naïve Indian. Study of the Basque *chalupa* and its evolution into the Canadian *chaloupe* open a window to this early middle ground and its role of enabling and structuring the colonization of New France.

The Basque Whaling *Chalupa*

During the 1560s, the *chalupa* became the principal small craft in the Basque whale hunt in southern Labrador. Prior to this date, two other boats, also powered by sail and by oar, were mentioned in a whale-hunting role: the *galion* and the *pinaça*. The Labrador whale hunt peaked in the 1560s and 1570s and many practices concerning the shallop became customary during these decades (Proulx 2007). Whaling captains required each harpooner to be the owner of a shallop. In practice however, the captain paid for the shallop's construction in the Basque Country, then sold the craft to the harpooner and charged him for its transport to Labrador. Due to its size, the boat was shipped across the Atlantic in disassembled parts and the captain paid for its completion in Labrador, for which he also charged the harpooner. The latter's share of the expedition's profit was prorated according to the number of whales he killed and according to his shallop's depreciation over 3 years. The harpooner chose his tillerman but the captain designated the four oarsmen. In the shore-based hunt, harpooners headed out daily into the Strait of Belle Isle to cruise for whales. They did not lash their harpoon line to the shallop to remain in contact with the stricken whale, but rather to wooden crosspieces, or "drogues," that slowed the whale and tired it out. Once killed, the whale was towed to the shore station to be flensed and rendered into oil.

The whalers' annual rhythm saw crews leave the Basque Country in June, spend about 120 days in Labrador from August to December and return to their home ports in early January. Whale oil had many uses, including lighting, caulking and soap-making, but most went to textile centers in the Low Countries Normandy and England where it was used to lubricate wool after it had been washed and dyed, in order to facilitate its weaving. In this larger picture of the whale oil industry, the shallop represented a small investment but it was crucial to the most specialized step in the entire process, that of procuring whales (Barkham 1988; Loewen 1999; Proulx 1995, 2007).

In fall, when the whaling crew cleaned up and departed, the *chalupa* entered a tenebrous time of its existence. Rare was the shallop that ever returned to Europe. Most were submerged in a pond or a lagoon, to keep their joints tight over winter. Archives in the Basque Country contain notarized contracts between shallop owners, when a non-returning harpooner sold or leased his boat to a colleague. These contracts, and the lawsuits involving them when they were not respected, contain many details of whaling customs regarding *chalupas*. In 1551, Esteban de Arriçaga of Orio left a *chalupa* in Brest (Vieux-Fort) and sold it for eight ducats to Joan de Aguirre, who died without paying. Ten years later, Arriçaga summoned Aguirre's widow to pay for the *chalupa* and for his accumulated legal costs (5 November 1561, AHPG, II, 1785:142; Barkham 1988:110). In 1572, Blasio de Echabe of Orio claimed 6 ducats for his *chalupa* that had been moved without his authorization from Gradun (Middle Bay) to Samodet (West St. Modeste) (8 June 1572, AHPG, II, 1798: 37; Barkham 1988:113). In 1549, Andrés de Armencha sold the *pinaças* he had left at Blanc-Sablon the previous season (3 May 1549, AHPG, III, 2575, 10:25v; Barkham 1988:72). These examples reveal a lively trade in used shallops in the Strait of Belle Isle, not only during the fishing season but also in winter when the harpooners were in the Basque Country and the boats remained in the New World.

In the ordered world of the sixteenth century fisheries, even abandoned boats carried a customary status. Basque judicial archives contain two civil suits involving "unhanded" whaleboats (*chalupas desmanparadas*), that is, they were left in Labrador by harpooners who did not expect to return (AGDG, civiles Elorza, 65:22r-v). In both cases, the boats were ultimately reused and the documents reveal the customs that guided whalers in these circumstances. In the first suit, the harpooner Lorenço de Mutio signed on with a whaler bound for Samodet (West St. Modeste) in 1565, and planned on using an old boat that he had left unhanded at Buitres (Red Bay) the previous fall (AGDG, civiles Elorza, 65:2r-22v). On arrival at Samodet, a crew was dispatched to Buitres, about 25 km distant, to fetch the *chalupa* which was found to be leaky and unseaworthy. Mutio hired a carpenter to make the boat navigable and soon he was out hunting; that year, he killed "*muchas bayllenas*." At the end of the season, he was paid as if he had used an old shallop and was left hanging for the cost of refurbishing his boat. In his lawsuit, he claimed (and obtained) a salary commensurate with the use of a new *chalupa*.

In the second lawsuit, the harpooner Joseffe de Echaniz of Orio left two *chalupas* in Terranova in 1578, one old and one new. He did not return the following year however, and instead authorized his colleague Joanes de Galbet to use them for the 1579 season. Their agreement was thwarted when Sebastian de Labastida, the captain of the 1579 voyage and a harpooner in his own right, took the *chalupas* for himself, promising to pay Echaniz. Complications soon arose. At one point in the season, Labastida lent the old *chalupa* for a *barrica* of wine to French cod-fishermen who used the boat for 2 months and then returned it to Labastida. To further confuse matters, Labastida tried to tow the new *chalupa* back to Europe, but lost it during the voyage. The court was asked to determine Echaniz's due, first for the use of the two *chalupas* and, second, for the loss of the new boat, in accordance with the "customs of the Province of Terranova" (1581, AGDG, civiles Elorza, 445: n.f.; 4 March

1581, AGDG, Pleitos civiles, 446: n.f.; Huxley 1988:116). One witness testified it was general knowledge among whalers on the "*carrera de Terranova*" that anyone might use an unhanded *chalupa*, with or without the owner's formal consent. Payment for using someone else's *chalupa* was agreed upon with the owner or his procurer, and was calculated at one *barrica* of whale oil per year. Another witness, an old hand who had been on 20 whaling voyages, stated that it was customary for those taking an unhanded *chalupa* to leave it in Terranova (and not tow it back to Europe). If by mishap a borrowed *chalupa* was lost during the fishery, its original owner bore the cost since he had been paid rent. In the case of Echaniz versus Labastida, the court records do not include the judgment, so the validity of these testimonies cannot be fully known.

The whale hunt declined into a sporadic activity after 1580, but Basque cod fishing steadily grew throughout the sixteenth and seventeenth centuries (Loewen and Delmas 2012). Fishermen embraced the *chalupa* and diffused it throughout the Gulf of St. Lawrence and the Atlantic coast. Seasonal crews set up shore stations from which fishermen set out in shallops to jig for cod, while a shore crew cleaned, salted and dried the fish. Fishermen used the shallops, but ship captains retained ownership and identified their boats with a personal mark. As with the harpooners, the cod fishing captains left their craft in the New World over winter and sometimes sold them to other captains. In one off-season sale of ten shallops left at Trepassey in southern Newfoundland, the owner stated that his mark appeared three times on each shallop, on the stem, the tack cleat and the inner planking (13 April 1606, AHPG, III, 2598, 3:50). He described the mark as "two fishhooks and a bar on the side" (*dos aguxeros con varreno del costado*), and then indicated where he had sunk the boats:

> … in the port of Trespas in the province of Terranova, on the bottom of the lagoon found in the said port, in front of the beach called *barrachoa*; and the said lagoon is on the left hand when going toward the said *barrachoa* …

Contracts and lawsuits involving the sixteenth century Basque fishery include many references to the "customs of Terranova," as established from the testimonies of veteran fishermen and whalers. This body of customary practice and jurisprudence shows the orderliness and pragmatism of the sixteenth century fishery, of which several aspects were transmitted to Natives and helped to structure the earliest North American middle ground. Concerning the *chalupas*, the customs of Terranova covered many aspects of shallop ownership, including overwintering, abandonment, their use as professional instruments, and their subjection to a captain's prerogative. "Unhanded" craft were commonplace, since turnover was high and the average life of a shallop was about 3 years. Many moderately used craft were left unhanded after each season. Especially in the cod fishery, the number of boats abandoned each year in the New World must have numbered several hundred. This practice made it easy for sixteenth century Natives to acquire shallops, as well as the customs that governed their use. Natives observed the socioeconomic relations that regulated shallop ownership and they adapted both the craft and its social significance to their own cultural context.

The Shallop in the Native Sphere

From the early sixteenth century, Natives in the Gulf of St. Lawrence and further west were attracted to European ships, boats, and fishing activities on the Atlantic seaboard. In September 1534, the explorer Jacques Cartier met a boatload of Iroquois-speaking men on the north shore of the Gulf near Natashquan, heading westward. They climbed aboard Cartier's ship as readily as if they were European sailors and informed the explorer that the European fishing fleet in the Strait of Belle Isle, from whence they came, had already departed for the winter. They too were on their way home, in the St. Lawrence Valley (Cartier 1981). In 1542, another boatload of Iroquois-speakers from the Québec City area arrived in the Strait to trade with Basque fishermen (Biggar 1930:462–463). Later sixteenth century sources reveal that Basques hired Natives to help with fishing activities (Barkham 1988). Not all Euro-Amerindian contacts involving boats were amicable. In 1543, Basque carpenters assembling shallops were ambushed and killed by unknown Natives, who took the shallops and sailed away in them (Biggar 1930:456).

A century later, a prominent Acadian fishing outfitter, Nicolas Denys, commented on the shallop's popularity among Micmac sailors:

> ... the Natives of the coast use canoes only for rivers, and they all have shallops for the sea, which they sometimes buy from Captains who are about to leave after completing their fishery; but most often, they take them from where the Captains hide them on the coast or in ponds, so as to use them on a later voyage. When the owners ... recognize [the shallops], they make no more ceremony of taking them back than do the Natives in making use of them (Denys 1672: 180).

As a colonial seigneur interested in developing his territory, Denys took a dim view of such practices founded in the pre-colonial "customs of Terranova." In his day, many practices remained from the sixteenth century and survived as innumerable acts of understanding between Micmac and French or Basques. One example comes from a contract written in 1608 at Le Havre (France), when the fishing captain Jehan Vincent sold two *barques* that he had left at Île Percée (Gaspé) to Estienne Lemercyer. Vincent specified that the craft had been left over winter in the hands of two Natives named Rougefort and Jouanis, who would identify them to Lemercyer. The *barques* were identified by a mark on the stern, reproduced in the act (28 February 1608, ADSM, 2E 70/119). Both Vincent and Lemercyer seem to have known Rougefort and Jouanis; the French and Basque consonance of the Natives' names is striking and leads one to wonder whether they were baptized or of partial European ancestry.

Transfers of shallops and their associated technology to Natives were common in the sixteenth century. When colonists arrived in the early seventeenth century, they encountered Natives in shallops busily crisscrossing the Gulf and following the Atlantic coast, attending to matters that often had little to do with fishing. An English ship making way from Sable Island to Maine in 1607 was accosted by a "Biscayne shallop" manned by eight Native sailors and a boy, wishing to trade (Burrage 1906:81–83). They spoke in French of their "Cheef Comander" Messamouet, an influential Micmac chief at La Hève

(Nova Scotia). Messamouet is known to have been an accomplished shallop builder for French fishermen and Native navigators, and he spent a winter in France about 1580 (Whitehead 1991:28–29). He was not the only Micmac chief whose prestige was based on the adoption of European boats. In 1607, the fiery Membertou summoned a fleet of 50–70 shallops, manned by 400 warriors, and sent them from Port Royal (Nova Scotia) across the Gulf of Maine to avenge the death of his compatriot, Panounias, at the hands of the Armouchiquois of Saco (Maine). When the fleet returned, Membertou recounted the exploit in a song to the French colonists at Port Royal, to ensure that they were duly aware of his military prowess (Whitehead 1986:227–232).

Natural sailors, the Micmac, seem to have found European shallops and fishermen to be a welcome addition to their maritime world, to the point of ascribing totemic qualities to their craft. In 1608, the explorer Marc Lescarbot described a Native shallop, sailing alongside St. Malo fishermen based at Canso (Nova Scotia), whose sail was painted with the effigy of a moose (Lescarbot 1609:577). The Jesuit missionary Jerôme Lalemant marveled at the skill of Native mariners who traversed broad stretches of open sea without a compass and often losing sight of the sun, relying only on their "imagination" to guide them (Thwaites 1959b:65). Astonished English colonists also penned accounts of Native sailing abilities (Whitehead 1986:224–232). In 1602, Native traders hallooed English colonists off the coast of Maine:

> Came towards us a Biscay shallop, with sail and oars, having eight persons in it whom we supposed at first to be Christians distressed. But approaching us nearer, we perceived them to be savages. These coming within call, hailed us, and we answered. Than after signs of peace, and a long speech by one of them made, they came boldly aboard …; these with a piece of chalk described the coast thereabouts, and could name Placentia of the Newfoundland; they spoke divers Christian words, and seemed to understand much more than we, for want of language, could comprehend (Archer 1843:73; Burrage 1906:330; Whitehead 1991:21–22).

French colonists were just as unprepared to see the Micmac's enthusiasm for the Biscayan shallop. At Port Royal in 1609, colonists awaiting provisions from Europe were disappointed time and again when the *chaloupes* they espied on the horizon turned out to be Native craft (Lescarbot 1609:628). In contrast to cod fishermen's silence on the subject, colonists' frequently mentioned Natives in shallops, suggesting that fishing crews were used to the sight. Although the colonists' neat distinction of "Savages" and "Christians" was shaken by the bicultural habits of these Middle Grounders, they were happy, when the fishermen left in fall, to have Europeanized Natives as neighbors over winter.

Shallops and Early "Canadians"

As the Biscayan shallop penetrated Native societies and became a focus of new social relations, it also acquired new cultural meaning within a "middle ground" context. Early French colonial texts associated *chaloupes* with Europeanized Natives and attributed a specific identity to these people, calling them *Canadiens*.

These texts contain some of the earliest known occurrences of this name. Gervais Carpin (1995), in tracing the history of the name *Canadien*, has shown that the word originally designated a Native group living in the St. Lawrence lowlands in the 1530s and 1540s. Jacques Cartier associated it with Iroquois-speaking, maize-growing people living in villages west of the Île aux Coudres, especially around Québec City. These *Canadiens* led a migratory lifestyle, growing maize and over-wintering near Québec City and travelling east toward the Gulf in summer to fish and hunt seal (Plourde 2011). Cartier encountered one boatload of *Canadiens* who were returning to the St. Lawrence from the Strait of Belle Isle in the fall of 1534, and another group from Québec City who were fishing in a summer camp near Gaspé. A boatload of Iroquois speakers from Québec City sailed to the Strait in 1541 to trade with Basque fishermen (Biggar 1930:462–463). These *Canadiens* and their villages disappeared some time between 1543 and about 1585, for reasons that remain incompletely understood (Tremblay 2006).

The name *Canadien* nevertheless survived. It appears next in the early seventeenth century, when it designated a Christianized Native community on Chaleur Bay, at Percé and Miscou, who maintained a close relationship with French and Basque fishermen and whalers. Carpin argues that *Canadien* signified the distinctive cultural identity of these Micmac, based on their Europeanized lifestyle and, not least, on their use of Biscayan shallops (Carpin 1995:76–80). He cites Samuel de Champlain, who wrote in 1624 at Québec City:

> On the 10th of the said month, the Natives came to build their dwellings near the Habitation. The next day, [Guillaume] de Caen arrived with two *barques* loaded with trade goods; the next day, trade with the Natives began; other *Canadiens* arrived at this time in several *chaloupes* (de Champlain 1973:1064).

Champlain's choice of specific nautical terms—*barques* for French colonists, *chaloupes* for *Canadien* traders—is noteworthy given the explorer's concise maritime vocabulary in all his writings.

Gervais Carpin explores the unique blend of Basque, French, and Micmac languages of these intermediate Canadians. He relates the case of a certain Juanchou, "*capitaine sauvage*" and a leader among the *Canadiens* at Percé and Miscou (Carpin 1995:80–81). The name of this "Native captain" is a typical Basque diminutive of the surname Juan (*Juan-txu*). Basque whalers from Saint-Jean-de-Luz entrusted him and his *Canadiens* with their supplies over the winter. When a Basque whaler went up in flames in 1624, Juanchou arranged for the crew's return home (de Champlain 1973:1104). Four years later, after an English corsair attack on Québec City, Champlain prevailed upon Jaunchou to shelter twenty colonists for the winter or, failing resources, to send them to France on Basque fishing and whaling ships departing from Gaspé (de Champlain 1973:1186–1187, 1206). In other words, this Native chief was an intermediary between Basque fishermen and French colonists. In 1638, Juanchou's son crossed the Atlantic to meet Louis XIII, who called the young man a son of the Empire and gave him gifts in the name of his people (Thwaites 1959a:222). In 1639, when "captain" Juanchou and his *Canadiens* travelled to Tadoussac in a flotilla of barques and shallops, the Ursuline nun Cécile de

Sainte-Croix found them to be more "polite" than the local Montagnais, and related that their children sang Christian songs in their native tongue (Campeau 1967:747).

On Chaleur Bay, contact between Micmac and Basques continued for nearly two centuries. Cartographic evidence indicates that Basques were fishing on the Gaspé Peninsula by about 1580 and Nicolas Denys, writing in 1672, provides details of their whaling at Miscou (Egaña Goya 1992, 1995; Denys 1672). Basque fishing in this region continued until the end of the French Regime in 1759 (Mimeault 1987; Turgeon 2000; Nadon 2004:1–16). Such seasonal contact left a durable cultural heritage among the local Micmac: the linguist Peter Bakker (1988, 1991) has found many Basque loanwords in this region. The *Canadien* cultural identity of these people lasted several generations. Possibly of mixed ancestry, they were at ease in the fishermen's world of boats, contractual agreements, and seasonal migrations, and they acted as a welcoming committee to French colonists at Québec City. After about 1640, as Carpin shows, the name *Canadien* came to designate the country-born descendants of French colonists, a meaning that still remains in use, although the word has also gained a larger significance.

Carpin indicates that the *Canadiens* lived at Percé and Miscou, and mapmakers from the seventeenth century provide other clues. In 1612, Samuel de Champlain (1973:114) marked *Canadiens* in the southeastern part of the Gaspé Peninsula, between the towns of Gaspé and Maria. In 1633, the Flemish geographer Joannes de Laet (1633:30–31) also indicated this region, and in 1689 the Basque mapmaker Pierre Detcheverry showed Native place-names around Chaleur Bay (Fig. 10.7).

The Chaleur Bay *Canadiens* seem to have attained the height of their cultural identity between 1580 and 1640, with the shallop as an identifying trait. They come

Fig. 10.7 Detail of a 1689 map by Pierre Detcheverry, showing the Gaspé Peninsula and surrounding areas. Habitacion Pichiguy is at bottom center. *Source*: Gallica

into focus as Middle Grounders, but they also raise questions about Native power shifts in the St. Lawrence basin during the sixteenth century. It is unknown when or how they inherited the *Canadien* identity from the St. Lawrence Iroquois. Access to European nautical technology and commercial networks may have allowed the Micmac to take over the route formerly used by Iroquoians between the St. Lawrence Valley and the Gulf. The shallop and familiarity with European customs gave the Micmac a strategic advantage during the same period, between 1543 and 1585, when Iroquois villages disappeared from the St. Lawrence Valley. While power shifts in the sixteenth century middle ground remain dimly understood, the shallop clearly played a role in the cultural ascendancy of the Micmac and the *Canadiens*. The decade of the 1580s seems important, as by this date the Micmac had fully acquired the shallop technology, Basque traders had moved into the St. Lawrence estuary, the Iroquois villages had disappeared and European trade goods began appearing on Native sites in the lower Great Lakes. In the early seventeenth century, a specific middle ground context, characterized by Micmac ascendancy, Iroquois retreat, and *Canadien* assistance, enabled French colonization in the St. Lawrence Valley.

Further east, seventeenth century maps identify another place that may refer to a community of Europeanized Natives. This is Pequeño Canada or Petit Canada in the northeastern Gulf, about 100 km west of the Strait of Belle Isle (Fig. 10.8). Basques whaled here about 1590 and called the place Canada El Pequeño (Azkarate et al. 1992:88–90), but the place-name survived long after intensive whaling in the Strait came to an end, and throughout the period when Basque cod-fishermen occupied the northeastern Gulf (Loewen and Delmas 2012). It appears on a Basque map made in 1674 by Denis de Rotis, showing fishing stations in western Newfoundland and the northeastern shore of the Gulf of St. Lawrence (Fig. 10.9). Its location is in the vicinity of St. Augustine and the present Innu community of Pakuashipi. Basque relations with the Innu are poorly understood, although it is known that Basques traded with Natives at Brest (St. Paul) in 1537 and found them to speak English, French, and Gascon. Natives worked on Basque fishing and whaling stations in this region during the 1570s (Biggar 1930:453–454; Barkham 1988; Litalien et al. 2004:101). There is no direct evidence that they adopted the Biscayan shallop, but in the light of research on the *Canadiens* of the Chaleur Bay, Pequeño Canada may also be interpreted as a middle ground community that grew out of contact between Basque fishermen and Innu sailors.

The Colonial *Chaloupe* in Québec City, 1663–1714

The final link in the cultural chain linking the Red Bay *chalupa* and the Québec City *chaloupe* appears in colonial notary records, preserved from 1663 onward. A study of boat and shipbuilding at Québec City also reveals a precise nautical vocabulary (Brisson 1983). Among the earliest craft built were the *barque* and the *chaloupe*, which had distinct sizes and functions. The *chaloupe* had a burthen of 3–10 t, measured from 33 to 35 French feet in length (10.7–11.4 m) and had no deck. The *barque*

Fig. 10.8 Detail of a 1674 map by Denis de Rotis, showing the Strait of Belle Isle and surrounding regions. Petit Canada is the 5th name from the upper left. *Source*: Gallica

surpassed 15 t and could reach 120 t, and had one deck. As Québec City grew and its economy diversified, *chaloupes* became more numerous and dominated medium-range fluvial cabotage. Between 1663 and 1700, *chaloupes* were most frequently mentioned private boat, eclipsed only by the army's mass construction of *bateaux plats* (Dagneau 2004).

In analyzing the number of boat and ship types built in Québec City, it is possible to deduce that the *chaloupe* was already at the height of its popularity when records began in 1663. Decadal construction of small and medium craft at Quèbec City shows that the classic "battleship curve" of popularity for the *chaloupe* is truncated by the absence of records before 1663 (Fig. 10.9; Table 10.1). By projecting the curve backward in time, it appears that the shallop's popularity began in the 1630s (Deetz 1977:121). Its adoption coincided with the time when contacts with the Chaleur Bay *Canadiens* were frequent

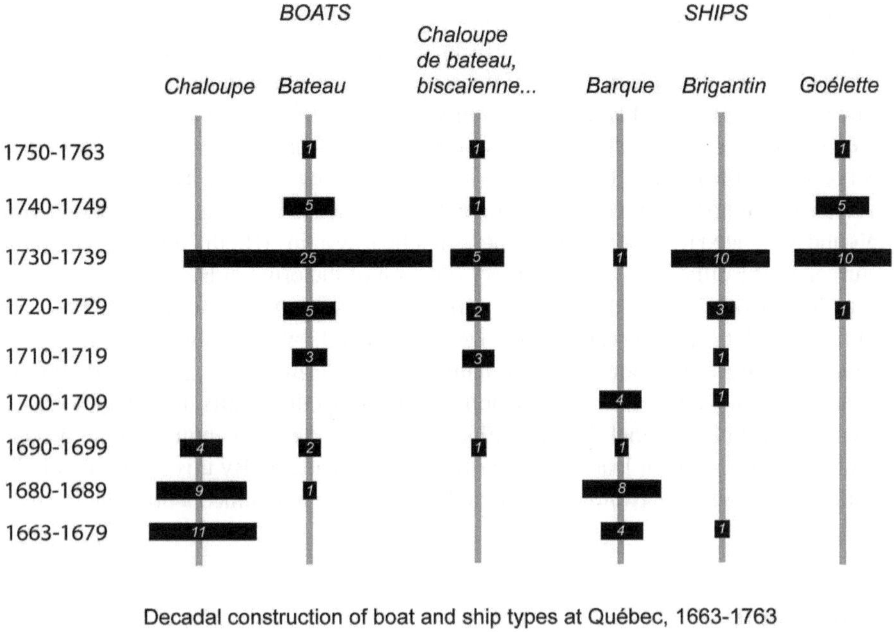

Fig. 10.9 Seriation of boat and ship types built at Québec City, based on notary records as published in Brisson 1983

Table 10.1 Decadal construction of boats and small ships at Québec City by type, 1663–1763 (Brisson 1983)

	Chaloupe	*Bateau*	*Chaloupe de bateau...*	*Barque*	*Brigantin*	*Goélette*
1750–1763		1	1			1
1740–1749		5	1			5
1730–1739		25	5	1	10	10
1720–1729		5	2		3	1
1710–1719		3	3		1	
1700–1709				4	1	
1690–1699	4	2	1	1		
1680–1689	9	1		8		
1663–1679	11			4	1	

and when the name *Canadien* began to be applied to the country-born children of French colonists. Québec City colonists thus appear to have adopted the shallop as part of their adaptation to their new cultural and environmental context, rather than bringing it with them from Europe. While no evidence can be advanced to show that colonial builders acquired the shallop from the Chaleur Bay *Canadiens*, this hypothesis fits with available data and with the idea of the middle ground as a condition for colonial success.

The graph also reveals that the names *barque* and *chaloupe* disappeared from the Québec City nautical vocabulary early in the eighteenth century, and other names emerged to take their place. Boats of the same size and function as the *chaloupe* were henceforth simply called *bateaux*. *Chaloupe*, usually qualified as *chaloupe biscayenne* or *chaloupe de bateau*, henceforth designated a somewhat smaller boat of 25 French feet or less (8.1 m), which functioned as a tender for a larger vessel. This new boat type, previously unknown at Québec City, reflected an increase in the number of larger vessels and their need for tenders. In this new "nautical system" (Rieth 2003), the *chaloupe* returned to the dimensions and auxiliary function of the original Basque whaling *chalupa*. In a parallel development, the *barque* also disappeared from the Québec City nautical vocabulary. In its category of size and function, two new ships appeared: *brigantin* and *goélette* (schooner). These names echoed terminology that was current in France and in the Anglo-American colonies. Taken together, shifts in nautical technology and vocabulary signal that the influence of the sixteenth century middle ground had drawn to a close, at least in its maritime manifestations. By this time, the middle ground had moved westward to the Great Lakes, where it assumed a different form.

Conclusion

The problems that are raised when attempting to explain apparent affinities between boat types are highly contextual, and call for a detailed knowledge of historical and cultural factors. The discovery of two boats with similar characteristics, the 1565 Basque *chalupa* from Red Bay and the 1751 Canadian *chaloupe* from Québec City, seemed to have no explanation when a direct, European transmission path was sought between the two craft. By enlarging the context to the Gulf of St. Lawrence, an indirect path appeared, covering two centuries of maritime cultural history. This path led through a pre-colonial "middle ground" of Native and European contact that the shallop's history reveals. In the Basque fishery, shallops were abandoned and taken by Natives; as well, alliances between fishermen and Micmac provided a systematic context for the transfer of nautical technology and customs to Native society. In some regions, shallop use and construction among Natives was very widespread and helped to forge a new cultural identity. Christianized, shallop-sailing Natives on Chaleur Bay, possibly of mixed European and Native parenthood, were known as *Canadiens* before the country-born children of French colonists acquired this name. They aided the Québec City colonists during the 1620s and 1630s, at a time when colonial boat builders appear to have adopted the St. Lawrence shallop. This complex path of cultural transmission brings a broader understanding not only to the shallop but also to the pre-colonial history of the St. Lawrence basin.

Acknowledgements This chapter grew out of research at Parks Canada under the direction of Robert Grenier. Others who played important roles in this research are Daniel LaRoche, Miren Egaña Goya, Charles Moore, Anne-Marie Larochelle, Marcel Moussette, Laurier Turgeon, and Françoise Niellon. My thanks go to each of them.

References

Archer, G. (1843). The relation of Captain Gosnold's voyage to the north part of Virginia, 1602. In *Collections of the Massachusetts historical society* (Vol. 3, 33rd series, p. 73).

ADSM (Archives départementales de Seine-Maritime), Rouen, France.

AGDG (Archivos General de la Diputación de Guipúzcoa, Tolosa, Spain.

AHPG (Archivos Historicos de los Protocolos de Guipúzcoa), Oñati, Spain.

Azkarate, A., Hernández, J. A., & Nuñez, J. (1992). *Balleneros vascos del siglo XVI (Chateau Bay, Labrador, Canada). Estudio arqueológico y context historico*. Vitoria-Gasteiz, Spain: Servicio Central de Publicaciones del Gobierno Vasco.

Bakker, P. (1988). Two Basque loanwords in Mikmac. *International Journal of American Linguistics, 55*, 258–261.

Bakker, P. (1991). Trade languages in the Strait of Belle Isle. *Linguistica Atlantica, 13*, 1–20.

Barker, R., Loewen, B., & Dobbs, C. (2009). Hull design of the Mary Rose. In P. Marsden (Ed.), *Your Noblest Shippe: Anatomy of a Tudor warship. The archaeology of the Mary Rose* (Vol. 2, pp. 34–65). Portsmouth, England: The Mary Rose Trust.

Barkham, S. H. (1988). Los Vascos y las pesquerías transatlánticas, 1517–1713. In S. Huxley Barkham & E. Ayerbe (Eds.), *Itsasoa 3: Los Vascos en el marco Atlántico Norte, Siglos XVI y XVII* (pp. 26–164). San Sebastián, Spain: ETOR.

Bélanger, R. (1971). *Les Basques dans l'estuaire du Saint-Laurent, 1535–1635*. Montréal, Quebec, Canada: Les Presses de l'Université du Québec.

Biggar, H. P. (1930). *A collection of documents relating to Jacques Cartier* (Vol. 14, p. 212). Ottawa, Ontario, Canada: Public Archives of Canada.

Brisson, R. (1983). *Les 100 premières années de la charpenterie navale à Québec: 1663–1763*. Québec City, Québec, Canada: Institut Québécois de Recherche sur la Culture.

Brown, J. S. H. (1980). *Strangers in blood: Fur trade company families in Indian country*. Vancouver, British Columbia, Canada: University of British Columbia Press.

Burrage, H. S. (Ed.). (1906). *Early English and French Voyages, Chiefly from Hakluyt, 1534–1608*. New York: Scribner.

Campeau, L. (1967). *Monumenta Novæ Franciæ* (Vol. 4). Montréal, Quebec, Canada: Éditions Bellarmin.

Carpin, G. (1995). *Histoire d'un mot: l'ethnonym «canadien» de 1535 à 1691*. Québec City, Québec, Canada: Septentrion.

Cartier, J. (1981). *Voyages au Canada*. Paris: Maspero.

Chapelle, H. (1980). *The history of American sailing ships*. New York: Bonanza Books.

Dagneau, C. (2004). The 'Batteaux Plats' of New France. *The International Journal of Nautical Archaeology, 33*(2), 281–296.

Dagneau, C. (2009). *La culture matérielle des épaves françaises en Atlantique nord et l'économie-monde capitaliste, 1700–1760*. Doctoral dissertation, Anthropology Department, Université de Montréal, Montréal, QC.

de Castro, F. V. (2005). *The Pepper Wreck: A Portuguese Indiaman at the mouth of the Tagus river*. College Station, TX: Texas A&M University Press.

de Champlain, S. (1973). *Œuvres, in Charles-Honoré Laverdière* (Œuvres de Champlain, Vol. 3). Montréal, Québec, Canada: Éditions du Jour.

de Gandía, E. (1942). *Primitivos Navegantes Vascos*. Buenos Aires, Argentina: Editoriál Vasca Ekin.

de Laet, J. (1633). *Americae utriusque descriptio*. Lugduni Batavorum (Leiden): Apud Elzevirios.

Deetz, J. (1977). *In small things forgotten: The archaeology of early American life*. New York: Anchor Books.

Denys, N. (1672). *Description geographique et historique des costes de l'Amerique septentrionale avec l'Histoire naturelle du Païs* (Vols. 2). Paris: Claude Barbin.

Egaña Goya, M. (1992). Basque toponymy in Canada. *Onomastica Canadiana, 74*(2), 53–74.

Egaña Goya, M. (1995). Les ports des Basques dans les routiers et la cartographie, XVIe et XVIIe siècles. In *L'aventure maritime, du golfe de Gascogne à Terre-Neuve* (pp. 55–62). Paris: CTHS.

Grenier, R. (1985). Basque Whalers in the new world: The Red Bay Wrecks. I. In G. Bass (Ed.), *Ships and shipwrecks of the Americas* (pp. 69–84). London: Thames & Hudson.

Grenier, R., Bernier, M.-A., & Stevens, W. (Eds.). (2007). *The underwater archaeology of Red Bay. Basque shipbuilding and whaling in the 16th century* (Vols. 5). Ottawa, Ontario, Canada: Parks Canada.

Harris, R., & Loewen, B. (2007). A Basque whaleboat: chalupa no. 1. In R. Grenier, M.-A. Bernier, & W. Stevens (Eds.), *The underwater archaeology of Red Bay. Basque shipbuilding and whaling in the 16th century* (Rigging, vessel use and related studies, Vol. 4, pp. 309–380). Ottawa, Ontario, Canada: Parks Canada.

Havard, G. (2003). *Empire et métissages. Indiens et Français dans le Pays d'en Haut, 1660–1715.* Paris: Septentrion and Presses de l'Université de Paris-Sorbonne.

L'Hour, M., & Veyrat, E. (2003). Analyser la culture matérielle maritime d'époque moderne: la contribution des épaves de la Natière (Saint-Malo). In C. Roy, J. Bélisle, M.-A. Bernier, & B. Loewen (Eds.), *Mer et Monde: Questions d'archéologie maritime.* Montréal, Québec, Canada: Association des Archéologues du Québec.

LaRoche, D. (1985). *Rapport d'étude préliminaire suite à la fouille de l'embarcation située dans la cour de la maison Estèbe à Québec (CeEt-7).* Montréal, Québec, Canada: Report to Ministry of Culture.

LaRoche, D. (1986). *La surveillance et le sauvetage de vestiges archéologiques au Musée de la Civilisation à Québec.* Montréal, Québec, Canada: Report to Ministry of Culture.

LaRoche, D. (1987). The small boats finds at the Musée de la Civilisation in Québec City. In A. B. Albright (Ed.), *Underwater archaeology proceedings from the Society for Historical Archaeology Conference* (pp. 108–112). Ann Arbor, MI: Society for Historical Archaeology.

LaRoche, D. (2009). Sailboats used on the St Lawrence river in Quebec City, Canada, in the 18th century: Tradition and adaptation. *The International Journal of Nautical Archaeology, 38*(1), 21–37.

Larochelle, A.-M. (2009). *La chaloupe à quille en Nouvelle-France. Une embarcation et son milieu.* Master's thesis, Anthropology Department, Université de Montréal, Montréal, Quebec, Canada.

Lescarbot, M. (1609). *Histoire de la Nouvelle France.* Paris: Jean Milot.

Litalien, R., Roth, K., & Vaugeois, D. (Eds.). (2004). *Champlain: the birth of French America.* Québec, QC: Septentrion.

Loewen, B. (1999). *Les barriques de Red Bay et l'espace atlantique septentrionale, vers 1565.* Doctoral dissertation, History Department, Université Laval, Québec, QC.

Loewen, B. (2007). The Basque shipbuilding trades: design, forestry and carpentry. In R. Grenier, M.-A. Bernier, & W. Stevens (Eds.), *The underwater archaeology of Red Bay. Basque shipbuilding and whaling in the 16th century* (The 24M Hull, Vol. 3, pp. 253–298). Ottawa, Ontario, Canada: Parks Canada.

Loewen, B., & Cloutier, C. (2003). Le chantier naval royal à Québec et le savoir maritime au XVIIIe siècle. In C. Roy, J. Bélisle, M.-A. Bernier, & B. Loewen (Eds.), *Mer et Monde : Questions d'archéologie maritime.* Montréal, Québec, Canada: Association des Archéologues du Québec.

Loewen B., & Demas, V. (2012). The Basques in the Gulf of St. Lawrence and adjacent shores. *Canadian Journal of Archaeology, 36*, 351–404.

Mimeault, M. (1987). *Destins de pêcheurs: les Basques en Nouvelle-France. Une étude de la présence basque en Nouvelle-France et de son implication dans les pêches en Amérique sous le Régime français.* Master's thesis, History Department, Université Laval, Québec City, Québec, Canada.

Moore, C. (1993). *Salmon fishing boats of the North American Pacific Coast in the era of oar and sail.* Master's thesis, Department of Anthropology, Texas A&M University, College Station, Texas, USA.

Moussette, M. (2002). Archéologie d'une rencontre. Les univers dualistes français et amérindiens dans l'Amérique septentrionale des XVIIe et XVIIIe siècles. *Recherches amérindiennes au Québec, 32*(1), 12–27.

Moussette, M. (2003). An encounter in the Baroque age: French and Amerindians in North America. *Historical Archaeology, 37*(4), 29–39.

Nadon, P. (2004). *La baie du Grand Pabos: une seigneurie gaspésienne en Nouvelle-France au XVIIIe siècle.* Montréal, Québec, Canada: Association des Archéologues du Québec.

Plourde, M. (2011). *L'exploitation du phoque dans le secteur de l'embouchure du Saguenay (Québec, Canada) par les Iroquoiens au Sylvicole supérieur (1000–1534 de notre ère).* Doctoral dissertation, Anthropology Department, Université de Montréal, Montréal, Quebec, Canada.

Proulx, J.-P. (1995). *Les Basques et la pêche à la baleine au Labrador au XVIe siècle.* Ottawa, Ontario, Canada: Parks Canada.

Proulx, J.-P. (2007). Basque whaling methods, technology and organization in the 16th century. In R. Grenier, M.-A. Bernier, & W. Stevens (Eds.), *The underwater archaeology of Red Bay. Basque shipbuilding and whaling in the 16th century* (Archaeology underwater: The project, Vol. 1, pp. 42–96). Ottawa, Ontario, Canada: Parks Canada.

Rieth, E. (1988). L'arquitectura naval. In H. Palou (Ed.), *Excavacions arqueologiques subaquatiques a Cala Culip: Vol. 2. Culip VI*. Girona, Spain: Museo d'arqueologia de Catalunya.

Rieth, E. (2003). La pirogue 2 de Mortefon (Charente-Maritime): remarques sur l'architecture monoxyle et le "système nautique" du bassin de la Charente au Moyen Âge. In C. Roy, J. Bélisle, M.-A. Bernier, & B. Loewen (Eds.), *Mer et Monde: Questions d'archéologie maritime*. Montréal, Québec, Canada: Association des Archéologues du Québec.

Thwaites, R. G. (Ed.). (1959a). *The Jesuit relations and allied documents* (Vol. 15). New York: Pageant.

Thwaites, R. G. (Ed.). (1959b). *The Jesuit relations and allied documents* (Vol. 65). New York: Pageant.

Tremblay, R. (2006). *Les Iroquoiens du Saint-Laurent. Peuple du Maïs*. Montréal, Québec, Canada: Éditions de l'Homme.

Turgeon, L. (1995). Pêcheurs basques du Labourd dans le golfe et l'estuaire du Saint-Laurent au XVIe siècle. In *L'aventure maritime, du golfe de Gascogne à Terre-Neuve* (pp. 213–234). Paris: CTHS.

Turgeon, L. (1996). Le chaudron de cuivre en Amérique: parcours historique d'un objet interculturel. *Ethnologie Française, 26*(1), 58–73.

Turgeon, L. (2000). Pêches basques du Labourd en Atlantique nord (XVIe–XVIIIe siècle): ports, routes et trafics. In *Itsas memoria, Revista de estudios maritimos del Pais Vasco 3* (pp. 163–178). Donostia-San Sebastián, Spain: Untzi Museoa-Museo Naval.

Turgeon, L., Fitzhugh, W., & Auger, R. (1992). Les objets des échanges entre Français et Amérindiens au XVIe siècle. *Recherches amérindiennes au Québec, 22*(2–3), 152–167.

Van Kirk, S. (1980). *Many tender ties: Women in fur-trade society, 1670–1870*. Winnipeg, Manitoba, Canada: Watson & Dwyer.

Vaugeois, D. (2002). *The last French and Indian war*. Montréal, Québec, Canada: McGill-Queen's University Press.

Vaugeois, D. (2005). *America. The Lewis and Clark expedition and the dawn of a new power*. Montréal, Québec, Canada: Vehicule Press.

White, R. (1991). *The middle ground: Indians, empires and republics in the Great Lakes region*. New York: Cambridge University Press.

Whitehead, R. H. (1986). Navigation des Micmacs le long de la côte de l'Atlantique. In C. A. Martijn (Ed.), *Les Mikmacs et la Mer* (pp. 224–232). Montréal, Québec, Canada: Recherches amérindiennes au Québec.

Whitehead, R. H. (1991). *The old man told us: Excerpts from Micmac history, 1500–1950*. Halifax, NS: Nimbus.

Chapter 11
The *Batteau Plat* of New France: Its Origin, Construction, and Design

Charles Dagneau

Introduction

Although French and other European sailors began fishing for cod in North America by the mid-sixteenth century, they did not establish a permanent base in the St. Lawrence valley (Eastern Canada) until the beginning of the seventeenth century. From this time, the French rapidly penetrated the continent along the waterway systems of Northeastern America. In order to navigate and to supply the trading posts and forts built deep inland, the French needed flat-bottomed crafts that were stronger than birch-bark canoes but lighter than most European boats (Fig. 11.1). Since it was necessary to navigate shallow waterways, cover great distances, and transport heavy loads, the *batteaux plats* were built by the hundreds to carry troops and supplies from Québec City to fortified places far upstream.[1]

French economic and military expansion in the New World soon put the French in conflict with Native peoples and New Englanders. During the first half of the eighteenth century, French colonizers took control of a large portion of the North American continent and established inland domination that finally ended with a decisive British victory during the Seven Years' War (1755–1763). In this context of semi-permanent war, the *batteau plat* came to be an essential part of every military operation in the New World for both sides of the colonial conflict. As J. Gardner (1987:20) wrote, "Perhaps more than any other single factor, it was the

[1] In French, the modern spelling of "batteau plat" is *bateau plat*, or the plural *bateaux plats*, which means flat boat. Both spellings, with one or two 't's, were used in seventeenth and eighteenth century. *Batteau plat* is the boat-type.

This text is partly derived from my Master's dissertation, *Les "batteaux plats" en Nouvelle-France* (Dagneau 2002).

C. Dagneau (✉)
Underwater Archaeology Service, Parks Canada,
1800 Walkley Road, Ottawa, ON K1H8K3, Canada
e-mail: Charles.Dagneau@pc.gc.ca

A.M. Evans (ed.), *The Archaeology of Vernacular Watercraft*,
When the Land Meets the Sea, DOI 10.1007/978-1-4939-3563-5_11

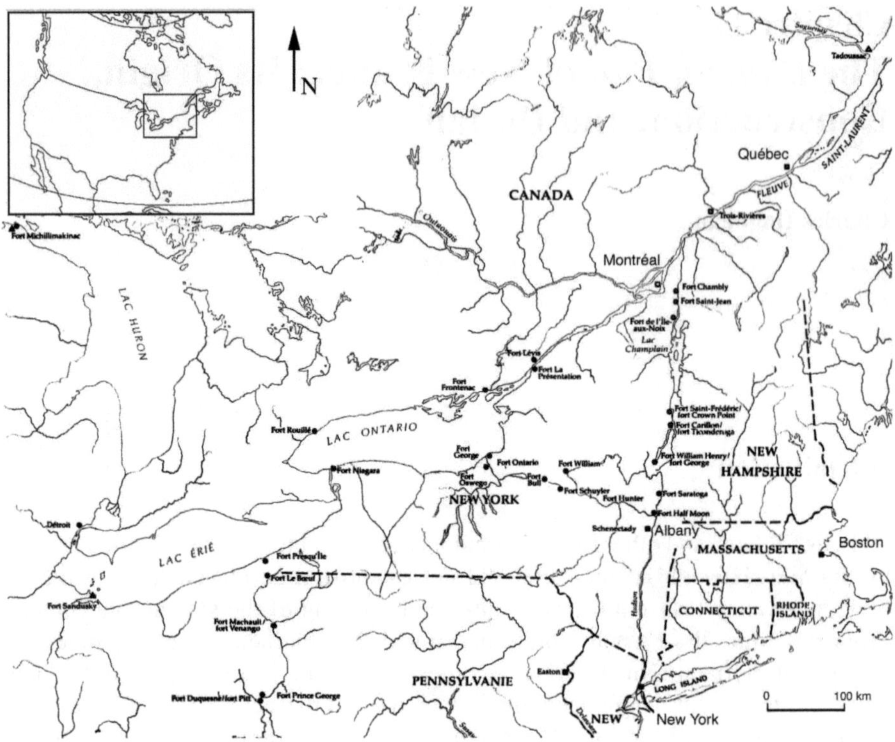

Fig. 11.1 Portion of the waterway network of Northeastern America (C. Dagneau, after G. J. Matthews, in Harris 1987–1993 (1):42)

batteau that not only built the power and wealth of New France, but also contributed to her downfall at the end [...]. The logistics that defeated the French was a logistics based upon batteaux."

The remains of four *batteaux plats* were found in 1985–1986 in the harbor of Québec City (Canada), during a rescue excavation conducted by archaeologist Daniel Laroche (1985, 1986, 2009) (Borden codes: CeEt-7 and CeEt-143). These small crafts, named *Batteau-1* to *4*, represented the only positive examples of this boat-type built by the French in the New World at the time and were the subject of a MA dissertation by the author (Dagneau 2002; Laroche 1986, 1987). Another *batteau plat* documented more recently in Kingston may well represent a fifth example, as it is very similar in construction (Dagneau 2009). It has been argued that the *batteau plat* industry arose and developed to such an extent in New France because the boat's design and construction were inherently efficient and that strong state policies were in place to develop maritime industries, including boat construction (Dagneau 2004).

On the other hand, John Gardner (1987:11) has suggested that the quickly and easily built "batoes" of eighteenth century New England emerged along with the power sawmill. Indeed, this new technology could have been introduced into the shipbuilding process to provide cheap wide boards for the construction of small flat-bottomed 'batoes.' We may enquire whether the *batteau plat*'s development in New France was also linked to the introduction of the power sawmill in a similar way, and if its construction relied on a standard design.

By addressing questions of origin, design, and technology, it is possible to understand *batteau plat* production in New France and to situate this boat-type in a long-term evolutionary perspective. An examination of written sources and archaeological remains suggest that the boat-type was first introduced in North America by the French in 1665 and evolved through time in various forms. A detailed architectural analysis of the Québec City remains indicates that the construction process relied on a standard design method. This unusual early instance of maritime boat production was also quite distinct, because unlike the case of New England, it was not related to power mill development.

Building Contracts and Literature

From the establishment of a Royal government in Quebec by 1665 to the fall of the colony in 1763, evidence of more than 1000 *batteaux plats* can be found in Québec City archives, mostly in notarial records prior to 1702. In general, contracts specified a length ranging from 23 to 25 *pieds* (7.47–8.12 m), a few others being 25–26 *pieds* long (8.12–8.44 m). Mentioned more rarely, the amidships breadth was 4–4.5 *pieds* wide. All these boats were to be invariably 22–24 *pouces* (60–65 cm) high (Table 11.1) (Dagneau 2004).[2]

Although building contracts were not used after 1702, colonial accounts available between 1715 and 1753 show consistent spending for the "construction, caulking, and purchase of batteaux and canoes for the service of the King"—that was probably enough to maintain a fleet of 200 *batteaux plats* or more (Centre d'archives d'outre-mer, France: COL C11A 115). Indeed, at least 400–500 *batteaux plats* were in service during the Seven Year's War (de Bougainville 2003:219–221). Moreover, one can also grasp the importance of this craft for military operations and the daily life of the colony when reading some contemporary writers (for example: de Baugy 1883:70–76; de Bougainville 2003:219–221; Kalm 1977:f.563, 693, 783; Dagneau 2004).

A similar phenomenon occurred in the New England colonies at the end of the eighteenth century, when the British began to build a similar craft known as "batoes". During the Seven Years' War, a "batoe service" in Albany provided craft for the British army. In 1758 alone, this service under the command of Colonel Bradstreet built no fewer than 1500 "batoes" (Meany 1998). Many examples were found in Lake George (New York State) in 1960, and described by J. Gardner (1967a, b, c, d, 1987), K. Crisman (1988) and J. Zarzinski (Pepe

---

[2] 1 *pied* (foot) = 12 *pouces* (inches) = 32.48 cm.

Table 11.1 *Batteaux plats* built between 1665 and 1760

Year	Owner/investor	Boat	Source
1665	The State	152 batteaux plats	General correspondence
1670	Toussain Toupin	1 bateau plat	Notarial records
1671	Abel Turquot	1 bateau	Notarial records
1673	The State	2 large batteaux plats	Notarial records
1684	The State	25 batteaux plats	Notarial records
1685	The State	75 batteaux plats, pine planks	Notarial records
1686	The State	120 batteaux plats	Notarial records
1687	The State	150 batteaux plats	Notarial records
1689	The State	40 batteaux plats	Notarial records
1691	The State	2 large heavy-load bateaux double-ended	1702 inventory
1691	The State	1 large bateau grenadier	1702 inventory
1692	Léonard Paillé René Drouillard	1 bateau	Notarial records
1694	The State	16 batteaux double-ended	1702 inventory
1695	The State	100 batteaux plats	Notarial records
1696	The State	40 bateaux double-ended	Notarial records
1693	The State	30 bateaux plats	Notarial records
1700	The State	40 bateaux double-ended	Notarial records
1701	The State	50 batteaux plats double-ended	Notarial records
1702	The State	1 bateau grenadier	1702 inventory
1759	The State	250 batteaux plats	General correspondence

Sources: Drolet-Dubé and Lacombe (1976), Dagneau (2004), Centre des archives d'outre-mer, France, COL C11A 20/fol.182-182v

Productions 2005). Their overall construction was comparable to that of the Québec City boats, though they differ in many details, such as their dimensions and the wood types used (Dagneau 2002).

The *Batteau Plat* Remains

The four crafts found in Québec City's harbor were not complete, nor were they in very good condition when uncovered. Many of the timbers were eroded or twisted, if not absent. The crafts were situated on the original shoreline of the St. Lawrence River, under a stone wharf built in 1752 (Laroche 1987:123, 303). It is believed that they date from between 1740 and 1751 (Dagneau 2002).

The description that follows is reconstructed from observations of the four excavated crafts, since each was incomplete (Fig. 11.2). The architecture of the *batteau plat* is quite simple. It has a flat bottom (also called a *sole*) made from four lengthwise planks, carvel laid. It is double-ended and the posts are notched so as to fit over the end of the bottom, and to protect the vulnerable front edge of the sole planks. The boat's sides consist of three carvel planks nailed at each end onto the curved posts, with no rabbet. Each craft has between 19 and 21 frames, all made in the same way: a flat floor timber associated with two knee-shaped futtocks. There is a sharp chine between the sole and the sides.

The floor timbers are fixed to the bottom planks, and the futtocks are nailed to both the bottom and the sides. On the other hand, the floors and futtocks are not fixed to each other. Interestingly, the outboard edges of the bottom are beveled in order to mate with the lowest side strakes. These strakes are in turn nailed to the edges of the sole for more strength. Remnants of inside planking were also found in one example.[3]

The *batteau plat* ranges from 9.90 to 10.05 m in total length. The breadth amidships is between 1.70 and 1.90 m at the gunwale. The height is 60–65 cm. According to these dimensions, the length to breadth ratio is between 1:5 and 1:6. These remains are longer and wider than what is generally known from written sources, but the height corresponds exactly to that given in period building contracts (Table 11.2). In fact, the Québec City *batteaux* seem to correspond to one of the larger types of

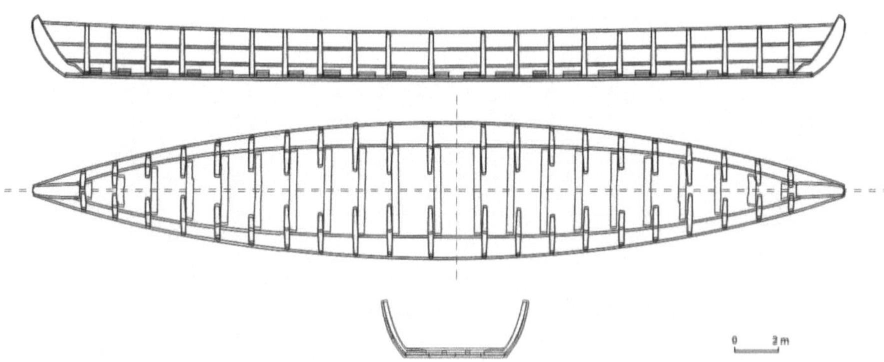

Fig. 11.2 Reconstruction of *Batteau 4* (C. Dagneau)

Table 11.2 Dimensions from remains and notarial records (cm)

	Length	Width	Height	Ratio length:breadth
Remains (total)	9.90–10.50	1.70–1.90	0.60–0.65	5.72
Remains (bottom)	*9.10–9.50*	*1.22–1.30*	–	–
Archives	7.47–8.44	1.30–1.46	0.60–0.65	5.76

Source: Dagneau (2002), Laroche (1986)

[3] For more detailed information on the construction and dimensions of the craft, see Dagneau (2004).

batteau that were built in the colony, known as a heavy-load batteau (*batteau de charge*) (Centre des archives d'outre-mer, France, COL C11A 20/fol.182-182v).

The wood species study shows a fairly constant picture of timber use (Fig. 11.3). Though many of the planks are missing, in all four *batteaux*, white pine (*Pinus strobus*) was used for the bottom planks and the sides, except for the lower strakes and the lateral bottom planks, which were made of oak (*Quercus* sp.). The floor timbers were all made of white pine, except for five oak floors in *Batteau 1* and maybe another in *Batteau 3*. The posts and all the futtocks are taken from naturally curved pieces of Eastern white cedar (*Thuja occidentalis*).

The Québec City architectural remains provide no information on the boat's fittings since they were all badly preserved above the lower strake. A towing ring, however, was found on one of the stem-posts. On the other hand, historical references state that the *batteaux plats* were propelled mostly by oars, but also towed and poled when necessary (de Baugy 1883:70–76; Kalm 1977:f.693, 841). They were fitted with a removable mast and a square sail, as also mentioned in some building contracts (Kalm 1977:f.821; Dagneau 2004). A draft of a 1776 "bateau" from the British Admiralty provides a good idea of the interior fittings of a *batteau plat*, with the location of the benches and the thole pins on the gunwale (Fig. 11.4) (Dagneau 2002).

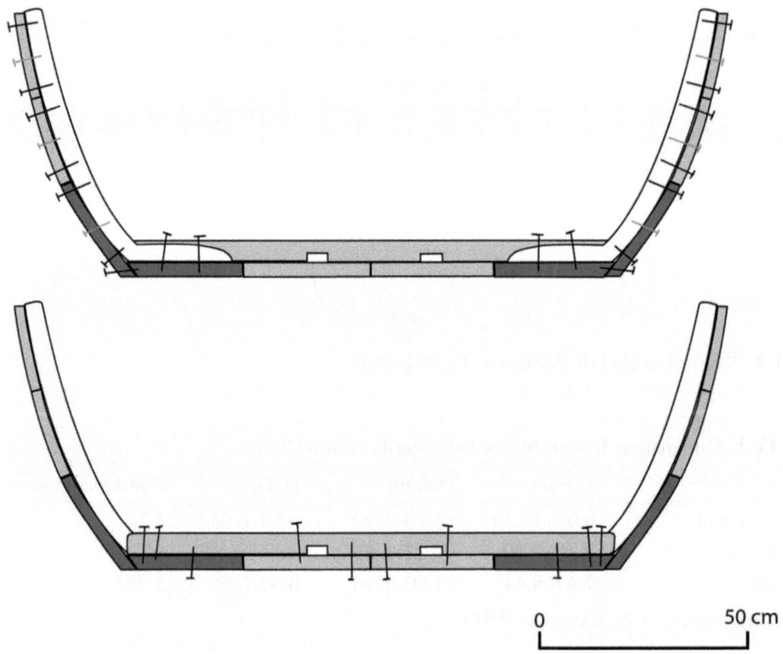

Fig. 11.3 Reconstructed master-frame section of *Batteau 4*. The *Batteaux plats* are made of pine (*light gray*), oak (*dark gray*), and white cedar (*white*) (C. Dagneau)

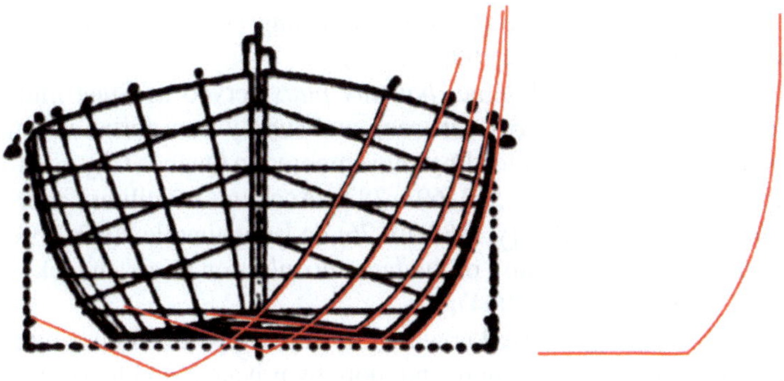

Fig. 11.4 A draft of a British "batoe" from 1776 (Chapelle 1951:35, after the Admiralty Collection of Draughts, National Maritime Museum, London)

Origin and Evolution of the *Batteau Plat*

Many scholars believe that the *batteau plat* is a boat-type derived from the small French "flats" (*plates*) used in the St. Lawrence estuary for cod-fishing in the early seventeenth century (Gardner 1987:6; Chapelle 1951:34). Some suggest the model could also come from the Dutch in New Amsterdam (Crisman 1988:131). However, it appears that the type could originate from elsewhere, since many traditional flat boats were used in Atlantic France and in the British Isles in the seventeenth century (Beaudouin 1985, 2000:32–34; Gardner 1987:10–11; Renault 1984). While there is doubt concerning the exact origin of the *batteau plat*, there is certainly little chance that this boat-type was a New World invention.

In fact, a French origin is the most probable if we consider the following. First, the name "batoe" used in English clearly derives from the French word "*bateau*", which basically means boat. Second, the *batteau plat* were first mentioned in New France in 1665, when 152 units were built in Québec City "to bring the troops to their barracks as well as to counter the Iroquois and attack them by way of the rivers" (Talon to Colbert, Nov 13, 1666. Archives Nationales du Québec à Québec, C11 A).[4] In New England, the earliest use of "batoes" by military is known to have been during Queen Anne's War, from 1702 to 1713 (Bellico 1992:15–16; Crisman 1988:138–139).

The 1665 Québec City *batteaux plats* were built by four royal shipwrights, all originating from the vicinity of La Rochelle (Brisson 1983:25–26; Pritchard 1971:9). Unfortunately, research has not yet produced any information on these shipwrights' activities in France before they moved to Canada. There is, however, a good chance that this type of boat was already used around La Rochelle at the time,

[4] Archives Nationales du Québec à Québec, hereafter ANQ-Q.

as such "flatties" were common vernacular fishing boats in South-Ouest France (Beaudouin 2000:32–34; Millot 1999:11–12).

Throughout the French Regime, *bateaux plats* served as State multi-purpose crafts. They were used primarily for troop and supply transportation, but also for mail delivery, as lighters for unloading ships moored in Québec City and for general trade in the colony (Kalm 1977:563, 693, 783). Accounts of military expeditions by contemporary writers (de Baugy 1883:70–76; de Bougainville 2003:219–221) provide vivid descriptions of the use of the *batteaux plats* on rivers and lakes in a wartime context (Dagneau 2002, 2004).

After the French Regime, we know that this boat-type was in use by the British military, but it came to be used more and more by private individuals. By the end of the nineteenth century, this flat-bottomed craft was widely found in North America in different forms. The colonial *batteau* is generally considered to be the direct ancestor of the Banks dory (Gardner 1987:10; Chapelle 1951:34). In fact, there are many parallels between the two types. The *batteau plat* also evolved into a form of log-driving boat called a "pointer," used by the lumbermen in different regions of Canada, and diverse flat-bottomed crafts were used as multi-purpose rowing skiffs by inhabitants along the St. Lawrence River until the early twentieth century (Chapelle 1951:80–84). On the Atlantic coast of Canada and in the United States, it evolved into a number of skiffs, wherries, and other small boats, in forms and sizes so numerous that it is hardly possible to consider them all (Chapelle 1951). Obviously, the development of many of these crafts was also influenced by boat-types other than the *batteaux plats*.

The *Batteau Plat* Design

Based on the large scale of *batteau plat* production in New France and the relatively standard shape of the Québec City remains, it appears that construction relied on standard design methods. Several pieces of evidence point in this direction, apart from the great number of crafts that were built. One revealing document from 1693 concerns the supply of 1000 "moulded" futtocks for the construction of the "King's batteaux plats". Another building contract mentions a "model" to help for the construction of the craft (ANQ-Q, notary Génaple, 26 Oct 1686 and 28 Jan 1693).

A 1776 "batoe" draft from the British Admiralty reproduced by H. Chapelle (1951:35) shows a moulding procedure known as hauling-down (Barker 1991), even though the French *batteau plat* might be more likely to show a variant of the *trébuchement* procedure as explained by Eric Rieth (1996). An analysis of the *Batteau-4* remains, which exhibit better preservation, provides the best evidence of the hull construction. Only half of this boat was found, but its futtocks seemed in better condition than those of the others.

Based on initial drawings prepared by Daniel Laroche (1986), who conducted the excavation, archaeologists tried to set up the frames in their original position, once the bottom was reconstructed. The corresponding nail holes were very helpful. The next step was to find the right position of the frames in elevation, as well as to discover the method used to create their shape. Was there a single mould consistent to all the frames as for the 1776 British "batoe," or were there several moulds used in conjunction with each other? Was the hauling-down procedure or the *trébuchement* procedure used? Was the boat built with only a master-frame plus one pre-erected frame at each end, then projected by placing ribbands on these reference frames? Generally, there are three or four basic modifications involved in the moulding process of a wooden craft, applied gradually from the master-frame to the ends. In the case of the *batteaux plats*, the process is quite simplified, because the boats are double-ended and thus identical forward and aft, without any rising of the floor.

Batteau-4 was selected for an in-depth architectural analysis attempting to understand the construction and conception of this boat-type. While most of the futtocks were too deformed to allow conclusive data on the complete design method, many futtock shapes clearly follow a circular arc section of 3-French foot radius (Fig. 11.5). The same was also found in the surviving stempost of *Batteau 4*. Any other arcs applied to the futtock lines were inadequate. An attempt was made to reconstruct the shape of *Batteau 4* based on this 3 ft arc, despite the fact that some distorted futtocks (nos. 13, 17, and 21) were clearly not fitting the model.

A few control parameters were used in order to smooth the lines of this reconstruction: (1) The position of the futtock and its angle (flat part horizontal); (2) The position of the circle center (height and distance from middle line); (3) The arc portion based on a 3 ft radius (observed and reconstructed); (4) The position of the "chine" or "*bouchain*" observed on remains; (5) The position of the gunwale (constant height); (6) The chord length; (7) The exterior angle; (8) A simple ratio for the master-frame. At the end of the process, some futtock lines were changed in order to smooth the *batteau*'s shape (Figs. 11.6 and 11.7). The reconstruction made in this manner is not satisfactory in all aspects, but represents a probable representation.

It is doubtful that a comprehensive moulding method was used throughout the hull. On the other hand, there is a good chance that the *batteau plat* was designed with a double pre-erected master-frame, or with four pre-erected frames: two in the middle and two toward the ends (Fig. 11.8). This is an easy method often used for small craft that could explain the problem of irregular futtock design encountered during analysis. If, however, small ribbands were used to guide the builder, they should have been nailed on the pre-erected frames showing characteristic nail hole marks. Unfortunately, an analysis of the nailing pattern could not demonstrate the existence of temporary ribbands as the nail holes are absent or no longer visible.

In the end, even though the architectural analysis was not completely conclusive in demonstrating the use of a particular method—whether it was the hauling-down, *trébuchement*, ribband, or a mix, the 3 ft circular arc observed in many cases may indeed be considered to be part of the *batteau plat* design method.

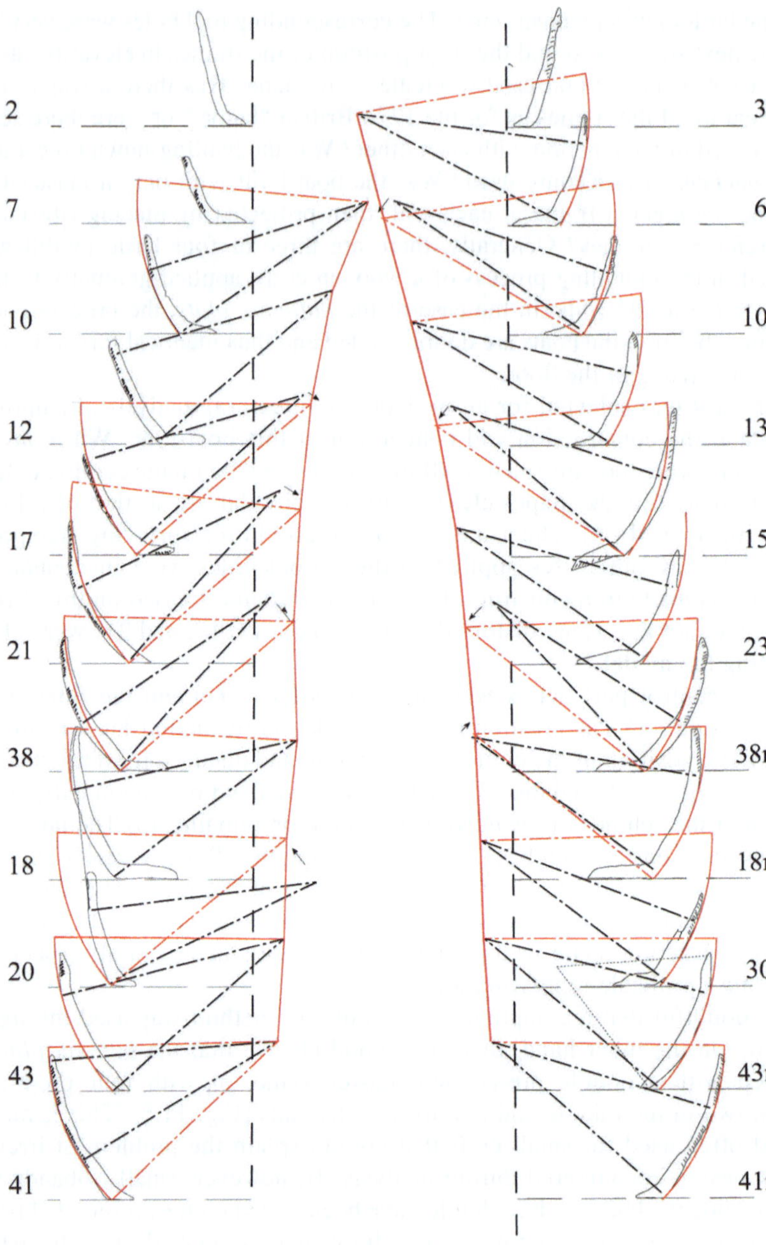

Fig. 11.5 Futtock drawings in their respective position. Some futtocks show a nice 3 ft radius circular arc section, others don't (C. Dagneau. Original drawings: D. Laroche)

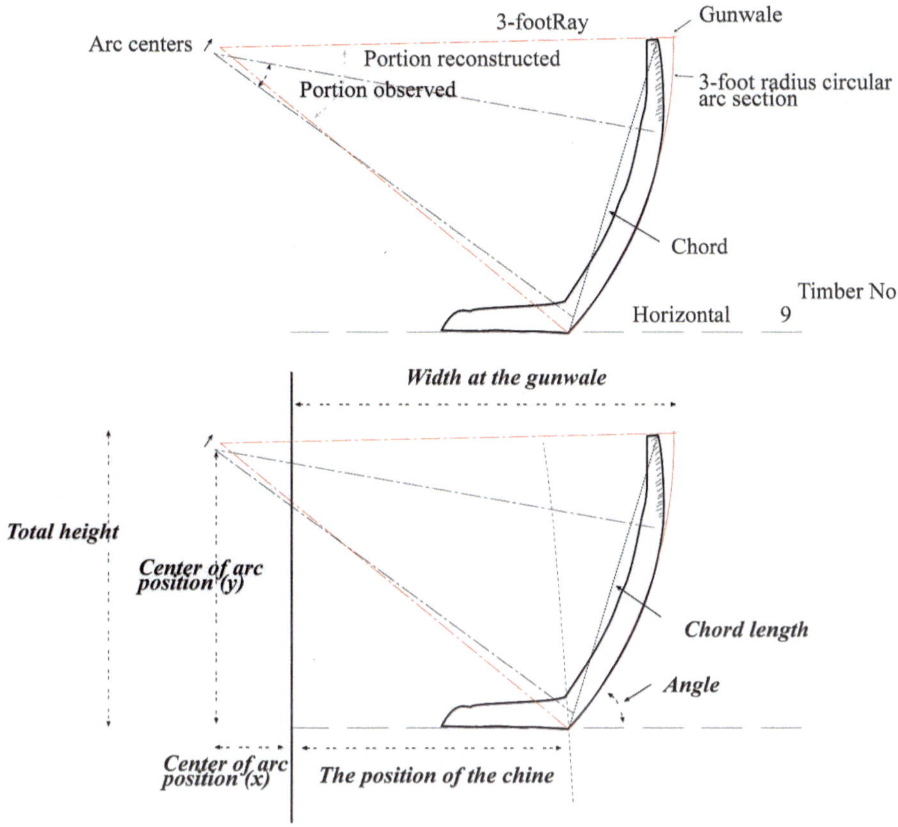

Fig. 11.6 Various control parameters used to verify and smooth the *Batteau 4* shape (C. Dagneau)

Wide Board Craft and Power Mill Development

For J. Gardner, the development of the "batoe" in America was closely linked to the development of the powered saw mill. "Wide lumber, sawn by machines," he writes, "may be what produced dory-type boats in Renaissance Europe as well as in maritime North America. Apparently cheap, quickly built, these were the 'plywood skiffs' of yesteryear" (Gardner 1987:15). Abundant wood resources for shipbuilding as well as the absence of pit sawyers' guild constraints found in Europe probably also contributed to the success of the *batteau plat* in America.

It is true that no guilds militated against sawmill development in New France, but the power mill was not as quick to develop as in New England. As a consequence, it is doubtful that the *batteau plat* industry established in Québec City as early as 1665 was made possible by power mill technology. During this period, there is no evidence of power saw mills near the Québec City shipyards (Fauteux 1928).

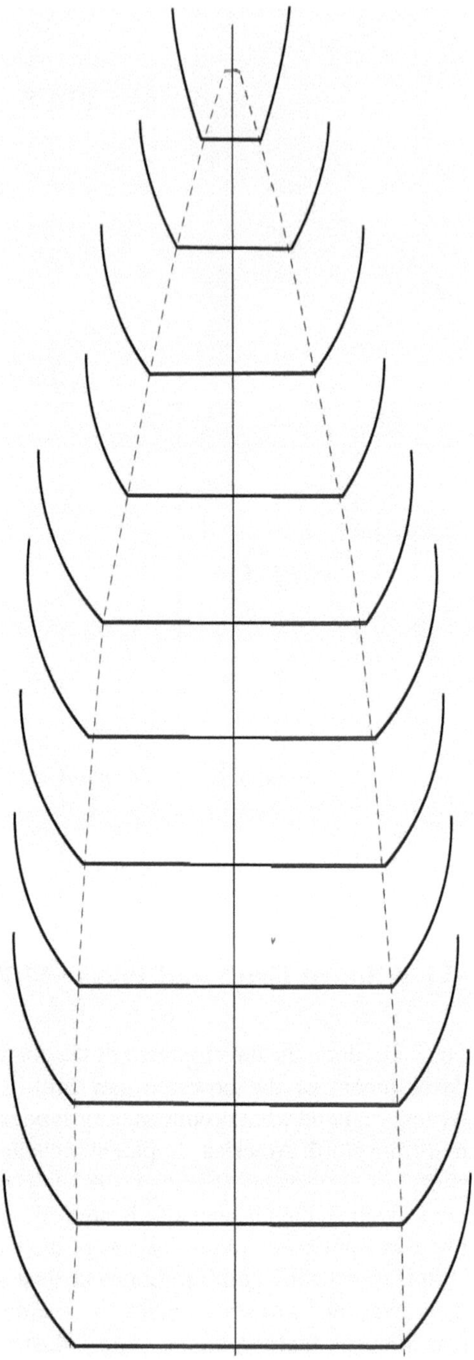

Fig. 11.7 Reconstructed futtocks shape of *Batteau 4* after smoothing process (C. Dagneau)

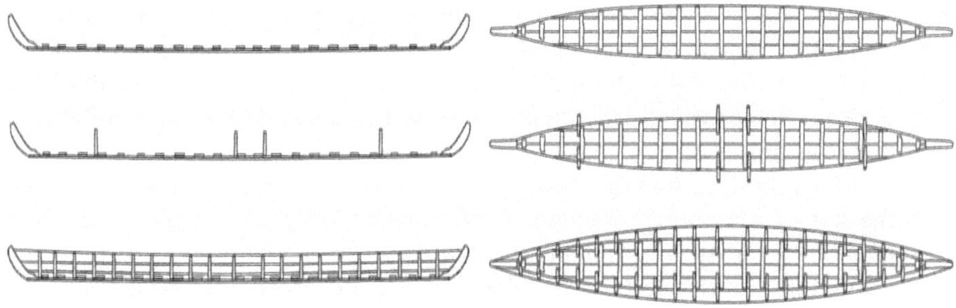

Fig. 11.8 A construction method using four pre-erected frames and ribbands to design the hull (C. Dagneau)

On the other hand, surviving contracts for the sale of planks (*bordages*) for the King's *batteaux plats* are signed by individuals who are often specifically identified as pit sawyers (*scieur de long*) and who worked in the forest during the winter (ANQ-Q, notary Rageot, 7 and 29 Jan 1686; notary Génaple, 9 Dec 1693). There is no mention of a power mill owner in these documents. Moreover, a close observation of the crafts' oak and pine boards has shown manual saw marks on multiple planks from *Batteau 3* and *4*. Thus, based on these various sources, it seems that the Quebec shipbuilders did not use the power mill for the construction of these crafts.

Apart from the relative backwardness of New France's economy in the seventeenth and eighteenth centuries, two reasons may help to absence of power mills in *batteau plat* construction. First, the French *batteau plat* was made primarily of softwoods (when not entirely), while the English "batoes" were made of both oak and pine. Softwood is more easily sawn than hardwood. This is certainly the result of the distinctive St. Lawrence valley woodland environment, where hardwood is scarcer (Farrar 1999). Second, French production never reached the level of English 'batoe' construction during the Seven Years' War (1755–1763) (Gardner 1987:22; Meany 1998). The greater efficiency of the New England batoe service is probably related to the use of the power saw mill.

Conclusion

From the thousands of *batteaux plats* built in New France, only four or five archaeological examples have been found to date. This early naval "mass-production" was an uncommon phenomenon in the seventeenth and eighteenth centuries, especially in the New World. The Quebec remains are believed to correspond to a large type of bateau called "heavy-load batteau." The colonial *batteau* seems to first occur in

North America in 1665, in Québec City. This boat-type may have derived from the cod-fishing flats used in the St. Lawrence valley in the seventeenth century but most probably from traditional craft used in South-Western France. Considered the Bank dory ancestor, the *batteau plat* is believed to have evolved into a variety of shapes and sizes over time.

It was hypothesized that this boat-type may have been built in a standard way, with the use of a moulding method. Unfortunately, detailed analysis of the best preserved Québec City *batteau plat* (*Batteau-4*) does not confirm the exact method of construction. Nevertheless, archaeologists were able to note the recurrence of a circular arc with a radius of 3 *pieds*. Finally, it seems that the *batteau plat* from New France was not made from mechanically sawn wide boards, as were their New England counterparts. Thus, the link between power sawmill technology and the dory boat-type development, as suggested by J. Gardner, needs to be reconsidered in the New France context.

Acknowledgments Thanks to Daniel Laroche who carried out the excavation of the *batteaux plats* remains and the preliminary analysis of the boat remains. Also thanks to Brad Loewen for comments on earlier versions of this paper. Any errors remaining in the text are mine. Special thanks to the Social Sciences and Humanities Research Council of Canada and the Society for Historical Archaeology for their financial support.

References

Barker, R. (1991). Design in the Dockyards, about 1600. In R. Reinders, & K. Paul (Eds.), *Carvel construction technique*. International Symposium on Boat and Ship Archaeology 5 (Amsterdam 1988) (pp. 62–63). Oxford, England: Oxbow Monograph 12.

Beaudouin, F. (1985). *Bateaux des fleuves de France*. Douarnenez: Éditions de l'Estran.

Beaudouin, F. (2000). *Les bateaux garonnais I. Les Cahiers du Musée de la batellerie* (Vol. 44). Conflans-Sainte-Honorine, France: Musée de la Batellerie.

Bellico, R. P. (1992). *Sails and steam in the mountains: A maritime and military history of Lake George and Lake Champlain*. New York: Purple Mountain Press.

Brisson, R. (1983). *Les 100 premières années de la charpenterie navale à Québec: 1663–1763*. Collection Edmond-de-Nevers No 2. Québec, QC: Institut Québécois de Recherche sur la Culture.

Chapelle, H. I. (1951). *American small sailing craft. Their design, development and construction*. New York: W.W. Norton.

Crisman, K. (1988). Struggle for a continent: Naval battles of the French and Indian wars. In G. F. Bass (Ed.), *Ships and shipwrecks of the Americas. A history based on underwater archaeology* (pp. 129–148). New York: Thames and Hudson.

Dagneau, C. (2002). *Les "batteaux plats" en Nouvelle-France*. Master's thesis, University of Paris 1, Panthéon—Sorbonne, Paris, France.

Dagneau, C. (2004). The 'Batteaux Plats' of New France. *International Journal of Nautical Archaeology, 33*(2), 281–296.

Dagneau, C. (2009). Batteau plat de Kingston, Navy Bay (Fieldnote 2010-11-79). Ottawa, Ontario, Canada: Underwater Archaeology Service, Parks Canada.

de Baugy, L.-H. C. (1883). *Journal d'une expédition contre les Iroquois en 1687: lettres et pièces relatives au fort Saint-Louis des Illinois*. Paris: E. Leroux.

de Bougainville, L.-A. (2003). *Écrits sur le Canada. Mémoires—Journal—Lettres (1756–1758)*. Québec, QC, Canada: Septentrion.

Drolet-Dubé, D., & Lacombe, M. (1976). Répertoire des contrats d'engagement, de marché de construction et autres types d'actes concernant les artisans du XVIIe siècle à Québec. Québec, QC, Canada: Parks Canada, unpublished.

Farrar, J. L. (1999). *Les arbres du Canada*. Ottawa, Ontario, Canada: Service Canadien des Forêts.

Fauteux, J.-N. (1928). *Essai sur l'industrie au Canada sous le Régime Français*. Québec, QC, Canada: Imprimeur du Roi.

Gardner, J. (1967a, April). Bateau played key role in American history. *National Fisherman*, pp. 8a, 9a, 19a.

Gardner, J. (1967b, August). Bateau reconstructed from remains. *National Fisherman*, p. 8a.

Gardner, J. (1967c, June). Construction details of old bateaux show basic design with variations. *National Fisherman*, pp. 8a, 9a.

Gardner, J. (1967d, May). Famous boat type in transitional stage. *National Fisherman*, pp. 8a, 9a, 28a.

Gardner, J. (1987). *The Dory book*. Mystic, CO: Mystic Seaport Museum.

Harris, R. C. (Ed.). (1987–1993). *Atlas historique du Canada* (Vol. 1). Montréal, Quebec City, QC, Canada: Les Presses de l'Université de Montréal.

Kalm, P. (1977). *Voyage de Pehr Kalm au Canada en 1749*. (J. Rousseau, G. Béthune, & P. Morisset, Trans.). Montréal, QC, Canada: Pierre Tisseyre.

Laroche, D. (1985). *Rapport d'étude préliminaire suite à la fouille de l'embarcation située dans la cour de la maison Estèbe à Québec (CeEt-7)*. Québec, QC: Report to the ministry of Culture.

Laroche, D. (1986). *La surveillance et le sauvetage de vestiges archéologiques au Musée de la Civilisation à Québec*. Québec, QC: Report to the Ministry of Culture.

Laroche, D. (1987). The small boats finds at the Musée de la Civilisation in Québec City. In A. B. Albright (Ed.), *Underwater archaeology proceedings from the Society for Historical Archaeology conference* (pp. 108–112). Ann Harbour, MI: Society for Historical Archaeology.

Laroche, D. (2009). Sailboats used on the St Lawrence river in Quebec City, Canada, in the 18th century: Tradition and adaptation. *International Journal of Nautical Archaeology, 38*(1), 21–37.

Meany, J. F. Jr. (1998). "Batteaux" and "Battoe Men": An American colonial response to the problem of logistics in mountain warfare. New York State Museum, New York. Page creation December 8, 1998, Retrieved February 22, 2006, from http://www.dmna.state.ny.us/historic/bateau.html.

Millot, G. (1999). Les gens de la rivière d'Etel. *Le Chasse-Marée, 126*, 2–15.

Pepe Productions. (2005). *Wooden bones: The Sunken Fleet of 1758 [DVD]*. New York: Glens Falls.

Pritchard, J. S. (1971). *Ships, men and commerce: A study of maritime activity in New France*. Doctoral dissertation, University of Toronto, Toronto.

Renault, F. (1984). *Bateaux de Normandie*. Rennes, France: Édition de l'Estran.

Rieth, E. (1996). *Le maître-gabarit, la tablette et le trébuchet. Essai sur la conception non-graphique des carenes du Moyen Age au XXe siècle*, Mémoires de la Section d'Histoire des Sciences et des Techniques (Vol. 9). Paris: CTHS.

Chapter 12
Vernacular Craft of the North American Great Lakes

Bradley A. Rodgers

Introduction, 30 Years Research in Retrospective

If "Vernacular" is defined in a maritime setting as a marine technology indigenous or characteristic of a particular locality then vernacular craft of the Great Lakes of North America must include virtually any vessel built there, particularly after nineteenth century circumstances dictated the development of new and unique vessel types. For until enlargement of the St. Lawrence Sea Way in 1895 and improvements to it in the 1950s, what was built on the lakes, generally speaking, stayed on the lakes. There were exceptions; certainly smaller ships built on the lakes did make their way into oceanic trade, especially after completion of the various locks and canals in the St. Lawrence River by 1848. Experiments in shipping directly from the lakes overseas did show promise. Yet access to the Inland Seas was still fairly restricted and movement of large or heavily laden ships with drafts that exceeded 8.5 ft (2.6 m) was out of the question (Musham 1948:57–58). The only other ocean access to the lakes was via the Illinois/Mississippi Rivers, and through the Erie Canal (1825) stretching from Lakes Erie and Ontario to the Hudson River over some 360 ft (109.7 m) of elevation. Nonetheless, canal and river access remained too restricted during much of the nineteenth century to allow passage of large, deep draft ships from the lakes to the ocean.

Over the course of the nineteenth century, tens of thousands of vernacular vessels were built on the Great Lakes ranging from wooden vessels of the largest classes, those that most influenced trade in the "Old Northwest," to craft such as tugs, ferries, pleasure craft, and small working craft like fishing tugs, bateaus, canoes, mackinaw boats, flats, and barges of all types. Of particular interest though are the large ships, the types of vessels that improved the lakes' trade and

B.A. Rodgers (✉)
Program in Maritime Studies, East Carolina University,
Admiral Ernest M. Eller House, Greenville, NC 27858-4353, USA
e-mail: rodgersb@ecu.edu

A.M. Evans (ed.), *The Archaeology of Vernacular Watercraft*,
When the Land Meets the Sea, DOI 10.1007/978-1-4939-3563-5_12

commerce during the nineteenth century and witnessed the transition from sail to steam. Sources of information previously underutilized in these types of studies include thesis sources and reports, which often include detailed vessel descriptions. It should be stressed that the following ship descriptions, historical or archaeological, are only brief summaries of these in-depth analyses. It should also be noted that the cited research should not be construed as the only archaeological work available concerning Great Lakes maritime topics. Texas A&M University, The Great Lakes Historical Society, and various other organizations have collaborated with East Carolina University over the years to produce a growing body of informative and professional reports and theses.

The Great Lakes research project began 30 years ago in 1985, as East Carolina University (ECU) in collaboration with the State Historical Society of Wisconsin (SHSW) began to conduct studies of nineteenth century ship classes. The initial survey of the schooner *Fleetwing* laid the groundwork for subsequent studies and collaborations (Cooper 1988). Since then ECU research has taken place on all of the Great Lakes, but has been focused primarily on sites in Lakes Michigan, Superior, and Huron. Research partners added since research commenced include the Federal NOAA—Thunder Bay Marine Preserve; the Michigan State Department of History, Arts, and Libraries; the Roger's Street Fishing Village Museum in Two Rivers, Wisconsin; and the Wisconsin Maritime Museum in Manitowoc, Wisconsin. Although by necessity the research originated as a series of site-specific surveys, it has evolved beyond this. Over the last three decades, researchers have identified and documented all of the major ship types that operated on the lakes in the nineteenth century along with their internal construction and design characteristics. This documentation includes full Phase II pre-disturbance archaeological recording of about 30 wrecked or abandoned ships and numerous scatter sites from known ship trap areas such as the Apostle Islands in Lake Superior, Death's Door Wisconsin, and the North Point Reef of Alpena, Michigan.

Perhaps the greatest value of this research, at least to potential researchers and or students of Great Lakes history and archaeology, is the publication of results. As with many vernacular watercraft, many of the references and records lie hidden in relatively unknown publications or remain unpublished in the form of Masters theses. The theses and reports in many instances are equivalent to well crafted and written primary source based books but are not as widely available or well known. The research conducted by ECU ranges from purely historical theses to, and including, remote sensing surveys using both magnetometer and side scan sonar. This body of research work is sometimes available through bookstores but can most often be acquired through interlibrary loans, direct contact, and PDFs through the ECU Maritime Studies Program, The State Historical Society of Wisconsin, and the NOAA Thunder Bay Marine Preserve. The research can be divided into published books and reports, MA theses, and funded research reports. The following list contains research work conducted in accordance with the original site-specific, and later expanded, research designs, and includes archaeological documentation conducted by and in cooperation with the other aforementioned sponsoring organizations:

Books and Published Reports

Bazzill, Dina M.
2007: *The Missing Link Between Sail and Steam: Steambarges and the Joys of Door County, Wisconsin*. ECU Report #19, PAST Foundation, Columbus, OH.
Cooper, David J. (editor)
1991: *By Fire, Storm, and Ice: Underwater Archaeological Investigations in the Apostle Islands*. State Historical Society of Wisconsin, Madison, WI, Sea Grant, NOAA, and U.S. Department of Commerce, USA.
Dappert, Claire P.
2006: *Oaken Whale with a Cast Iron Tail; Single Decked Wooden Bulk Carrier* Monohansett. Maritime Studies Program, ECU Report #13, Greenville, NC.
Rodgers, Bradley A.
1996: *Guardian of the Great Lakes; The U.S. Paddle Frigate Michigan*. University of Michigan Press, Ann Arbor, MI.

ECU Theses

Cantelas, Frank J.
1995: An Archaeological Investigation of the Steamboat *Maple Leaf*. Master's thesis, Department of History, East Carolina University, Greenville, NC.
Cooper, David J.
1988: 1986–1987 Archaeological Survey of the Schooner *Fleetwing* Site, 47 DR168, Garrett Bay, Wisconsin. Master's thesis, Department of History, East Carolina University, Greenville, NC.
Dappert, Claire P.
2005: Oaken Whale with a Cast Iron Tail: The Wooden Bulk Carrier *Monohansett*, A Reflection of Her Kind. Master's thesis, Department of History, East Carolina University, Greenville, NC.
Gray, Jefferson J.
1998: Fueling the Fire: An Underwater Archaeological Investigation of the Claflin Point Wreck in Little Sturgeon, Wisconsin. Master's thesis, Department of History, East Carolina University, Greenville, NC.
Hartmeyer, Phil
2014: Passengers, Packages, and Copper: The Steamer *Pewabic*, Its Archaeology, Management, Material Culture, and the Development of the Keweenaw Peninsula. Master's thesis, Department of History, East Carolina University, Greenville, NC.
Hoyt, Joseph C.
2008: The Transition from Wood to Iron in Great Lakes Bulk Carriers: An Historical and Archaeological Investigation of the Wooden Bulk Carrier *Continental*, and Early Iron Hulled Bulk Carriers. Master's thesis, Department of History, East Carolina University, Greenville, NC.

Jaeschke, Brian

2003: Black Angels of the Ice: The History of Chartered Ice Breaking on the Great Lakes by the Rail Ferries *Sainte Marie* and *Chief Wawatam*. Master's thesis, Department of History, East Carolina University, Greenville, NC.

Kerfoot, Sara C.

2015: Catastrophic Disaster in the Maritime Archaeological Record: Chasing the Great Storm of 1913. Master's thesis, Department of History, East Carolina University, Greenville, NC.

Knoerl, Kurt

1993: Beneath Niagara, Methodological Approach to an Inundated 18th Century Site, Fort Niagara. Master's thesis, Department of History, East Carolina University, Greenville, NC.

Kopp, Nadine

2012: The Influence of the War of 1812 on Great Lakes Shipbuilding. Master's thesis, Department of History, East Carolina University, Greenville, NC.

Monk, Kimberly E.

2003: A Great Lakes Vessel Type: Archaeological and Historical Examination of the Welland Sailing Canal Ship, *Sligo*, Toronto Ontario. Master's thesis, Department of History, East Carolina University, Greenville, NC.

Moore, James D. III

2003: Return to the Stone Age: The Maritime History and Nautical Archaeology of Sturgeon Bay, Wisconsin's Dolomite Industry. Master's thesis, Department of History, East Carolina University, Greenville, NC.

Pecoraro, Tiffany A.

2007: Great Lakes Ship Traps and Salvage: A Regional Analysis of an Archaeological Phenomenon. Master's thesis, Department of History, East Carolina University, Greenville, NC.

Rodgers, Bradley A.

1985: The Iron Sentinel: The *USS Michigan*, 1844–1949. Master's thesis, Department of History, East Carolina University, Greenville, NC.

Ronca, Filippo

2006: The Historical and Archaeological Investigation of the *Selah Chamberlain*, a 19th Century Great Lakes Bulk Carrier. Master's thesis, Department of History, East Carolina University, Greenville, NC.

Weir, Andrew

2007: A Historical and Archaeological Analysis of the Middle Island Life-Saving Station: Applying Site Formation Theory to Coastal Maritime Infrastructure Sites. Master's thesis, Department of History, East Carolina University, Greenville, NC.

Whitesides, Scott M.

2003: Spatial Patterning Aboard the Millecoquins Wreck: Interpreting Shipboard Life and Functional Use of an Early 19th Century Great Lakes Sailing Vessel. Master's thesis, Department of History, East Carolina University, Greenville, NC.

Zant, Caitlin N.
2015: Unloading History: Historical and Archaeological Investigations of the Self-Unloading Schooner-Barge, *Adriatic*. Master's thesis, Department of History, East Carolina University, Greenville, NC.

ECU and Sponsor Supported Collaborative Reports

Rodgers, Bradley A.
1996: *The 1995 Pre-disturbance Wreck Site Investigation at Claflin Point, Little Sturgeon Bay, Wisconsin*. Research Report #10, Department of History, East Carolina University, Greenville, NC.
Rodgers, Bradley A. and Annalies Corbin
2003: Mud Box—Filled with Stone: the Wreck of the Scow Schooner *Dan Hayes*. *International Journal of Nautical Archaeology* 32:210–224.
Rodgers, Bradley A., James D. Moore III, Annalies Corbin, Jacqueline D. Piero, and Andrew Pietruszka
2006: *From Quarry to Quay: Shipwrecks at McCracken's Cove*. Research Report # 17, Department of History, East Carolina University, Greenville, NC.
Rodgers, Bradley A., Russell T. Green, and Annalies Corbin (editor)
2003: *Of Limestone and Labor: Shipwrecks of the Stone Trade, The 1999 Bullhead Point Stone Barge Investigation*. Research Report #11, Department of History, East Carolina University, Greenville, NC.
Rodgers, Bradley A., Samuel M. Blake, Brian S. Jaeschke, James D. Moore III, and Annalies Corbin (editor)
2003: *The Bones of a Bulk Carrier: The History and Archaeology of the Wooden Bulk Carrier/Stone Barge,* City of Glasgow. Research Report # 12, Department of History, East Carolina University, Greenville, NC.

Wisconsin Historical Society Reports (Partial Listing)

Cooper, David J.
1989: *Survey of Submerged Cultural Resources in Northern Door County: 1988 Field Season Report*. State Historical Society of Wisconsin, Madison, WI.
Cooper, David J. and Bradley A. Rodgers
1989: *Report on Phase I Marine Magnetometer Survey in Death's Door Passage, Door County Wisconsin*. State Historical Society of Wisconsin, Madison, WI.
Cooper, David J. and John O. Jensen
1995: *Davidson's Goliaths: Underwater Archaeological Investigations of the Steamer* Frank O' Connor *and the Schooner-Barge* Pretoria. State Historical Society of Wisconsin, Madison, WI.

Jensen, John O.
1994: Oak Trees and Balance Sheets: James Davidson, Great lakes Shipbuilder and Entrepreneur. *American Neptune* 54(2):99–114.
Jensen, John O., David J. Cooper, Frank J. Cantelas, and David V. Beard
1995: Archaeological Assessment of Historic Great Lakes Shipwrecks: Surveys of the Steamers *Niagara* and *Francis Hinton*. State Historical Society of Wisconsin, Madison, WI.
Meverden, Keith N. and Tamara L. Thomsen
2005: *Wisconsin's Cross-Planked Mosquito Fleet: Underwater Archaeological Investigations of the Scow Schooners* Iris, Ocean Wave, *and* Tennie and Laura. Technical Report, State Archaeology and Maritime Preservation Program, Wisconsin Historical Society, Madison, WI.
2006: *Wheat Chaff and Coal Dust: Underwater Archaeological Investigations of the Grain Schooners* Daniel Lyons *and* Kate Kelly. Technical Report, State Archaeology and Maritime Preservation Program, Wisconsin Historical Society, Madison, WI.
2010: *Small Boats on a Big Lake: Underwater Archaeological Investigations of Wisconsin's Trading Fleet 2007–2009*. Wisconsin Historical Society, Madison, WI.

The initial research design for the Great Lakes region was very simple, it called for area-wide remote sensing surveys in areas of known ship traps to identify cultural resources along with site-specific recording of as many different sorts of lakes vessels as could be worked given time and monetary constraints. Collaboration between government cultural resource concerns and ECU's academic program in the end proved immensely successful. Within a decade researchers had recorded a sufficient number of vessels to dovetail an archaeological understanding with in-depth, historical research. In doing this, discernable economic and trade patterns emerged along with a clearer understanding of nineteenth century marine technical design changes and improvements over time to the various ship classes. The research design at that point was modified to include a search for those vessel types that had not previously been studied and for which little knowledge existed. It also included in-depth historical analysis of the people and economic factors that led to construction of these vessels. In time historical vessel classification was combined with archaeologically recorded data to identify and detail most of the larger nineteenth century Great Lakes vessels, and in so doing, answer questions that were out of analytical reach two decades before. Over the course of this project, the research focus advanced from simply asking, "what is that wreckage lying on the bottom," to proposing overarching questions concerning how and in what manner marine technology advanced during the nineteenth century and how it aided society in exploiting its resources. Now with the data collected researchers can begin to ask questions concerning how marine technology and change influenced nineteenth century culture, sociology, history, and the historical landscape of all historic ports located along the lakeshores.

Historical and Geographical Background

For background to this research, it should be noted that the Great Lakes of North America comprise nearly 100,000 mile^2 of fresh water sea and thousands of miles of shoreline. From the US east coast it is possible to travel nearly 1000 miles (1609 km) into the heart of North America via the lakes. The economics of the US westward migration in the nineteenth century, therefore, made these waterways (along with the western rivers) the commercial super highways of the day. Unofficial estimates put the number of shipwrecks on the lakes in the tens of thousands. Any estimate of numbers, however, is difficult as ships that wreck or become "partial wrecks" are often refloated, rebuilt, and put back into service, either under their old name or a new one (Pecoraro 2007:62). By any estimate the archaeological resources lying in these fresh water seas are enormous and must also include intentional abandonment sites. It should also be noted that the preservation qualities of the Great Lakes' cold fresh water cannot be overstated and although new threats to sites seem to crop up at times, from pollution to exotic species infestations, the sites have remained remarkably preserved, arguably the best underwater archaeological classrooms in the world.

The largest deepest and coldest of the Great Lakes is Superior. Operations in this area were conducted out of Bayfield, Wisconsin, while recording a dozen wrecks in the Apostle Islands (Cooper 1991). These ships were eventually identified as representatives of the predominant nineteenth century ship classes.

Proceeding from Lake Superior east, or "down bound," as it is called in lake's jargon through the Sault Locks (the highest volume lock system in the world, first opened in 1856) is Lake Huron and further, the northeast coast of the State of Michigan, home of the NOAA Thunder Bay Marine Preserve. Thunder Bay is perhaps the premier sanctuary for surviving cultural material and wrecks on the lakes and a preeminent archaeological repository. From the sanctuary heading west through the Straights of Mackinac is Lake Michigan where the preponderance of ECU's research has taken place in Door County, Wisconsin, an area of extreme importance in the trade routes of the lakes, and which offers over 150 miles of rugged coastline, bays, islands, reefs, and inlets. Below Lake Huron the southernmost lakes of Erie and Ontario have only been visited infrequently during ECU surveys at such places as Buffalo, Erie, Toronto, and Fort Niagara.

Great Lakes' Ship Types

The following descriptions of vessel classes are arranged, more or less, in the order that they appear historically in the nineteenth century on the Great Lakes and they are described through their chief features and uses. It should be remembered that these descriptions are only summaries. Once again, researchers in need of in-depth inquiry are encouraged to seek out theses and field reports that pertain to their interest. One rule of thumb, as initially set by the parameters of archaeological study, was that for each hour a research team spent in the field, four were needed to collect historical and ancillary data. If anything, the heavy field schedule over the decades has shown that the ratio of archival research to field research is actually much higher than four to one.

Generic early century schooners and brigs of the Great Lakes (late eighteenth century to 1835) are for the most part not included in this discussion and are represented by only one archaeological example from the ECU research, the Millecoquins wreck in Michigan's Upper Peninsula (Whitesides 2003; Michigan History Museum 2004). A comparative study of early nineteenth century vessels, including War of 1812 merchant vessels, is now underway and should be completed as an East Carolina University thesis in 2012 (Kopp 2010).

In any discussion of early nineteenth century ship construction on the Great Lakes, the Millecoquins wreck can serve as a baseline, a standard reference from which to measure the divergence of ship construction on the lakes from traditional to vernacular. Therefore, some details of this wreck deserve telling here. The Millecoquins wreck was excavated and documented by ECU for two seasons in 1991 and 1994 with the research report pending additional field and archival study. The Michigan Historical Museum however, carries both photos and a description of the site (Michigan History Museum 2004). In overview, the Millecoquins vessel is approximately 65 ft (19.8 m) in length and 22 ft (6.7 m) in beam with a 5 ft depth of hold, likely a two-masted schooner or brig. It contained no centerboard but had an American penny in the mast step dated to 1833. As a generic cargo vessel of the early nineteenth century the Millecoquins shipwreck could not be termed vernacular as it looked and was built in a similar manner to the multitude of shallow draft vessels plying the Atlantic coastal and Caribbean trade. Its only allowance for the differences in geography and geology encountered on the lakes is its shallow draft for negotiating the Lake St. Claire mud flats near Detroit. But a shallow draft is not unusual for any coastal vessel in need of navigating the sand bar sills of unimproved rivers. The Millecoquins vessel was not unlike many east coast craft in this regard. It can be seen that at this point early or perhaps proceeding 30 years into the nineteenth century Great Lakes ship types had not greatly changed from their coastal progenitors. Although the Millecoquins wreck is very historically and archaeologically important, particularly for the area in which it was wrecked, its most important use here is to demonstrate that early Great Lakes ships were not vernacular but continued the western or European tradition of ship construction imported from the east coast. This would change and change quickly (Fig. 12.6).

From the 1830s onward ship types manufactured on the Great Lakes began to diverge considerably from their oceanic cousins. The new classes of ships developed on the lakes began to specialize according to economic and industrial influences. Gone were the days of generic cargo vessels like Millecoquins. Major ship types identified during lake surveys and dated from the mid to late nineteenth century fall into seven different general classes of vessel (with some exceptions); Steamers, Grain Schooners, Canal Schooners, Scow Schooners, Steam Barges, Wooden Bulk Carriers, and Passenger/Freight Propellers. The first class Side-wheel steamers, or what are known on the lakes as "Steamers," were large, elegantly appointed and fast passenger liners ascendant from the 1830s on, chiefly to move vast numbers of migrants and entrepreneurs east to west. Grain Schooners,

Canal Schooners, and Scow Schooners formed the backbone of bulk cargo commerce during the mid-nineteenth century but continued in service until well into the twentieth century. Schooners on the lakes successfully transitioned into the steam age through the consort system, whereby they were towed several at a time to their destination in line astern by a schooner converted to steamship, known as a Steam Barge. This freighting system eventually gave way to larger purpose built Bulk Carriers by the last quarter of the nineteenth century, whose single ship capacities exceeded the multi-ship consort system, while providing greater safety and less overhead expense in sailors and rigging. Rounding out the seven classes are Passenger/Freight Propellers, a vessel class that extended rail service to unimproved backwater harbors during the second half of the nineteenth century (Fig. 12.7).

Some of these ship types have ocean-going equivalents, but the actual vessels developed and designed for the environmental circumstances of the lakes (with shallow drafts, long narrow hulls, and some unusual wind and steaming characteristics) are unique in overall maritime history. For example, it is very unusual for an oceanic Bulk Carrier built of wood to exhibit a seven to one length to beam ratio (except perhaps extreme clippers), but this is not an unusual ratio on the lakes. Other traits such as the plumb bow, used extensively for oceanic and lake steamers in the late nineteenth and early twentieth centuries, may have originated on the Great Lakes in response to locks, where a vessel simply needed a boxy shape in order to fit within the confines of the lock. In all probability, this feature was seen by oceanic builders and passed on in their new construction, oddly enough, for esthetic purposes.

Steamers

Steamers, or the great passenger paddle wheelers that moved vast numbers of immigrants across the lakes to new homes in the west, do have oceanic equivalents in the form of the transoceanic liners, such as the *Great Western* that crossed the Atlantic in 1837 (Fig. 12.1). Previous to this, however, a good many steamers plied the passenger trade in the Great Lakes. *Walk-in-the-Water* was the first steamer on the upper lakes in 1818, a vessel built to ferry passengers in Lake Erie but unremarkable in size at 135 ft (41.1 m) in length by 32 ft (9.8 m) in beam and a depth of 8.6 ft (2.6 m) (Mansfield 1899:593). Between 1844 and 1857, however, these humble steam vessels morphed into "palace steamers" with 25 of them built at over 1000 t old measure (Labadie 1989:22). *Niagara* is an excellent representative of her class at nearly 245 ft (74.7 m) in length, with a 33 ft beam (9.1–10.7 m) and a 14 ft (4.3 m) depth of hold; 1084 t old measure (SHSW 2000; Jensen et al. 1995:6). These vessels could carry several hundred passengers and were known for their opulence and speed. The route for immigrants heading west often started in New York, proceeded up the Hudson, and into the Great Lakes via canal boat on the Erie Canal. From Buffalo on Lake Erie passengers could sojourn hundreds of miles to Chicago via these large upper lakes steamers, unrestricted in their navigation of Lakes Erie, Huron, and Michigan. Should they choose, immigrants could move

a

b

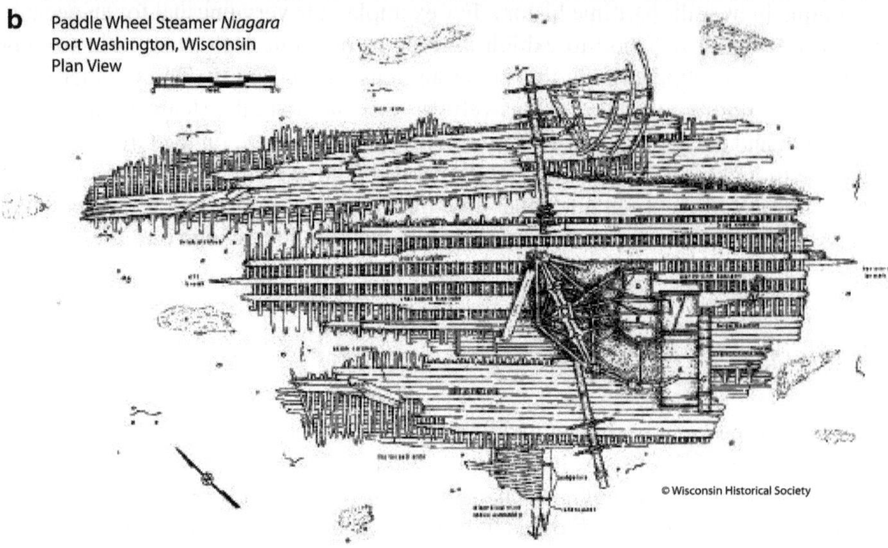

Fig. 12.1 *Above*: Steamer *Western World* is an excellent example of a Great Lakes Palace Steamer of the mid-nineteenth century. *Below*: *Niagara* has given us our first comprehensive view of the inner workings of one of these giants (*Top*: Courtesy of Historical Collections of the Great Lakes, Bowling Green State University. *Bottom*: Courtesy of Wisconsin Historical Society-WHS)

further west via canal and river transportation out of Chicago to the Illinois River, down the Mississippi, and on to the Missouri, Platt or Red Rivers.

Of the steamers examined by ECU and SHSW teams, *Alabama* near Buffalo, and *Niagara* near Milwaukee, only the *Niagara* has been documented in great detail. *Niagara* caught fire off Milwaukee on 24 September 1856, burned and sank taking over 60 people with it (Jensen et al. 1995:6, 15). Its wreckage shows the typical heavy bilge keelsons (5 in number) used to support the weight of the walking beam engine while giving the hull longitudinal support. It also still contains the paddle shaft and part of one paddlewheel. The large low-pressure steam engine (6 ft in diameter (1.8 m) and 14 ft long (4.3 m)) dominates the center of the wreck site, along with the walking beam. The boiler lies some distance from the main wreck site (Jensen et al. 1995:20–29; Jensen 1999:216). Even in its present condition the size and dominance of the vessel's engines and machinery tell investigators that this was an expensive ship to build and operate, and was not intended to handle large amounts of cargo. *Niagara* demonstrates all of the attributes of a passenger vessel built for speed with passenger accommodation above the main deck. *Niagara* represents a ship class that has only a few representatives in the archaeological record, increasing her importance both archaeologically as well as historically.

Grain Schooners

Of the schooner classes on the lakes, the best documented of the Grain Schooners is *Lucerne* along with the newly discovered and recorded *Daniel Lyons* and *Kate Kelly* (Cooper 1991; Meverden and Thomsen 2006). When archaeologists documented *Lucerne* in June and July of 1990 in the Apostle Islands of Lake Superior, they were unprepared for the sheer size of the vessel and the fact that it still contained much of its cargo (Fig. 12.2). The bow of the wreck still sported her green and white colors, prominent after 100 years in the cold fresh water. *Lucerne* broke its back approaching the shallows of Long Island and sank after having been separated from its escort during a blizzard in 1886. Since the vessel sank in shallow water, some of the eight-man crew were able to briefly escape into the rigging, but as fate would have it, they survived only minutes longer than those washed overboard, their frozen bodies were found in the rigging days after the wintry gale. So ended the life and career of the *Lucerne*, as so often is the case, a historical tragedy turned archaeological boon (Cooper 1991:31–58).

Grain schooners are known for their large size and double decks. Most bulk carriers are also historically listed as carrying two decks, a tween deck and a weather deck dividing the cargo hold. The mystery concerning whether the tween deck was ever fully planked or simply left open to better facilitate loading and unloading has not been solved archaeologically, and could not be answered at the *Lucerne* site.

Lucerne was 195 ft (59.4 m) by 34 ft (10.4 m), with a 14 ft (4.3 m) depth of hold and calculated to be over 900 t old measure but 651 gross tons. These ships were sharply built for speed and carried cargoes of grain, coal, and iron ore. *Lucerne* was unhindered by the locks at Sault Ste. Marie (exiting Lake Superior to the other lakes) in 1873, when she was built, as they could accommodate vessels 350 ft (106.7 m) long by 35 ft (10.7 m) in beam and 16 ft (4.9 m) deep. By the time of her demise in 1886 the locks were further enlarged to accommodate mammoth steel vessels up to 800 ft (243.8 m) by 50 ft (15.2 m) and 21 ft (6.4 m) depth. As mentioned there is no evidence of the second tween deck on *Lucerne* and it seems likely that a second deck in this instance meant that the vessel was outfitted with deck beams for lateral hull support but no actual deck, which would have interfered with the loading and unloading of bulk commodities. When *Lucerne* was destroyed she was carrying a cargo of high-grade iron ore, much of which is still spread across the bottom near the wreck site (Cooper 1991:31–58).

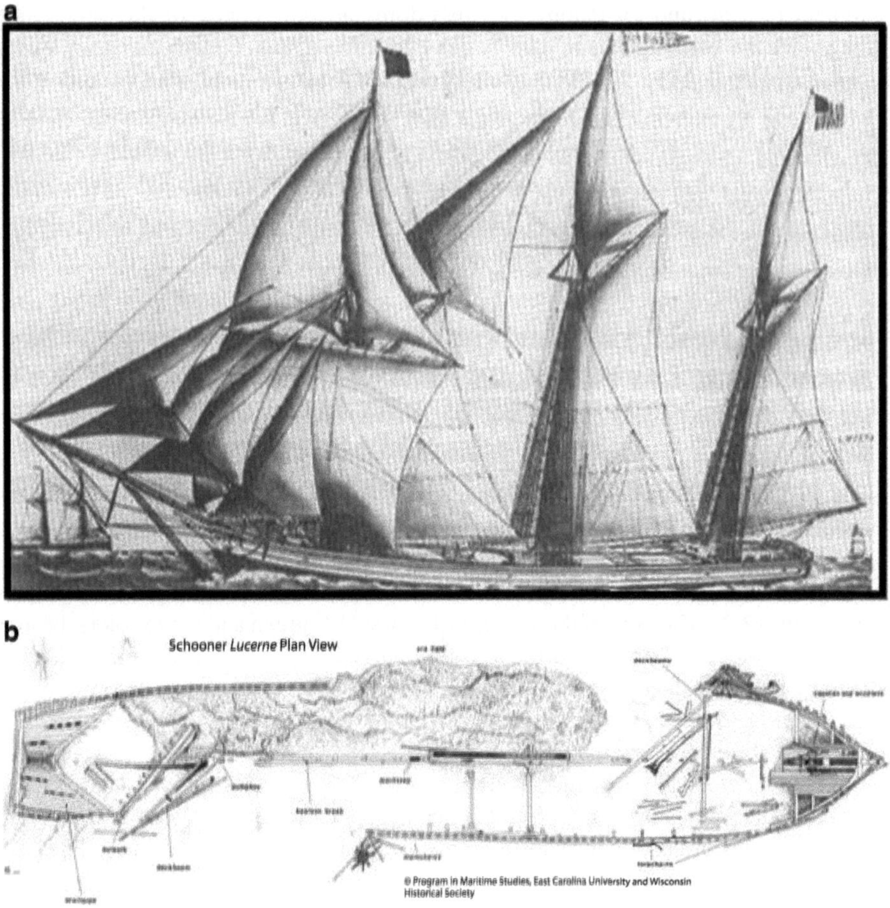

Fig. 12.2 Painting of the Grain Schooner *Lucerne* under full sail. Grain Schooners typically exhibited an Atlantic or clipper bow, popular in the mid-nineteenth century, combined with a traditional early century transom stern. The triangular fore topsail is known on the lakes as a Rafee. The wreck of *Lucerne* still contains large quantities of iron ore (*Top*: Image Courtesy Marine Collection, Milwaukee Public Library. *Bottom*: Courtesy the Program in Maritime Studies, East Carolina University and the Wisconsin Historical Society)

Canal Schooners

There are many examples of canal schooners in the archaeological record of the lake bottoms. At least half a dozen of these schooners have been the focus of projects including the Death's Door remote sensing survey in 1988 and 1989 and ECU's 1999 and 2002 surveys of Bullhead Point and McKracken's Cove in Sturgeon Bay, Wisconsin. One of the most spectacular examples of this type of vessel, *Bermuda*, lies near Grand Island in Lake Superior and is fully intact up to the deck. Ironically, for an archaeologist, the more disarticulated wrecks actually offer a better vantage from which to view the internal structure of any vessel, intact examples are rare and can be difficult to document in three dimensions.

Canal Schooners or canallers were specifically built to navigate the Welland Canal between Lakes Erie and Ontario. Welland was necessary to bypass Niagara Falls, a 350 ft (106.7 m) dolomite precipice between the upper lakes and Lake Ontario. As built in 1829, the Canadian canal's smallest locks could accommodate vessels 110 ft (33.5 m) long and 22 ft (6.7 m) in beam with a depth of 8.5 ft (2.6 m). Improvements were ongoing from the late 1830s, but by 1845 the canal's smallest locks could accommodate ships 142.5 ft (43.4 m) long and 26.25 (8 m) feet in beam with 10 ft (3.0 m) drafts (Mansfield 1895:229–236). Ships built to navigate the enlarged locks of 1845, like the canaller *Sligo* and *Fleetwing*, were boxy in cross section, had a high length to beam ratio over 5:1, had virtually no dead-rise, and contained plumb, very bluff bows, and square transom sterns designed to fit inside the Welland lock doors (Fig. 12.3). These ships were also invariably designed with loading ports cut into the sides of their hulls to accommodate the loading of lumber. In extreme cases, these ships passed with only inches to spare and could get stuck even if a rope was caught between the hull and the lock (Cooper 1988:67; Monk 2003:45). Other vessels, like the schooners *Oak Leaf* and *Ida Corning* located at Bullhead point in Sturgeon Bay, Wisconsin, were built to some of the box like dimensions of canallers but modified (lengthened) over time so they could not have fit the second 1845 Welland locks.

Great Lakes Sailing schooners built by the late 1830s were invariably outfitted with centerboards, originally offset to the side of the keel/keelson structure, but modified after 1860 so that the centerboard passed directly through the keel/keelson. Though not a canaller, *Ida Corning* exhibits the Grand Haven or jackass rig of the latter part of the century carried by many canallers with the main mast removed to accommodate cargo handling (Rodgers and Green 2003:36–41). Oddly, the two-masted rig configuration was said to be favored by lakes sailors for its good sailing attributes (Martin 1995:40–41). The wreck of *Ida Corning* contains an iron hogging strap on the bilge ceiling for longitudinal support, a somewhat unusual feature of this schooner and a characteristic that helped in the vessel's identification and probably an attribute used by canallers to offset their extreme length to beam ratio (Rodgers and Green 2003:36).

Canal Schooners, like all other vessels used on the Great Lakes, were subject to improvement at any time both in hull and sail configuration, and toward the end of their useful lives as cargo vessels or barges they were often adaptively reused to extend the deep face of various wharfs, such as that at Bullhead Point for loading stone. Abandonments such as these in Sturgeon Bay, Wisconsin, are ideal for study since they are located in shallow water and are often abandoned in a grouping or used as cribs for wharfs and wharf extensions (Rodgers and Green 2003:45–47; Rodgers et al. 2006:39).

a

b

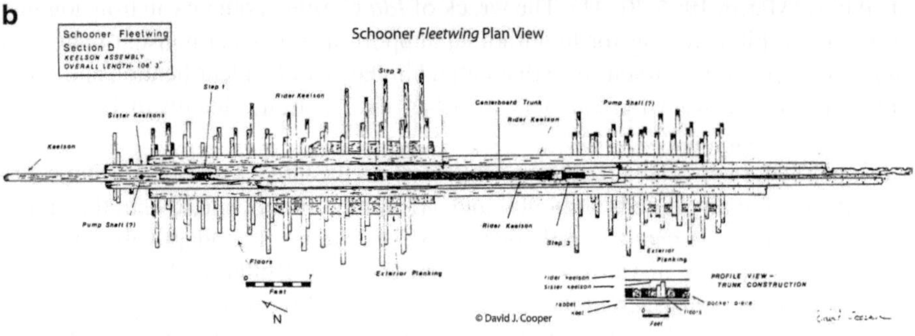

Fig. 12.3 *Above*: Schooner *Fleetwing* stripped of her running rig for the winter. The *Fleetwing* survey by Cooper and Rodgers in 1985 initiated the ECU/WHS collaboration that successfully recorded major nineteenth century vessel classes. The identification of *Fleet Wing* as a likely Canal Schooner became apparent only after many other schooners had been recorded (*Top*: Courtesy the Iconographic Collection, Wisconsin Historical Society. *Bottom*: Courtesy of David J. Cooper)

Scow Schooners

In September of 2001 and 2002, archaeologists were given the opportunity to study in detail the third type of schooner represented in this archaeological sample, a scow schooner. The scow schooner *Dan Hayes* is conveniently located on the bottom of Sturgeon Bay, Wisconsin, at a site across from Bullhead Point (Fig. 12.4). It should be mentioned that Sturgeon Bay represents a gold mine of abandoned vessels from the mid to late nineteenth century brought in for wharf cribbing and to act as barges for the limestone trade. Harbors such as Sturgeon Bay offer an archaeologist the resources to study various ship types in a relatively benign environment, allowing documentation at a much faster pace than can be achieved in an open water project that is invariably complicated by boats and equipment needed for the rigors of deeper diving.

Dan Hayes represents the low budget version of a cargo schooner on the Great Lakes. Though first introduced in 1827, these schooners saw widespread use in the second half of the nineteenth century. They look, quite literally, like a box with sails, a vessel only their captains could love. The bow in the case of *Dan Hayes* was V-shaped, and the stern was a shallow sloped apron. In cross section *Dan Hayes* is rectangular, slab sided, and has no dead-rise. Stringer and keel/keelson internally offer longitudinal strength while the side planking is edge fastened. Internal and external bottom and ceiling planking lay athwartship for easy repair and low construction cost. These ships could be built in sections and the bottom of *Dan Hayes* at least, looks to have been built inverted before being righted for the addition of the sides. Despite their unusual design, these small ships served unimproved hinterland ports with distinction. Their extremely shallow draft and lack of external keel allowed them to pass into water depths unapproachable to traditional rigs. Unloaded and sailing before the wind there was reportedly not another type of schooner that could keep pace with these ships. Fully loaded, however, these vessels had to be a handfull to sail. Recently, the State Historical Society of Wisconsin recorded another of these ships, the *Ocean Wave*, a wreck lying almost fully intact off Door County (Rodgers and Corbin 2003; Meverden and Thomsen 2005).

Steam Barges

Some of the more recent work for this project took place in September of 2005 and involved the documentation of a Steam Barge located near Sunset Park, also in Sturgeon Bay. This project was set up to verify the results of an earlier survey conducted at McKracken's Cove in September of 2002, in which a schooner style vessel hull was recorded that contained fittings and engine supports, perhaps representing the first close look at a Steam Barge (Rodgers et al. 2006:30). Steam Barges represent a transition between sail and steam on the Great Lakes. Steam Barges are described in historical literature as schooners outfitted with steam engines. Our mission in this instance, therefore, was to document archaeologically

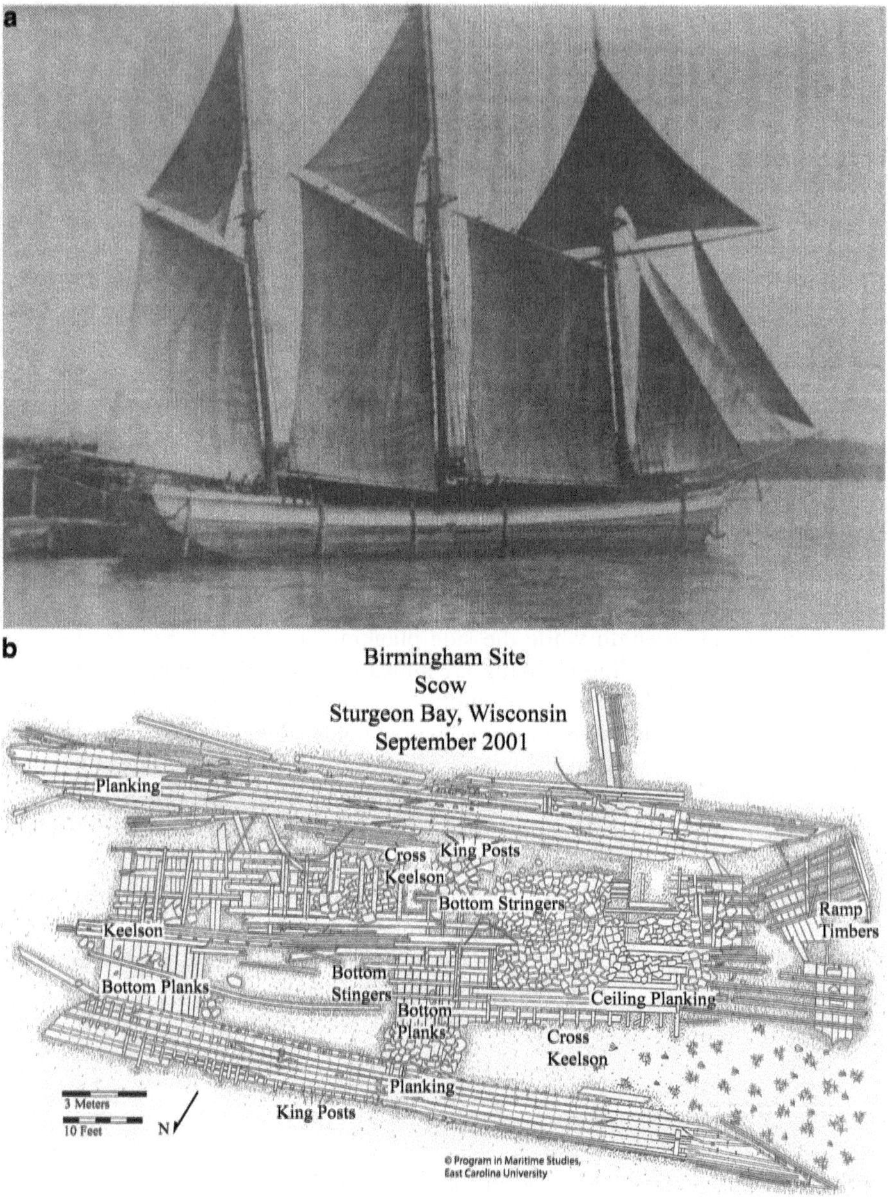

Fig. 12.4 *Above*: The *Dan Hayes* loading lumber at a mill wharf while drying her sails. Scow Schooners are lightly built with no projecting keel structure ensuring their shoal operational draft. A centerboard would be a necessity to sail such a box like craft (*Top*: Courtesy of Manistee County Museum. *Bottom*: Courtesy the Program in Maritime Studies, East Carolina University)

just how shipbuilders were able to put a steam engine into a schooner and how these steam powered vessels differed from their later, larger, and more sophisticated brethren, the Bulk Carrier. Historic and photographic evidence concerning the differences are vague and the two vessels, aside from size, look outwardly at any rate, to be almost identical.

Few Steam Barges have been studied archaeologically but based on these examples it can be concluded that these vessels, perhaps more properly known by their west coast terminology, "Steam Schooners," are truly THE missing link between sail and steam on the Great Lakes (Bazzill 2007:90). Surprisingly enough considering the dearth of knowledge concerning these vessels for many years, two recent reports have opened the way to further research (Bazzill 2007; Labadie and Herdendorf 2004a). First seen on the lakes in the late 1840s to carry lumber, the Steam Barge rose to prominence during the depression of 1857. These ships combine a schooner hull with a steam engine, usually of the inverted direct acting single cylinder type. Later the efficiency of the compound engine was harnessed through use of tandem compound engines of the Holt variety, oddly known on the lakes as a "steeple compound" (Bazzill 2007:29–32; Gardiner 1993:157).

Illustrations and photographs of Steam Barges generally show a one-decked vessel with a fore and aft deck configuration (pilot house foreword and crew quarters aft). This represents a deck arrangement change from their earliest incarnations showing both pilothouse and engineering aft (Labadie 1989:24; Rodgers et al. 2006:19; Bazzill 2007:33). With fore and aft configuration these ships look very much like the later wooden Bulk Carriers except that Steam Barges, like schooners, rarely exceeded 160 ft in length while Bulk Carriers approached 300 ft in length (Rodgers 2003:7). Internally, however, these are two completely different ships and their internal support structures easily demonstrate why there is a size difference (Fig. 12.5). Steam Barges differ very little from schooners, they are planked and configured internally in the same manner as a schooner. Steam barges contain one deck, with the earlier and schooner converted versions generally sporting centerboards in the cargo hold (Rodgers et al. 2006:30–31; Labadie and Herdendorf 2004a:7, 32–33). The use of centerboards in steamships (generally considered anti-leeway devices in fore and aft rigged sailing ships such as schooners) remains somewhat of a mystery. It can be speculated that the centerboard helped stabilize these vessels in a crosswind or kept them from rolling in a beam sea since they had limited engine power and a very shallow draft. The centerboard may also have been used to stabilize the ship when it was using sail assist while steaming. Internally steam barges also contain bilge keelsons (heavy longitudinal supports), presumably for internal engine and boiler support. One of the only differences between schooners and Steam Barges internally is additional floors near the stern of a Steam Barge to help support the weight of engines and boilers (Bazzill 2007:59; Rodgers et al. 2006:30).

Steam Barges were also known on the Great Lakes as "lumber hookers," and were designed to carry cut timber or lumber in the hold and stacked on deck as high as 15 ft (4.6 m) (Bazzill 2007:1–2). They could also carry, as need be, any

a

b

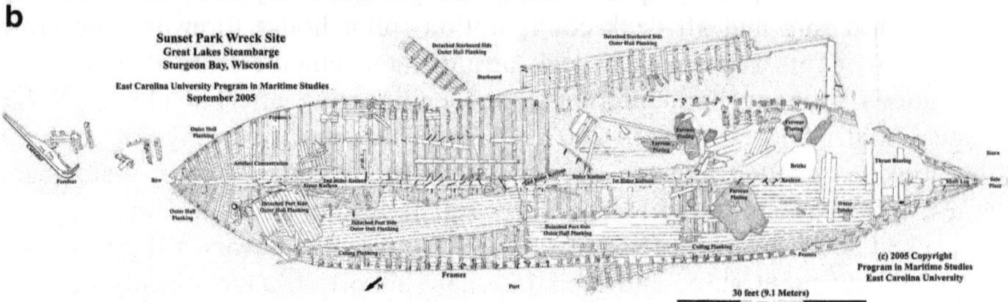

Fig. 12.5 The *Joys* exhibits all of the typical characteristics of a Great Lakes Steam Barge. The fore and aft deck arrangement became typical of both Steam Barges and Bulk Carriers in the 1860s. Many steam barges retained a centerboard internally, a reminder of their sailing schooner heritage. *Below*: The *Joys* allowed researchers our first detailed internal glimpse into this vessel class, indeed a schooner outfitted with an engine (*Top*: Courtesy Historical Collections of the Great Lakes, Bowling Green State University. *Bottom*: Courtesy of the Program in Maritime Studies, East Carolina University)

other bulk commodity from grain to coal and rock. Generally, however, their ceilings were not configured to take the abuse doled out to the specialized bulk carriers. Repair and replacement of the ceiling in a Steam Barge and its expensive longitudinally placed ceiling would have been expensive compared to maintenance of a Bulk Carrier's hold covered by cheap, short, and easily replaced athwartship planking.

Steam Barges on the Great Lakes were generally seen to be towing a number of schooners in line astern. This system was known as the "consort system," a freight hauling method by which one powered schooner (Steam Barge) tows several sailing schooners (up to seven) in line astern. This made good sense on the Great Lakes since sailing ships were often hampered by lack of sea room as well as inadequate sailing conditions in confined ports, rivers, and canals. The consort system greatly increased both the speed and the amount of cargo that could be carried by vessels on the lakes while lowering the overhead costs (Bazzill 2007:2). Each schooner towed by a Steam Barge contained only a skeleton crew while steaming from point a to point b without time wasted sailing. Unfortunately, the consort system led to many shipping losses and it can only be imagined the danger faced by other vessels encountering an escort and consorts in dense fog. The consort system was eventually overtaken by a safer and more profitable system for moving bulk commodities, the wooden, and later steel versions of the Bulk Carrier.

Bulk Carriers

Few, if any, Great Lakes Bulk Carriers had been archaeologically recorded previous to ECU's June/July 1990 Apostle Islands project in which archaeologists recorded the remains of the steel hulled bulk carrier *Sevona* (1890) and the wooden Bulk Carriers *Fedora* (1888) and *HD Coffinberry* (1874) (Cooper et al. 1991:83–93, 93–102, 117–131). Since the original Great Lakes research design called for the recording of vessels of all types, archaeologists dutifully recorded these very large wrecks, mostly unaware of their true significance. Little did researchers know at the time that they had opened the door to further study of this unusual and extremely important vessel type, initiating a study of perhaps the only ship design other than the steam barge that successfully transitioned completely from wood to iron and then steel (Hoyt 2008).

By September 2000, it was readily apparent that archaeologists needed to know more about the history and archaeology of Bulk Carriers, so they set their sights on the wreck of *City of Glasgow* (1891) located in Lake Michigan just off the mouth of the Sturgeon Bay Ship Canal. Bulk Carriers are the largest ships that project divers attempted to record underwater and their massive size was daunting from a logistical standpoint. Fortunately, *City of Glasgow* was a cut down version of a Bulk Carrier that lays nearly on shore, and it was a doable project. The survey revealed *City of Glasgow* to be of such an unusual construction and design compared to traditional wooden ships and the recordings made earlier of the Apostle island Bulk Carriers, that archaeologists had to confirm their findings. The *Selah Chamberlain* Project conducted by the State Historic Society of Wisconsin, the *Monohanset* (1872) Project in Lake Huron in June 2004 conducted within the NOAA Marine Sanctuary at Thunder Bay, and the June 2006 recording of the wooden bulk carrier *Continental*, corroborated the *City of Glasgow* findings and revealed the unusual nature of Great Lakes Bulk Carriers, a ship type that still exists and is responsible for much of the economic prosperity brought about by cheap transportation of raw mineralogical materials to the industrial centers located on the Great Lakes (Fig. 12.6).

Fig. 12.6 *Above*: Bulk Carrier *Margaret Olwill* loading from a gravity feed hopper loader or pocket dock. This vessel class, with the loading and unloading equipment created for them, forms an integrated cargo system whereby these very large ships could be loaded and unloaded in a matter of hours. *Below*: *Monohansett*'s robust internal design demonstrates how Bulk Carriers can be built to immense size while retaining the strength and durability to carry mountains of unpackaged cargo cost effectively (*Top*: Courtesy Historical Collections of the Great Lakes, Bowling Green State University. *Bottom*: Courtesy the Program in Maritime Studies, East Carolina University)

Within a generation after the development of the Steam Barge, lakes builders began to modify the Steam Barge into a much larger freighter known as a Bulk Carrier. As mentioned, Wooden Bulk Carriers could look outwardly very much like a Steam Barge with fore and aft deck house configuration. Internally, however, Bulk Carriers are designed and built much differently, usually with two decks and many longitudinal stringers in their hold sitting atop floors, giving a solid bed to a double set of relatively short athwartships ceiling planks unheard of in any other type of wooden ship construction with the exception of ferries and flats. With this internal build the ships became much longer and larger than Steam Barges and were able to be maintained and repaired internally at nominal expense since the doubled short-planked ceiling was easily and routinely replaced (Rodgers 2003:34). Elihu M. Peck is credited with the first wooden bulk carrier, *R. J. Hackett* in 1869 (Mills 2002:1). Bulk Carriers carried rough bulk commodities like crushed and quarried rock, lime, iron ore, and coal. Since these minerals were loaded through gravity feed pocket docks from great heights, the ceiling took a good deal of abuse.

Examples of Wooden Bulk Carriers examined archaeologically by ECU researchers demonstrate classic characteristics of Wooden Bulk Carriers including multiple sets of longitudinal bilge stringers set on heavily constructed doubled and tripled floors supported with iron basket trusses and massive chine planks. These were extremely large ships, *City of Glasgow* for instance was 297 ft (90.5 m) in length by 41 ft (12.5 m) in beam, with a molded draft of 20.42 ft (6.2 m) for a gross tonnage slightly in excess of 2002 t (Rodgers 2003:8). These numbers push the limits for what is possible in wooden ship construction and indicate the ingenuity of the design as well as the amount of iron that went into it for support. They also push the limits for length to beam ratio in wooden ships at 7:1 an incredible number for a giant wooden ship intended to carry bulk commodities. Generally long narrow ships are the strict purview of iron and steel hulled liners and warships, both intended for speed; lakes carriers, however, were forced into these dimensions to fit the locks at Sault Ste. Marie.

The engineering and machinery fitted for use in these vessels was fairly standard, with compound and tandem compound engines in extended use along with scotch boilers. These single cylinder and stacked cylinder engines were cheaper and took up less space than the highly efficient triple expansion engine that came into use in the 1870s. Evidence for how these engines were fitted and how the through hull stern gland was arranged is evident on *Monohansett*. This ship is also outfitted with a standard four-blade propeller, a device that sacrificed efficiency for smooth power application, a necessity for wooden hulled steamers prone to leaking with prolonged hull stress, particularly propeller vibration (Dappert 2005:80–83).

Passenger/Freight Propeller

As mentioned a "steamer" in lakes' parlance refers specifically to side-wheeled paddle steamers. Propeller driven ships were called, appropriately enough, "propellers." The last of the seven great ship classes in ECU's nineteenth century ship study on the Great Lakes is called a Passenger/Freight Propeller. For this research two field projects were conducted, the first in September 1995, at Claflin Point, Little Sturgeon Bay, Wisconsin, on an unknown vessel (Fig. 12.7). The second project took place in September 1999, at Bullhead Point in Sturgeon Bay, on a ship called *Empire State* (1862). Passenger/Freight propellers are rather unusual vessel types in the archaeological record and archaeologists were fortunate to find two such vessels to study.

a

b

Fig. 12.7 *Above*: The *Quebec* and *Ontario* set good examples of mid to late nineteenth century Passenger/Freight Propellers. These ships often extended package delivery and passenger service to local ports beyond the reach of railroads on the Inland Seas. *Below*: The interior of the Claflin Point vessel demonstrates it is built with a shallow draft for speed on a very high length to beam ratio (*Top*: Courtesy the Wisconsin Maritime Museum, Carus Collection. *Bottom*: Courtesy the Program in Maritime Studies, East Carolina University)

Passenger/Freight propellers became a popular vessel type as railroads extended their passenger monopolies into the lakes for service to communities that lay beyond the reach of iron rails. They carried package freight and deck cabin passengers earning a premium for speed and service but not intended to carry large quantities of cargo. These were fast ships with moderate to extreme dead-rise, a high length to beam ratio, and a fine entry at the bow and lots of tuck in the stern for good hydrodynamics, maneuverability, and speed enough to attract passengers. *Empire State* was 212 ft (64.6 m) long by 32.7 ft (9.9 m) in beam with a 12.2 ft (3.7 m) depth of hold (Rodgers and Green 2003:27). It seems likely, however, that their extremely sharp hull design made them less suitable for barge conversion. Many were apparently scrapped rather than converted to barges, hence their scarcity in the archaeological record.

The Claflin Point vessel demonstrates the classic characteristics of these ships with its extreme dead-rise, high length to beam, and tall slab sides. Claflin Point was also outfitted with an internal hogging arch laid on the ceiling. Photographs and illustrations of other vessels in this class often show external hogging arches rising above the superstructure and deck cabins. Internally the ships are built in conventional manner much like a schooner (except for the dead-rise). Claflin point had only one keelson and two bilge keelsons to support her machinery, but builders greatly compressed the framing and floor spaces below the engines and boilers virtually making a platform of solid wood to support these heavy devices within the hull. One other longitudinal support, that is hard to see or document, is massive chock timbers placed between the floors under the keelson. These added stiffness to the ship's backbone (keel/keelson) without taking up valuable space inside the hold, as would rider or sister keelsons (Rodgers et al. 1995:22–24).

During analysis it was seen that Claflin Point represents a barge conversion that sank at its wharf late in the nineteenth century. It was obviously carrying a load of stone to be used to shore up the wharf and the cargo is still lined up neatly in two rows on either side of the keelson. Other evidence of the barge conversion comes in the form of tie rods and turn buckles used to hold in the two sides of the ship in the absence of deck beams, possibly removed to make loading and unloading that much easier during barge conversion (Rodgers et al. 1995:23).

Conclusion

In all, ECU's search for vernacular craft of the Great Lakes has yielded a tremendous amount of information concerning the ships that made commerce flourish on the Inland Seas of the nineteenth century. From simple and rather generic schooners and brigs of the early nineteenth century such as the Millecoquins wreck, vessel types became ever more sophisticated and specialized with the use of centerboards and schooner sailing rigs optimized for conditions on the lakes. On and below decks they were also forced to keep pace with developments in faster loading and unloading technologies. The introduction of steam witnessed an explosion in the development and assortment of ship types designed to take advantage of each economic and environmental niche offered by the Great Lakes. These vessels were particularly influenced in shape and construction by the intervening lake's canals, locks, and unique navigation conditions. Loading and unloading of bulk commodities progressed to fever pitch efficiency on the Great Lakes, so that portside loading and unloading went from many days (if not weeks) for the small vessels at the beginning of the century, to a matter of hours. This progress is all the more impressive considering the immense size of the cargoes carried by the gigantic vessels of the end of the century. In fact, the switch to iron and steel construction on the lakes toward the end of the nineteenth century was a direct reflection of the profitability and economics of size more than any other factor and although most of the archaeological work carried out to date has focused on wooden ship classes, it ultimately sheds more light on later technological developments such as the transition to steel (Hoyt 2008).

Finally, though archaeologists are researching and documenting the remains of large and elaborately constructed vessels in order to analyze and add detail to historical economic trends, it should never be forgotten that these ships also represent the lives of people who served or took passage on them. Personal items found amid the wreckage are a profound reminder of this fact (always left in place during Phase II, pre-disturbance documentation). Yet in archaeology there is always a temptation to read more into individual artifacts or even artifact collections than can often be justified under the circumstance, and a careful balance must be kept in archaeological analysis between the data and imaginative speculation. There are limits to archaeological inquiry. ECU's 30 year database of Great Lakes ship wrecks is not yet sophisticated or extensive enough to reflect or answer some broad sociological questions. For instance, there may be a temptation to interpret the apparent trend toward fewer and fewer personal belongings found on wreck sites near the end of the nineteenth century compared to earlier sites, to reflect an ever declining value placed on individual sailors as the nineteenth century progressed and the ships became less of a home and more of a place of labor. So many other variables dictate artifact survivability in the archaeological record, however, that questions of this sort are more properly answered through historical sources, where it is clearly seen that the industrialization of shipping became an increasingly dehumanizing process during the nineteenth century (Thompson 2000:331).

Archaeology can, however, in this long-term study, point out HOW technology (in the form of ship design change) increased efficiency and cost effectiveness in cargo ships. Archaeology combined with history and economics can also provide a better understanding of how traditional ship design can be converted or is purposefully evolved over time into vernacular, or regionally specialized watercraft, allowing entrepreneurs on the lakes in this instance, to better cope with the vagaries of a region's geography and geology. In an archaeological sense, industrialization profoundly affected the design, proliferation, and use of vernacular ships on the Great Lakes during the nineteenth century, a trend that is certainly reflected in the archaeological record lying on the bottom of the Great Lakes and reflecting a nineteenth century historical trend of increased industrialization in the USA as a whole.

References

Bazzill, D. M. (2007). *The missing link between sail and steam; Steambarges and the joys of door county, Wisconsin* (Research Report #19). Greenville, NC: Department of History, East Carolina University and Columbus, OH: PAST Foundation.

Cooper, D. J. (1988). *1986–1987 Archaeological survey of the Schooner Fleetwing Site, 47 DR168, Garrett Bay, Wisconsin*. Master's thesis, Department of History, East Carolina University, Greenville, NC.

Cooper, D. J., Partlow, M., Rodgers, B. A., Smith, G. T., & Watts, G. (1991). *By fire, storm, and ice: Underwater archaeological investigations in the Apostle islands*. Madison, WI: State Historical Society of Wisconsin.

Dappert, C. P. (2005). *Oaken whale with a caste iron tail: The single-decked wooden bulk carrier Monohansett*. Master's thesis, Department of History, East Carolina University, Greenville, NC.

Gardiner, R. (Ed.). (1993). *The advent of steam: The merchant steamship before 1900 (Conway's history of the ship)*. London: Conway Maritime Press.

Hoyt, J. C. (2008). *The transition from wood to iron in Great Lakes bulk carriers: An historical and archaeological investigation of the wooden bulk carrier Continental, and early iron hulled bulk carriers*. Master's thesis, Department of History, East Carolina University, Greenville, NC.

Jensen, J. O. (1995). Oak trees and balance sheets: James Davidson, Great lakes shipbuilder and entrepreneur. *American Neptune, 54*(2), 99–114.

Jensen, J. O. (1999). The history and archaeology of the Great Lakes steamboat *Niagara*. *Wisconsin Magazine of History, 82*(3), 198–230.

Jensen, J. O., Cooper, D. J., Cantelas, F. J., & Beard, D. V. (1995). *Archaeological assessment of historic Great Lakes shipwrecks: Surveys of the steamer Frank O' Connor and the schooner-barge Pretoria*. Madison, WI: State Historic Society of Wisconsin.

Kopp, N. (2010). *The Navy Bay Wreck*. Thesis prospectus, Department of History, East Carolina University, Greenville, NC.

Labadie, P. C. (1989). *Submerged cultural resources study: Pictured rocks National Lakeshore*. Southwest Cultural Resources Center Professional Paper #22. National Park Service, Department of the Interior, Washington, DC.

Labadie, P. C., & Herdendorf, C. E. (2004a). *Wreck of the steam barge* adventure*: An archaeological investigation in Lake Erie at Kelleys Island, Ohio* (Tech. Rep. #1). Vermilion, OH: Great Lakes Historical Society.

Labadie, P. C., & Herdendorf, C. E. (2004b). *Wreck of the Scow Schooner* W. R. Hanna*: An archaeological investigation in Lake Erie at Kelly's Island, Ohio* (Tech. Rep. #2). Vermilion, OH: Great Lakes Historical Society.

Mansfield, J. B. (1899). *History of the Great Lakes*. Chicago, IL: J. H. Beers.

Martin, J. C. (1995). Sailing the freshwater seas: A social history of life aboard the commercial sailing vessels of the United States and Canada on The Great Lakes, 1815–1930. Doctoral dissertation, Department of History, Bowling Green State University, Bowling Green, OH.

Meverden, K. N., & Thomsen, T. L. (2005). *Wisconsin's cross-planked mosquito fleet: Underwater archaeological investigations of the Scow Schooners Iris, Ocean Wave, and Tennie and Laura* (Tech. Rep.). Madison, WI: State Archaeology and Maritime Preservation Program, Wisconsin Historical Society.

Meverden, K. N., & Thomsen, T. L. (2006). *Wheat chaff and coal dust: Underwater archaeological investigations of the Grain Schooners Daniel Lyons and Kate Kelly* (Tech. Rep.). Madison, WI: State Archaeology and Maritime Preservation Program, Wisconsin Historical Society.

Michigan History Museum. (2004). Schooner in the sand exhibit, unlocking the secrets of a Great Lakes shipwreck. Lansing, MI: Michigan History Museum. Retrieved May 18, 2004, from hal.state.mi.us/mhc/museum/explore/museums/himus/special/schooner/questions.html.

Mills, R. H. (2002). *Wooden steamers on the Great Lakes*. Vermilion, OH: Great Lakes Historical Society.

Monk, K. E. (2003). *A Great Lakes vessel type: Archaeological and historical examination of the Welland Sailing Canal Ship, Sligo, Toronto, Ontario*. Master's thesis, Department of History, East Carolina University, Greenville, NC.

Musham, H. A. (1948). Early Great Lakes steamboats. The battle of the Windmill and afterward 1838–1842. *American Neptune, 8*(1), 37–60.

Pecoraro, T. A. (2007). *Great Lakes ship traps and salvage: A regional analysis of an archaeological phenomenon*. Master's thesis, Department of History, East Carolina University, Greenville, NC.

Rodgers, B. A. (2003). *The bones of a bulk carrier: The history and archaeology of the wooden bulk carrier/stone barge City of Glasgow* (Research Report #12). Greenville, NC: Department of History, East Carolina University.

Rodgers, B. A., Blake, S. M., Jaeschke, B. S., & Moore, J. D. (1995). *The 1995 predisturbance wreck site investigation at Claflin Point, Little Sturgeon Bay, Wisconsin* (Research Report #10). Greenville, NC: Department of History, East Carolina University.

Rodgers, B. A., & Corbin, A. (2003). Mud Box—Filled with stone: the wreck of the Scow Schooner *Dan Hayes. International Journal of Nautical Archaeology, 32*(2), 210–224.

Rodgers, B. A., & Green, R. T. (2003). *Of limestone and labor, shipwrecks of the stone trade: The 1999 Bullhead Point stone barge investigation* (Research Report #11). Greenville, NC: Department of History, East Carolina University.

Rodgers, B. A., Moore, J. D. III, Corbin, A., Piero, J. D., & Pietruszka, A. (2006). *From quarry to quay: Shipwrecks of McCracken's Cove: The 2001–2002 Sturgeon Bay Wreck and Wharf investigation at the Birmingham site* (Research Report #17). Greenville, NC: Department of History, East Carolina University.

State Historical Society of Wisconsin (SHSW). (2000). *Cultural resource diving cue, Niagara.* Madison, WI: University of Wisconsin Sea Grant and State Historical Society of Wisconsin.

Thompson, M. L. (2000). *Graveyard of the lakes*. Detroit, MI: Wayne State University Press.

Whitesides, S. M. (2003). *Spatial patterning aboard the Millecoquins wreck: Interpreting shipboard life and functional use of an early 19th century Great Lakes sailing vessel.* Master's thesis, Department of Anthropology, East Carolina University, Greenville, NC.

Chapter 13
The Support System for Riverine Trade in the United States

Steve J. Dasovich

Introduction

"There remains much to uncover concerning the hidden maritime history of America's inland rivers, especially the history that sank with the hundreds of steamboat wrecks and related sites on the upper Missouri River between 1819 and 1920" (Corbin 2006: xv). This quote is from a maritime history book and typifies the approach used to understand how the era of steam helped colonize the western half of the United States. History, as the quote regales, lies with the steamboats, or in this case, steamboat wrecks. Certainly, steamboats are the most significant single element of westward expansion until the railroad reached the Great Plains. Steamboats, however, could not operate by themselves. They had to be repaired, kept from harm, and their contents, mostly goods but also many passengers (and a few tourists) had to be cared for. All of these concerns had to be addressed through a network of buildings, structures, and locally appropriate vernacular watercraft. Still, steamboats overshadow their ancillary support systems, and deservedly so. It is no wonder then that when considering maritime archaeology along the inland rivers of the United States, steamboat wrecks are usually the first site-type that springs to mind.

In the Midwestern United States, the two most recognized steamboat wrecks are those of *Bertrand* and *Arabia*. Both steamboats sank in the mid-1860s and both have excellent museums dedicated to them, the former in Iowa and the latter in Missouri. These two wrecks yielded a significant number of artifacts, items which could have been found in any general store or mining supply store circa 1870. These boats were of "normal" size for western steamboats (Petsche 1974; Hawley 1998).

S.J. Dasovich, Ph.D. (✉)
Anthropology & Sociology Department, Lindenwood University,
Young Hall 108B, St. Charles, MO 63301, USA
e-mail: SDasovich@lindenwood.edu

A.M. Evans (ed.), *The Archaeology of Vernacular Watercraft*,
When the Land Meets the Sea, DOI 10.1007/978-1-4939-3563-5_13

These two wrecks were essentially salvage projects. Archaeological control was minimal for *Bertrand* and essentially nonexistent for *Arabia*, and the data generated is focused on the material culture of the goods carried. Nautical architecture was not of primary importance. Despite the lack of archaeological control during the salvage efforts, these two vessels form the base level of knowledge for Midwestern steamboat wreck sites in terms of site formation processes, material culture, and western river nautical architecture. Other wrecks have more recently been investigated, including work on *Heroine* in the Red River in Oklahoma (Crisman 2005), *Maple Leaf* in the St. Johns River in Florida (Holland et al. 1993; Cantelas and Bradley 1994), and *Montana* (Corbin and Rodgers 2004, 2008), and these are adding more significant information to what is known about steamboat archaeology in the inland waterways of the United States. Numerous general descriptions and measurements of steamboats show, at least, the general trends of steamboat construction styles and methods. Unfortunately, more detailed plans and written descriptions of steamboat construction are still lacking (Corbin and Rodgers 2008: 56). Data is still best gathered through archaeological work, and such projects will contribute to a more complete picture of riverine transportation.

Steamboats were not the only form of water craft in use during the 1800s as part of the inland waterway maritime trade. Bull boats, canoes, pirogues, the bateau, flatboats, keelboats, Mackinaws, barges, and rafts were actually more common forms and predate the steamboat. Of these, flatboats may have had the greatest numbers overall on the rivers, especially in the downstream trade. All of these craft were much less expensive to construct than steamboats, most especially rafts and bull boats. Rafts as well as some of the other types were usually meant for one-way transportation of goods on the larger rivers. Upon arrival at their destination, they were sometimes taken apart and the lumber then sold or perhaps traded. Such vessels were often the first to carry Europeans into river drainage systems in the interior of North America. Further, these boats often carried the first permanent settlers to locations that went on to become large cities. Outside of written, period descriptions of these small boats/craft, little is known about them because they were fragile (like bull boats and birch bark canoes) or, like rafts, were taken apart upon reaching their destinations and therefore are very difficult to find archaeologically.

Despite the predominance of steamboats and other watercraft these craft were only one part, albeit a significant part, of riverine trade. While there were many types of significant watercraft, they could not have succeeded to the level they did without a vast support network on land and water that included other forms of transportation and a significant variety of buildings and structures. These buildings and structures were specially designed and constructed for the maritime trade they supported. The full story of inland waterways cannot be told without discussing the support system for these vernacular vessels.

The land-based support system depended upon environmental and geographic conditions, and changed over time. Consider, for instance, Ice Harbors, which were first built by the US government in 1896 (Bradley 1993). In the waterways of the Deep South, such highly specialized facilities were not necessary. In the northern Great Plains however, where they were necessary due to extremely cold winters, they were used to great success (Bradley 1993). Ice Harbors had a significant number of specialized onshore structures and artifacts and at some of these locations some examples of these specialized components are still extant (Dasovich 1993). Specialty facilities such as these were not uncommon, no matter their purpose. The most ubiquitous of the maritime trade support structures, warehouses, and storage buildings were constructed wherever and whenever necessary or profitable. Even more specialized structures that supported water-borne tourism like picnic shelters, bandstands, and pavilions were constructed in park-like grounds that were sometimes only accessible by water. Wharf-boats, quarter-boats, construction barges, shoreline erosion mats (mattresses), piers, docks, and levees are representative of a longer list of support craft and structures that can still be seen in use today.

The vast support system for, and different uses of, the riverine trade is not commonly considered during historical research. Even less frequently are they considered by maritime archaeologists. Vernacular riverine features include different types of watercraft, littoral support structures, and the steamboat excursion trade. In the end, perhaps, a better understanding of such topics will make it easier to identify and interpret littoral cultural remains archaeologically. This can be accomplished by reviewing the general demographic and economic growth patterns along the western interior waterways, documenting specific examples of several growth patterns, and exploring some of the diverse ancillary byproducts of the riverine trade. Maritime history is not exclusive to the steamboat wreck.

Economic and Demographic Development Along the Inland Waterways of the Midwest

In order to understand why standing architecture and archaeological sites related to the riverine trade are located where they are, it is necessary to understand the process that necessitated their construction. Usually, such support structures and buildings were able to be constructed due to economics. Development of economic and demographic infrastructure on land was often accomplished through use of the river and lake "highway" system. West of the Appalachian Mountains and south of the Great Lakes, development was usually based along river systems. Water offered a convenient mode of transportation through the use of various, sometimes highly specialized watercraft. As mentioned earlier, many types of vernacular watercraft were developed for use in the interior of the United States, first by Native Americans, then by Europeans. These craft were specialized for the role they were to perform. River commerce in the Colonial Period was probably started by traders, both Indian and European, and river transportation was quickly used for both commerce and settlement. In the Midwestern United States, most of the early towns were settled along major rivers for this reason. The first towns in this region were in what is termed the Old Northwest, essentially along the Ohio River and its tributaries, and the southern shores of the Great Lakes. French speakers came to this region along the St. Lawrence River and Great Lakes. English speakers came through the

Appalachian Mountains on overland trails. Eventually, significant river cities like Pittsburgh and Cincinnati grew from small settlements at muddy boat landings to become hubs for commerce and boat construction for the inland waterways.

Boat traffic on the Mid-continent's rivers developed in rough stages based upon technology levels, water levels, and competing forms of transportation. Starting with dugout and other canoe types, as well as bull boats, these earliest forms of watercraft could travel most types of waterways. After Europeans established settlements in the region, craft of more complexity became more common. In general, the more complex the craft, the more capacity there was for cargo. This was driven by an ever increasing demand for goods from settlements, traders, and the various military establishments of the French, Spanish, British, and Americans. This demand for goods led Europeans and, later, Americans to establish most of their settlements along waterways because of the relative ease of the transportation of goods to and from these locations. Goods were usually offloaded along the muddy shores of any given river, and carried by boatmen, hackmen, slaves, and other laborers up to the higher ground for distribution to wagons. Wagons then carried the goods to stores or military posts in and around the town or further inland.

Boats would load and unload at points along the rivers where there were customers and where there was easy access to dry land. The issue of access to land is not as easy a problem to solve as one might first surmise. Access to land was and is of vital importance to river commerce. The earlier flatboats and keelboats usually preferred longer, flatter landing areas where they could partially "beach" themselves or at least tie up to larger trees to hold them against shorelines where there was also ample open land to unload, load, and bring customers aboard. Later, steamboats would partially solve the issue of having to "land" or beach themselves at specific locations (a plantation for instance) by having a device often called a "stage plank" on the bow of the boat that would swing out to reach higher shorelines in areas, where muddy flats and "beaches" were not available.

Locations along rivers called "landings" appear on many maps of Midwestern waterways in the 1800s. As the examples below demonstrate they are listed in several publications in large numbers after steamboats began regular service in any given area. Essentially, any point where a steamboat could reach shore with just the stage plank could be used as a landing, but the location also had to have access to customers and goods. Even still, the number of potential landings certainly numbers in the thousands as Hall (1884) suggests. Significant plantations and farms, small villages, and larger towns would all have landings. Some maps, particularly those made for steamboat travelers such as ribbon maps, which were long, narrow maps made to accommodate long stretches of rivers (Sevier 2010), show landings and their associated farms, towns, and roads. This information allows researchers to link a landing to specific families or business enterprises as well as gain more insight into the commercial structure of riverine trade. Landings are sometimes shown in association with only a farm field, suggesting a plantation or large farm/family estate, sometimes next to a platted town, and other times with only a road leading down to the landing. Usually, a landing has a town or family name to distinguish it from others nearby. In some cases, only a landing name is given without reference to a specific town, field, or road, and it is not possible to specifically locate the actual landing spot because only a name appears on the map without a corresponding location. This may indicate that the landing is only vaguely located in a certain area and that no visual signs of the landing could be seen when the river was surveyed. Landings could, by definition, be just about anywhere along a river. As discussed below, several resources have attempted to list all known landings along stretches of certain rivers. The numbers of landings were higher before railroads reached most of these areas, with some exceptions. The Gasconade River in Missouri, which is a smaller tributary of the Missouri River, is one example where railroad expansion did not significantly affect river trade because of the relatively small amount of goods and passengers being moved by boat and because the railroad corridors did not follow the river course but ran perpendicular.

The distance between landings both along rivers and lakes is measured in river or water miles. Landings were sometimes very close together, especially if they were operated by rival commercial interests (Lepley 2001). The numbers of landings also seem to have a direct relationship to population density; the higher the density, the more landings. A review of Giffen (2001) suggests that about 76 landings existed along the Missouri River between its mouth just north of St. Louis, Missouri and Council Bluffs, Iowa, a distance of 613 river miles, in 1846. This would average one landing every eight river miles. At the opposite end of the Missouri River (in an area of minimal population) using an estimated river mileage of 500, Lepley (2001) records 14 landings between Fort Union, Montana and Fort Benton, Montana, for an average of one landing every 36 miles. Most of these landings lasted only a few years each. On the Mississippi River, a river with a higher population density, the Mississippi River Commission, in an 1879 report to Congress (published in 1881) shows on one of its maps (Appendix J, Plate 1) that on an approximately 45 river mile stretch of the Mississippi River between French's Point (just north of New Madrid, Missouri) to Tiptonville, Missouri, 18 known landings existed, for an average of one landing every 2.5 miles (Burdick 1881). For the smaller Gasconade River in central Missouri, Shrader (1993) mentions 13 landings along a distance of 77 river miles, from its mouth to Vienna, Missouri, an average of one landing every six miles. Finally, Lingenfelter (1978) shows 34 landings during the 1860s and 1870s along approximately 380 river miles of the Colorado River between its mouth in the Gulf of California to Rioville, Nevada, averaging one landing per 11 river miles. Interestingly, Lingenfelter also shows the same map for the 1880s and 1890s, after the railroads reached the Southwestern United States. Twenty-two landings are shown for the same stretch of river, expanding the average miles per landing to 17, a loss of 35 % of the landings from the previous two decades when railroads were not present.

As the flow of goods increased to and from certain river landings, traders, merchants, and the military often constructed facilities for storage, preservation, and transportation of these goods. Sometimes, in areas where the growth was gradual or settlement occurred later, warehouses were not constructed until later in the 1800s. This was especially so on the smaller rivers and inland locations further west. Along the upper Ohio River, however, warehouses were constructed perhaps as early as the late-1700s for the purpose of storing goods in preparation for the Spring flood season when downstream transportation was generally easier and faster. Non-perishable goods were ported over the mountains from cities like Philadelphia and stored at warehouses near present-day Pittsburgh. The modern town of Elizabeth, Pennsylvania, for example, started as a warehouse location for non-perishable goods (Baldwin 1941: 187).

In the western rivers of the United States, most goods were simply stored without shelter, on shore while others were stored on the water. Through the early 1800s, especially along the Ohio and Mississippi Rivers, individual traders and merchants who took their goods downstream to sell, converted part of their flatboats into floating stores, complete with shelves and cabinets. They would stop at larger riverine farms, tiny hamlets, and villages did not have a general store, as well as river fords and crossings to sell their goods. These traveling salesmen were soon to be replaced by the terrestrial stores of merchants in the many small towns being settled after displacement of the Indians. Some flatboats were also used by skilled craftsmen such as blacksmiths and millers, some of whom incorporated current-powered grist-mills into their flatboats. Still other boats had libraries and various other forms of entertainment such as theaters for plays (Baldwin 1941).

After the end of the Revolutionary War, settlers began to cross the mountains in ever increasing numbers. In the drainages of the upper Ohio River, the first towns were settled via overland routes. The towns along the navigable rivers, however, became the most prosperous and were used as the starting places for the majority of western expansion expeditions. In these towns, the center of commerce and industry was along the river. Warehouses often covered the original, and still used landing areas. Eventually, these landings became crowded with boats, boatmen, and other laborers. To help regulate river traffic, towns began to develop maritime structures, such as docks and wharfs. There is little information as to when the first of these structures was constructed, but some of the towns of the upper Ohio reported docks and wharfs being washed away by floods in the late-1700s (Baldwin 1941: 188).

Along the Missouri River, the population did not increase significantly until after the War of 1812. The conclusion of this war saw a marked decrease in Indian hostility in what is now the state of Missouri, making it safer for families to immigrate into the still wild interior of Missouri. Like the Ohio River Valley, towns were usually established by settlers migrating overland first. Early settlements in Missouri often did not survive for long, but those that did rely on the Missouri River for their continuing success over a roughly 50-year period. It was not until steamboat travel fully commenced in the 1830s that the majority of settlements further north along the Missouri River Valley became well established.

Examples of River-Based Growth Patterns

Transportation has historically influenced settlement patterns and densities, but it has also shaped growth patterns at regional and national levels. Although the rivers offered relatively fast and convenient modes of travel, overland routes were also traveled by large numbers of people. Overland and river transportation were not mutually exclusive, because many overland trails ultimately crossed waterways. Smaller watercourses were crossed at fords while larger and deeper waterways were crossed via a ferry. The locations of fords and ferries often served as a catalyst for town growth. At locations where ferries existed, the probability of a town being established in the Midwest during the 1800s was higher than other locations. The busier the crossing, the more likely ancillary structures and buildings would be constructed. These crossings, and their associated structures, were strategic both militarily and economically and form significant, riverine, archaeological site types. The locations are usually noted on early maps and often mentioned in writings of travelers.

An example of one such town is Washington, Missouri, in Franklin County on the Missouri River. Washington is located approximately 60 river miles upstream of St. Louis and approximately 44 miles upstream of St. Charles. The town was one of several of the same name (a popular place name in the early 1800s) on or near the Missouri River in Missouri. Unlike the others, this Washington survived and thrived because it had a ferry crossing. The City of Washington is a river town. The original town was settled and later platted around a ferry crossing and boat landing on the Missouri River. The first record of the name Washington at this location is found in county court records, dated August 22, 1822, stating that a licensed ferry was operating at "Washington on the Missouri River." Apparently, several other ferries were in operation near Washington as well, but eventually most crossings were undertaken at this location. When the first ferry operation was started at what became known as Washington, is not known. The Washington Ferry was, in 1840, a "crank" ferry where up to four men operated a turning crank to bring the ferry across the river. Later, the ferry became a horse ferry where horses were used to pull the craft across the river (http://washingtonmo.com/history/). The town grew up around this ferry crossing and the later steamboat landing, called Washington Landing. The *Independence* became the first steamboat to take a trip up the Missouri River, in 1819, but it was not until 1829, after Washington was established, that regular steamboat service included this portion of the river.

Washington saw even greater growth once regular steamboat service reached the town. By the 1840s, the town's merchants and nearby farmers were prospering through the trade in tobacco. The demand for tobacco from Missouri was such that stores and warehouses were beginning to cover Washington's waterfront.

> We did a big business, and shipped a quantity of tobacco by boat—leaf tobacco, in hogsheads of 1,000 pounds each. At that time very little wheat was raised, and tobacco and corn were the principal staples. William Cowherd built a large tobacco warehouse, 35 by 80 feet, in the bottom below our store, and thus we had storage facilities. (Gregory 1981: 23)

This quote, apparently from the 1840s, shows the need to construct storage facilities along the river to keep goods preserved and safe while waiting for steamboat transportation. Even earlier, Gregory (1981) refers to records describing another tobacco warehouse and a display shed, possibly used to market and or sell tobacco. These warehouses built in the bottoms occasionally would be washed away by floods, sometimes with all the goods. In 1848, the railroad reached Washington, eventually changing the emphasis of the town from the river to the railroad.

Another settlement, called Missouri Town, originally platted by Daniel Boone in what is now Franklin County, Missouri, did not fare as well as Washington. Little is understood about this early town site (circa 1800) except that it may have had a post office, store, and several houses. Its location is known, though. It was on the Missouri River about two miles upstream from the current town of Defiance, Missouri. The town was placed on and around a natural levee, itself already the ancient home of Woodland and Mississippian Period peoples, in the broad floodplain of the Missouri. The town was reportedly abandoned after just a few years due to numerous episodes of flooding.

Most surviving river towns either were constructed on the bluffs or high terraces of river valleys or, like Ste. Genevieve, Missouri in 1785–1786, relocated from the floodplain to the terrace or bluff. River towns, both historic and present day, have had to balance the convenience of river access with the security of higher ground. The great Midwest flood of 1993, for example, forced several modern towns on the Mississippi River like Valmeyer, Illinois, to essentially relocate to higher ground or, as in the case of Hardin, Illinois, move vital services to the top of the bluff. Many of the early towns or platted towns (whether occupied or not) were placed in the bottoms in order to be near the river and therefore near the boat landings. If, however the river happened to flow against a bluff or high terrace, then towns might be originally placed on the higher ground. The idea was to make commerce easier and more convenient for traders and customers; the closer to the water, the better.

Some towns reversed the order of development moving down from the high ground to be closer to the water. Natchez, Mississippi was founded on a high bluff on the east bank of the Mississippi by the French (originally called Fort Rosalie) in 1716. This outpost eventually became a very important commercial town, but its location on the top of a steep bluff made access to the river difficult. About 1785, an area directly below the bluff and town began to see an increase in construction activity. Known as Natchez Under-The-Hill, this area was built around a river landing for flatboats and other water craft before the steamboat era. Eventually, Under-The-Hill flourished as a loading and unloading port of call for vessels. The town of Natchez (up on the bluff) became a wealthy city by the Civil War, due to the efforts of the lower classes who lived and worked in Under-The-Hill. Under-The-Hill had three streets, each progressively higher in elevation than the next, which ran parallel to the river. The lowest, Water Street, ran directly along the Mississippi River, but there were buildings on both sides of Water Street. These buildings were on pilings and stood at least partially if not fully over the water. During the late 1800s, a straight line railroad, called The Bluff City Railway, was constructed to ease the effort of hauling goods up the steep road and into Natchez. Buildings in Under-The-Hill included warehouses, offices, a few stores, and numerous saloons, brothels, and hotels (Moore 1958; Gandy 1987). Every building (and even the small railroad) in Under-The-Hill was constructed as part of the maritime trade.

River commerce made much of the Midwestern United States prosper during the first half of the 19th century. This trade necessitated a significant amount of physical property support to help it be successful, as can be seen at Natchez Under-The-Hill. Steamboats and flatboats by themselves could not ensure that property, people, and goods would reach their destinations safely and equitably. Support structures and buildings became necessary for the continued profitability of river commerce, and included warehouses, stores, hotels, and entertainment buildings. Even houses and offices were built in support of river trade. There are, however, other supporting structures and buildings that developed not directly for goods and passengers, but for safety, organization, river maintenance, and protection of property.

Support Structures

Between the 1850s and 1880s, the apex years of the steamboat, landings at major river cities could contain enough steamboats and flatboats, side to side, to cover many miles of shoreline. Sometimes, steamboats would be packed against the shore in two or three layers, unloading and loading from boat to boat until the goods reached the landing. This amount of river traffic had to be controlled. Some cities constructed docks that would allow for direct transport of goods from a boat to the landing without having to walk across other boat's decks. Sometimes, certain companies would construct private docks. Cities would also spend their own funds on improving landings to better serve maritime commerce. The major steamboat ports, like St. Louis, started this process with other, smaller towns quickly following suit. Towns would recoup these costs usually through the assessment of a wharfage fee for each boat landing, no matter the type or size. Fees would vary by town. Some towns even set a schedule of drayage fees for hackmen who usually carted goods to and from steamboats and warehouses (Lass 2008). Improvements to a landing might also include paving the usually collapsing sides of the bank of a landing with cobblestones, such as can still be seen at the St. Louis riverfront.

Wharfs were often added where unloading and loading needed to be completed faster to alleviate the boat congestion. Wharfs allowed temporary storage space, usually in the open air, so goods could be exchanged faster to allow the boat to leave sooner, alleviating some of the landing congestion. In support of this faster service, some individual steamboat lines, cities, or trading companies would construct wharf-boats. Wharf-boats were (and are) essentially floating buildings used as terminals for goods and company personnel. Hall (1884: 175) defines them as "a broad, flat-bottomed, strongly built barge with square ends, having an overhanging platform and large deck-house to protect the accumulations of freight and

Fig. 13.1 Steamboats along the Gasconade River in winter with a wharf-boat at far left

shelter the officers of the steamboat companies" (Fig. 13.1). They could be towed to any location on the water, as necessary. Hall also suggests that these vessels were expensive to construct and only were used, where there was a regular trade with steamboats. He further suggests that the main reason they were constructed in the first place was due to fluctuating water levels on the Missouri River. This would suggest that he thought wharf-boats were invented and used first on this river, spreading to other rivers later. These would give access only to company boats, giving them priority landing areas and again speeding up the transfer of goods. There is no indication of when wharf-boats were first used, but earlier versions were made from the hulls of old steamboats. These floating structures had multiple roles. According to a description from Gandy (1987: 24) of wharf-boat uses between 1870 and 1920, "the wharf-boat, floating up and down with the rise and fall of the water, made landing easier on this fluctuating and unpredictable river. Wharf-boats housed offices, storage areas, reception areas for boat crews, and sometimes lodging and dining facilities."

Another version of the Wharf-boat is the Quarter-boat (Fig. 13.2). The Quarter-boat apparently was a floating apartment for river workers. Photos of barges with buildings on them often appear in collections showing shoreline stabilization work

Fig. 13.2 Historic photograph of a Quarter-boat

using willow mats and rock along the Missouri and Mississippi Rivers (Fig. 13.3), for example, where large numbers of men were used for these large projects. Wharf-boats are still used today for the same purposes as described by Gandy.

Steamboating on the northern rivers had its own issues that brought about a late fix from the federal government. Northern rivers have a limited transportation season due to the rivers freezing. Sometimes, boats would be stranded by low water in the autumn and then fall prey to the crushing forces of ice. Higher expenses were incurred by boat owners when they had to relocate their steamers downriver to areas that did not freeze, and then bring them back upriver in the late spring floods. To help alleviate this situation, the federal government constructed ice harbors. These harbors were equipped with repair facilities, warehouses, ways (wood boatways for hauling steamboats out of the water), and other structures that would get steamboats out of the water, have space to store them through the winter, and repair and refit them before being put back in the water. The government operated ice harbors in many areas of the country. On the upper Missouri River in the Great Plains region, two such ice harbors were completed, the first, at Mandan, North Dakota called Rock Haven was completed in 1896. The second, completed in 1897, was located at the mouth of the Big Sioux River on the Missouri River at Sioux City, Iowa, and was called the Sioux Ice Harbor (Bradley 1993). Others were constructed through

Fig. 13.3 Rock and debris, including a steam engine boiler and concrete-filled wood casks, still in use for shoreline stabilization

the early 1900s. These later harbors appear to have mostly been actual in-water harbors protected by massive concrete walls, like one finished by 1910 just south of Cincinnati, Ohio. These later ice harbors could be used as repair docks during non-winter months (Popular Mechanics 1910).

Excursion (Tourist) Trade

From an anthropological perspective, tourism is not an unknown research area. The tourist trade on the Midwestern rivers played a significant role in the story of maritime trade as the economics of western waterways is not only based upon goods. Passengers were usually to be found on any steamboat plying the western waters. One type of passenger, however initiated a specific economic factor; the tourist. While passengers usually *traveled* on watercraft for emigration or a job, the *tourist* passenger was along for the ride, and as much entertainment as they could get. This distinction echoes with Crick's (1989) discussion of the same two types of people.

Crick distinguishes travelers as those who are moving about with the "notion of work (travail)" and the tourist "where passivity rather than activity reigns" (1989: 308). He goes on to say "Tourism is a … manufactured, trivial, inauthentic way of being … made safe by commercialism" (Crick 1989: 308). Tourism "is merely about a world discovered (or even created) by entrepreneurs, packaged and then marketed" (Crick 1989: 308), which well supports riverine excursion tourism as yet one more ancillary feature of riverine activity.

Steamboat lines began to see the opportunities in the tourist trade early in the 1800s. Usually labeled "excursion" trips, certain steamboat companies, as early as the 1830s, were advertising long excursion cruises/tours. These were offered during the summer months only, but it was not until the 1880s that the excursion market sometimes became the sole focus of certain steamboats.

An excellent example of this innovative type of maritime enterprise comes from Kankakee, Illinois. In 1884, a man by the name of W. T. Gougar bought a small steamboat, *Minnie Lillie*, with a capacity of 40 people, then operating on the Kankakee River. He also bought land that he converted into a private resort/park, named Gougar's Grove. Through the next five years, Gougar added another larger steamboat, with a capacity of 300 people, and a steam launch to his fleet. At the park, over a 15-year period, Gougar added a two-storey club house, numerous small buildings, a carousel, a toboggan slide (called Shoot the Chutes), a water tower, row boat rentals, electrical and water works buildings, picnic and dancing pavilions, restrooms (called public comfort buildings), and a large entrance gate with a ticket office in each of the two large pillars of the gate. Gougar's company continued for several years after his death until 1911 (Lindsay 1987).

Other steamboats on many of the Midwest's rivers used excursion cruises to supplement their income, especially in the more southern waters that did not ice over. Steamboats on the Colorado River participated in this type of effort starting in the 1890s (Lingenfelter 1978). Along the Missouri River, where the railroads ran essentially parallel to its course, the excursion business seems to have been less vigorous. Along the Osage and Gasconade Rivers in central Missouri, however, where railroads crossed perpendicularly, the steamboat industry had a longer life and excursion cruises lasted as long as the steamboats. During the 1850s in St. Louis, there was a thriving excursion industry on much larger boats than Gougar's largest, 40 years later (Meyer 1967). It must be assumed that the larger river cities all had large excursion trades. Indeed, today, most still do, though the boats are now usually decorative paddle wheelers with diesel engines.

The peak of the excursion cruises is suggested to be around 1917 (Meyer 1967: 43). Excursion cruising was a thriving industry from the 1830s through the 1960s. Meyer's research found the first advertisement for excursion cruises by steamboat captains in 1837. This advertising focused around long, vacation cruises to the Upper Mississippi River, with the cruise given the label of *The Fashionable Tour* by the famous western artist George Catlin. This tour developed its own marketing scheme complete with advertising, traveling salesmen, lecture tours, and displays of very long (up to 1200 yards and 12 ft high), hand-painted, panoramas consisting "of great unwinding rolls of painted canvas which artists exhibited in America and in Europe to the accompaniment of lectures" (Meyer 1967: 46). These tours were so popular, that even though the 1830s and 1840s were one of the United States' busiest immigrant periods, tourists sometimes outnumbered immigrants on these steamboats. Like Gougar decades later, park areas along the Mississippi River were developed for the early excursion industry. The St. Louis area's most popular was a 100 acre park at Kimmswick, Missouri, south of St. Louis, called Montesano. Another north of St. Louis was called Silver Island. The St. Louis area had at least eight such parks. These and others like them eventually boasted amenities like a scenic railway (at Montesano), dining halls, theaters, picnic facilities, refreshment stands, dancing platforms, baseball fields, and even runner's race tracks (Meyer 1967).

Archaeological Manifestations

What is the archaeological signature for a tourist venue like Montesano or Gougar's Grove? How does one distinguish between the material culture of a railroad warehouse versus a steamboat warehouse? Is the wreck in the river a wreck site of a steamboat or the scuttling location of a wharf-boat? Are those pilings the remains of a dock, wharf, or erosion control structure? These questions are just the tip of the iceberg in terms of archaeological identification of terrestrial and shoreline sites related to the maritime trade on inland waterways. Maritime archaeology on these waterways concentrates on water craft. This is not surprising, given most maritime archaeologists are quite interested in the nautical architecture of ships and boats and the goods they carried. Consider, for example, illustrations in several significant texts on maritime history topics. Illustrations are often used to help focus on significant issues being addressed in the text. In Hunter's *Steamboats on the Western Rivers: An Economic and Technological History* (1993), there are 29 illustrations. The majority of these show steamboats or their parts, with the only exceptions being five maps, one general view of a city riverfront (the perspective being too far to see much detail), and a picture of the falls of the Ohio River. In Bauer's (1988) *A Maritime History of the United States*, there are 37 illustrations, only four of which show any details of non-vessel structures and other buildings. These two books generally are devoted to maritime history, and gloss over things that are not directly associated with a vessel. There is little information on shore-based structures and buildings.

Two other general maritime history books show a few images of such things as wharf-boats and shore-based transportation. Lepley (2001), largely a picture book, has numerous images showing landings. Several of these have good views of warehouses and other commercial buildings. Lass' book, *Navigating the Missouri* (2008) has two photos showing what appear to be wharf-boats at the St. Louis landing in the 1850s. One of these wharf-boats (2008: 131) also shows the company specific use of such a vessel, complete with the company name, "Keokuk Daily Packet," and even an advertising banner. Such books appear to be slight exceptions to the rule.

Indeed, sometimes even survey reports completed by government agencies, such as the U.S. Army Corps of Engineers, have very little information about shore-based support structures and buildings. For example, Custer and Sandra (1997) completed a documentary survey of submerged cultural properties for an approximate 300 mile stretch of the Upper Mississippi River (between mile markers 300 and 614) and an approximately 250 mile stretch of the Illinois River (between mile markers 80 and 327), all within the jurisdiction of the Rock Island District. This study clearly shows the discrepancy between steamboats (and other watercraft) and their documentation, and support property documentation in the literature. This study located 131 vessel sites, but only seven structures or navigational markers (these markers are actually "famous" rocks). The study was meant to determine the locations of possible submerged cultural resources and did not attempt to document any maritime related sites which were not at least partially underwater. This certainly lessens the number of support structures that might be documented, but it also ignores such structures as wing dams, levees, dikes, and erosion control devices. Further, the short, textual content of the report has nothing to say about the role of support structures with the exception of three boatyards that are thought to be partially submerged in the study area. One must look to the nonacademic sources to find discussion and photos of support craft and structures. Early journals with issues devoted to the maritime trade often have many photos of shoreline scenes along the river systems. A good example is the 1970 issue of *The Palimpsest*, published by the Iowa Historical Society. There are several pages of excellent photos of facilities belonging to the Diamond Jo Line, a steamboat packet company operating mostly in the upper Mississippi watershed (Petersen 1970).

As the Custer and Sandra report documents, there is relatively little data to be found from archaeological investigations of shore-based maritime facilities. If surveys are completed, they are often looking for shipwrecks. In the Missouri River, a survey completed by the author covered a 15 river mile stretch between St. Charles, Missouri and ending just past Chesterfield, Missouri. Completed in 2003 during low water stages, the most visible traces of the maritime trade were the myriad channel control devices like wing dams and dikes, and the ubiquitous erosion control willow mats. Usually, the dikes and other channel control structures were only represented by the pilings sticking up out of the water. The willow mats had few if any rocks left on them, but the mats were firmly in the bank, still doing their job. This survey only located one piece of a wreck of a vessel, stuck in the middle of the river on a sand bar, which has doubtlessly washed further downstream by now.

River surveys by boat may document such structures and wrecks still in the water. Documenting the supporting shore infrastructure, however, is usually best done from land, unless the river has shifted or was dammed covering up the old landings/towns, like the Osage River forming the Lake of the Ozarks in central Missouri. The shoreline locations of such structures tend not to be surveyed with regularity. There are several reasons for this. First, as seen along any major river, the river shifts course, making former shoreline properties inland properties, and no longer immediately recognizable in their original role. Second, it often takes significant research time to identify a structure or building as originally playing a role in riverine activity. Most surveys undertaken through the Cultural Resource Management system do not have such time to conduct detailed research. Finally, if the property is in a valley, it is in a depositional environment and can often be in an area with a high water table, making it difficult to locate, or, as in the Primus Emerson site discussed below, very difficult to complete any fieldwork. Over the past ten years, and through at least 1000 compliance projects, the author has only been able to conduct significant research on four shoreline sites.

Of these four sites, two were chosen for detailed research not because of a contract, but because of research interest. The first site is the circa 1769 Louis Blanchette Homestead site (23SC2101) in St. Charles, Missouri, overlooking the Missouri River. Louis Blanchette was the founder of the city of St. Charles. He built a cabin and trading post on the Missouri River, along an old (even then) Indian trail now called the Booneslick Trail (after Daniel Boone and his sons who came to the area in 1799). The cabin faced the river because the river was the major route for transportation and because it was built before the implementation of the street grid in the 1790s. The site has at least three late 1700s structures, one partially extant (dating to at least 1793) but changed over the years, and two found archaeologically. This project was started not because of its possible maritime-related genesis, but because Blanchette is famous as the city founder. This site probably would not have attracted attention otherwise.

Much like Ford found in his survey of portions of Lake Ontario (2009), properties like this appear and disappear along waterways. This trading post could have died out quickly, but for whatever reason, it survived and thrived. At Blanchette's homestead site, there were no European-constructed roads when he settled there, and the river served as the main thoroughfare. For archaeologists, realization of the site's location and orientation are important. Sites along rivers and lakes that also have roads on the other side of the property can be more of a challenge to interpret. For the Blanchette site, the realization of the orientation of the still used, late-1700s building foundation oriented along the river and not the street grid, called for a completely different survey strategy. Today, the building has a front yard facing the street grid. In 1769, that same area was the backyard.

The artifact assemblage has yet to yield anything that is directly maritime in nature, but French colonial ceramics (Normandy Plain and St. Onge) have been found here, offering clues to early water transportation and maritime trade linking the Missouri River to the St. Lawrence River and the Atlantic Ocean beyond. This site cannot be fully interpreted without describing its reason for being, and that reason is the Missouri River. St. Charles is a good example of a town started by riverine trade. Blanchette and his fellow French-Canadians were hunters, trappers, and traders conducting riverine trade with their canoes. Today's city of St. Charles exists as a result of this pre-steamboat trade.

Switching to a more recent site component, the second site is the Primus Emerson site (23SL2292) located at the mouth of the River Des Peres overlooking the Mississippi River in the City of St. Louis. Due to possible future construction activity, the author directed a survey resulting in the location of this multicomponent site with various prehistoric and historic components. Of interest here is a linear structure encountered approximately 1.5 m below more recent fill that is interpreted by the author as a dock or wharf that ran parallel to the Mississippi River. It is made up of planks and square piles, making a structure that is approximately five meters wide and at least 30 m long based upon the locations of excavation trenches placed during the surveys, the most recent being in the Spring of 2010 (Dasovich 2009; Booth and Hajic 2010). This structure could have been part of the Eads Boat Works; the ways for this famous boat yard are still visible just downstream. Unfortunately, as soon as the structure would be uncovered, water started to fill the trenches due to being very near the Mississippi and the high water table. No detailed structural photos or drawings could be completed, but the boards/planks were mechanically sawed and were made from very large hardwood trees. No fasteners were observed. Primus Emerson was the owner of a marine railway shipyard through a company called the Sectional Docks Company. The location of this railway/shipyard is not precisely known, but site 23SL2292 was once his property. It is possible that the structure seen during recent survey work was part of the railway. The railway was used to haul large boats out of the river, sideways, for repair and refitting. To do this without uneven stress on chains and the boat, the railway was powered by steam engines inside a power house structure. A shaft, 390 ft long, ran the entire width of the rail line, and evenly pulled the chains attached to the boat and carriage. Such a device was very expensive. Hall (1884) writes that Emerson built this in 1856 or 1857. Fieldwork by the author did not locate any rails or the location of the power house, but the long structure could have been associated with the propulsion shaft and chain supports.

The third site was at the Selma Farm Landing on the Mississippi River, south of Herculaneum, Missouri. This property is privately owned today and was once owned by a wealthy family. Of particular interest is the fact that the property is today, still known as a "landing." A small Italianate, castle-like mansion was constructed in the 1800s and the size of the estate necessitated its own river landing. The landing was along a steep bank at the foot of a limestone bluff, upon which the castle/mansion still stands. Oral tradition placed several warehouses on top of the bank that were constructed for storage of goods to be shipped from the plantation. During the mid-1800s, a railroad was constructed along the top of the bank, possibly destroying much of the evidence for these warehouses. The project was necessary because a new flood-protection levee was to be constructed near the mouth of a small tributary. An initial survey of the top of the bank indicated the presence of nineteenth-century artifacts and the archaeological crew completed test excavations. The limestone foundation of one warehouse was partially uncovered during these excavations (Fig. 13.4). Since construction activity for a new levee was not going to impact the warehouse area, the Missouri State Historic Preservation Office did not require any further work in this area. Artifacts did suggest the foundation was from a mid- to late nineteenth century building. The crew conducted probing for the rest of the foundation and found the dimensions to be relatively small, at approximately 30 by 40 ft (Dasovich and Warner 2004).

The fourth location is along the Meramec River in St. Louis County. The remains of two boat hulls lie along the shore of this small river (a tributary of the Mississippi),

Fig. 13.4 The limestone foundation of a warehouse is uncovered during excavations at the Selma Farm Landing along the Mississippi River, south of Herculaneum, Missouri

and are close to the old Route 66 bridge, now included in a St. Louis County Park. The site of these two wrecks has variously been suggested as the location of a ferry crossing, construction barges for the bridge, or just the wreck site of two boats/barges. These hulls have not been studied in any detail and are often submerged. Is this site the location of a ferry crossing before the bridge was built and therefore the wrecks are remains of ferries? It may just be circumstantial that the wrecks are so close to the purported ferry location and the bridge that replaced it. This site suffers from a lack of further historical research about possible terrestrial components. The general assumption among local archaeologists and historians is that these are ferryboat hulls. To substantiate this claim, archaeological evidence of the ferry buildings/structures would be helpful. To date, no survey has been documented. Without such supporting data, the ferry hypothesis will not be very convincing. Even a significant survey of the wrecks may not be enough to support the ferry hypothesis. Regardless, this site serves to illustrate the difficulty in determining the link between wrecks and onshore facilities.

Across the United States, other projects have focused on shore installations, especially military fortifications. Most of these have been along saltwater or in the Great Lakes. Civilian property constructed for participation in maritime trade is not commonly investigated unless by accident through compliance projects such as the brief surveys mentioned above. It is probable that numerous reports exist on such structures and buildings in gray literature, but a more mainstream example of work on nonmilitary sites is found in Russell et al. (2010). This study is about sites related to the maritime/nautical tourism in Yellowstone National Park. Surveys conducted by the National Park Service located both underwater and onshore sites, usually nautical in nature, such as wrecks and docks.

Recently, McDonald (2011) discusses timber retaining structures in urban environments. Using examples from several east coast cities, she authored a detailed study on vernacular architectural styles and how they are linked to these maritime support structures. Mostly for land building, the structures McDonald discusses are part of the maritime world, though not always constructed by maritime specialists. Many of her examples are from riverine environments in the northeastern United States. There is no mention of similar sites in the trans-Appalachian west.

Another study, this one on a larger scale than most, is a dissertation project completed by Benjamin Ford while at Texas A&M University in 2009. His study of Lake Ontario and its shoreline looked at the littoral maritime cultural landscape. Farmsteads, industrial locations that he calls "a node of rural industry" (2009: 309), and other large-scale activity areas along the shore may include multiple, maritime-related structures and buildings such as boatyards and docks. While not discussed in Ford's conclusion, many of the residential and commercial buildings and structures that are necessary to run the shoreline industries, often mills and lumber yards, and house the workers/owners, are all in place as a direct result of the maritime trade on the Great Lakes. This same basic principle can be applied to the non-Great Lakes inland waterways.

Conclusion

The riverine trade support system of the nineteenth and early twentieth centuries was only slightly variable. Terrestrial, riverine sites vary in size and utility. From simple landings at the end of a road or at a ferry, to larger farms and plantations, to large cities with multiple wharves, warehouses, and docks, these are all maritime support locations. Riverine trade was the largest driving force in the westward expansion of the United States. For any of these support sites to be considered successful in their time, they needed to grow and last. Trading posts, military forts, and small villages located along navigable waterways tended to grow only if a combination of environmental, geological, and economic traits existed.

Geographically and archaeologically, these shoreline locations formed small nodes, each with their own reasons for existence, though the reasons may be similar. First, access to reliable transportation was a necessity. A location along a river would be suitable to meet this requirement, but unless the settlement/post was positioned out of the floodplain (like Ste. Genevieve, Missouri in the 18th century), flooding might curtail the chances for success. River traffic had to be able to reach the shore within a manageable distance for the effort to be cost effective. Stopping at the many landings along any given river system meant slower travel. If the stops were not profitable to the boat, then that landing would cease to exist as a port-of-call and any infrastructure or population center would have been left with serious issues. Indeed, many of the landings on the Upper Missouri River suffered this fate when the steamboat trade stopped calling. If the river changed course, towns may stop growing or even cease to exist. Vicksburg, Mississippi was originally settled as a river town on the Mississippi River. In 1876, the Mississippi River shifted course and left the city dry and financially devastated. Eventually, in 1903, after 25 years of effort, the Yazoo River (in a project known as the Yazoo River Diversion Project) was rerouted along the Vicksburg bluffs, restoring some of the town's river economy.

Good transportation, though, means very little if there is little to transport. The second trait seen in successful river landings and towns is therefore resources. Perhaps mining or farming was profitable in the area. If so, there was a means to support both a population and a riverine transportation system. Along the lower Colorado River, for example, mining was king. The riverine trade focused on bringing mining and living supplies to the miners and their support system and then carried out the spoils of mining (silver, gold, and various ores). After many of the mines were abandoned, the tourist trade helped some of the river trade continue through excursion cruises (Lingenfelter 1978). Along the lower Missouri River in Missouri during the 1840s and 1850s, tobacco was the chief cash crop and this helped the region in and around Franklin County become one of the most populous areas in Missouri by 1850.

Are there distinct riverine trade characteristics recognizable through archaeological investigations of buildings and structures? Few, if any, projects have specifically investigated the archaeological data searching for links between riverine trade and buildings and structures in the littoral zone. Perhaps personal items carried by rivermen being found in significant numbers in any given building or group of buildings would be a strong indicator that this building was used in the maritime trade. The real problem is that the material culture being transported on the river systems are the same or very similar to that of terrestrial material culture. For example, the vast majority of goods found and transported on the steamboats *Bertrand* and *Arabia* were meant for use on land. Archaeological fieldwork coupled with significant historical research is necessary to directly tie littoral sites to the riverine trade.

How would one describe, in a research report, the archaeological signature of Gougar's Grove? How would an archaeologist relate the socioeconomic variables seen at such a site to the riverine trade without significant historical research? If this location is ever investigated, what signature does a toboggan slide or picnic table leave and how does that signature get tied into a riverine economy? This archaeological site, if it still exists, would be considered eligible for listing on the National Register of Historic Places because of its significance to the local and regional economy of its time period, and it is doubtful that many such water-based tourist facilities still exist. As far as the small fleet of Gougar's boats is concerned, no records of the loss of the steamboat *Minnie Lillie* (Gougar's first boat) have been found, and the vessel's final disposition is unknown. If the wreck of this boat is found, the archaeological signature and interpretation would be that of a small stern wheel steamboat (approximately 55 ft long) that would look like a short-haul packet boat seen on many small rivers. Its small size would be attributed to the size of the river it cruised and the correspondingly small amount of goods and passengers it would have been asked to carry. Gougar's second steamboat, with a significantly increased capacity from *Minnie Lillie* (40–300 passengers), might surprise an archaeologist with its size for a small river.

In the Missouri River, in the Washington Marina (Washington, Missouri), two shipwrecks can be seen at low water. These wrecks appear to be wrecks of barges or small steamboats. No excavation has been completed and what can be seen of the vessels during low water episodes changes over time. The main local supposition for these wrecks being in their present location has been that they were used as storage/work boats for the Missouri River Bridge, just upstream. In view of the discussion of wharf-boats and quarter-boats, these wrecks may actually have been directly associated with the busy steamboat landing also immediately upstream of their resting place. Of course, these could also be barges or small steamboats that sank along the landing and settled in this location after a short drift. Consideration of the various recycled uses of old steamboat and barge hulls should always be considered for any wreck found in a river, especially near a landing.

The identification of riverine support structures and ancillary littoral nodes can be difficult. Often, the connection between a terrestrial, littoral site is not considered for its ties to riverine trade. Few projects such as Ford's (2009) dissertation specifically look into this connection. Steamboat and other watercraft wrecks are significant, among other things, because of their time-capsule quality. They can depict a moment in time, complete with all the trimmings of artifacts and nautical architectural advances. The shore-based support system, however, can reveal a long and much more detailed record of riverine trade both in its archaeology and historical documentation than only the consideration of a wreck. Consider the construction of a vessel and the support structures needed for lumber, metalwork, and workspace. Next, consider the finished vessel and the goods/people it will carry. Where were those goods stored before shipping? Where were people housed until the vessel arrived? After the vessel arrived, how did the cargo get to shore and where did they go? Were wharves and docks necessary? Were erosion control devices necessary to keep landings and wharves intact? What structures and buildings were constructed to make profit from the cargo, both goods and passengers? There are many more ancillary sites to consider as well. Historic archaeologists know the necessity of merging historic documentation with fieldwork. Historic period sites such as farmsteads, industrial buildings, parks, or modified landscapes which are located along a river are more than likely directly associated with the riverine trade. Much more of the United States' hidden maritime history can be found through a careful consideration of these associations.

References

Baldwin, L. D. (1941). *The keelboat age on the western waters*. Pittsburgh, PA: University of Pittsburgh Press.

Bauer, K. J. (1988). *A maritime history of the United States: The role of America's seas and waterways*. Columbia, South America: University of South Carolina Press.

Booth, D. L., & Hajic, E. (2010). *Addendum to cultural resource survey: Carondelet Coke Plant, St. Louis, Missouri*. Report. Jefferson City, MO: Missouri State Historic Preservation Office.

Bradley, L. E. (1993). Government Ice Harbors on the Upper Missouri. *North Dakota History: Journal of the Northern Plains, 60*(3), 28–37.
Burdick, U. (1881). *Reports of the Mississippi River Commission, 1881*. Washington, DC: US Government Printing Office.
Cantelas, F. J., & Bradley, A. R. (1994). The Maple Leaf: A case study of cost-effective zero-visibility riverine archaeology. *International Journal of Nautical Archaeology, 23*(4), 271–282.
Corbin, A. (2006). *The life and times of the steamboat Red Cloud or how merchants, mounties, and the Missouri transformed the West*. College Station: Texas A&M University Press.
Corbin, A., & Rodgers, B. A. (2004). *Leviathan of the plains: History, excavation, and architecture of the steamboat Montana*. Report. Jefferson City: Missouri State Historic Preservation Office.
Corbin, A., & Rodgers, B. A. (2008). *The steamboat Montana and the opening of the West: History, excavation and architecture*. Gainesville: University of Florida Press.
Crick, M. (1989). Representations of international tourism in the social sciences: Sun, sex, sights, savings, and servility. *Annual Review of Anthropology, 18*, 307–344.
Crisman, K. (2005). The heroine of the Red River. *The INA Quarterly, 32*(2), 3–10.
Custer, J. E., & Sandra, M. C. (1997). *An investigation of submerged historic properties in the Upper Mississippi River and the Illinois Waterway*. Rock Island, IL: US Army Corps of Engineers, Rock Island District.
Dasovich, S. J. (1993). Letter to the editor. *North Dakota History: Journal of the Northern Plains, 60*(4), 40.
Dasovich, S. J. (2009). Phase one archaeological survey: Carondelet Coke 12-Acre Addition Addendum, St. Louis County, Missouri. Report. Jefferson City: Missouri State Historic Preservation Office.
Dasovich, S. J., & Warner, K. A. (2004). Phase two national register of historic places eligibility testing: Selma Farm Levee, Sites 23JE813 and 23JE814, Jefferson County, Missouri. Report. Jefferson City: Missouri State Historic Preservation Office.
Ford, B. L. (2009). *Lake Ontario Maritime cultural landscape*. Doctoral dissertation, Department of Anthropology, Texas A&M University, College Station.
Gandy, J. W. (1987). *The Mississippi steamboat era in historic photographs: Natchez to New Orleans, 1870–1920*. New York: Dover.
Giffen, L. E. (2001). *Walks in water: The impact of steamboating on the lower Missouri River*. Jefferson City, MO: Giffen Enterprises.
Gregory, R. (1981). *The German-Americans in the Washington, Missouri area*. Report. Washington, MO: Missouri Historical Society.
Hall, H. (1884). *Ship-building industry in the United States*. Washington, DC: Department of the Interior, US Government Printing Office.
Hawley, G. (1998). *Treasure in a cornfield: The discovery and excavation of the steamboat Arabia*. Independence, MO: Paddle Wheel.
Holland, K. V., Manley, L. B., & Towart, J. W. (Eds.). (1993). *The Maple Leaf: An extraordinary American Civil War shipwreck*. Jacksonville, FL: St. Johns Archaeological Expeditions.
Hunter, L. C. (1993). *Steamboats on the Western rivers: An economic and technological history*. New York: Dover.
Lass, W. E. (2008). *Navigating the Missouri: Steamboating on nature's highway, 1819–1935*. Norman, OK: The Arthur H. Clark Company.
Lepley, J. G. (2001). *Packets to Paradise: Steamboating to Fort Benton*. Missoula, MT: Pictorial Histories.
Lindsay, T. J. (1987). *Captain Gougar and his steamboats: Excursions on the Kankakee River*. Kankakee, IL: Lindsay.
Lingenfelter, R. E. (1978). *Steamboats on the Colorado River, 1852–1916*. Tucson: University of Arizona Press.

McDonald, M. R. (2011). Wharves and waterfront retaining structures as vernacular architecture. *Historical Archaeology, 45*(2), 42–67.

Meyer, D. J. (1967). *Excursion steamboating on the Mississippi with Streckfus Steamers, Inc.* Master's thesis, Saint Louis University.

Moore, E. W. (1958). *Natchez Under-The-Hill*. Natchez, MS: Southern Historical.

Petersen, W. J. (1970). The Diamond Jo Line. *The Palimpsest, LI*(4).

Petsche, J. E. (1974). *The Steamboat Bertrand: History, excavation, and architecture*. Washington, DC: National Park Service and U.S. Department of the Interior.

Popular Mechanics. (1910). *Popular mechanics: An illustrated weekly review*. Vol. 14(5), p. 718.

Russell, M. A., Murphy, L. E., & Bradford, J. E. (2010). Maritime archeology of tourism in Yellowstone National Park. In A. Corbin & M. Russell (Eds.), *Historical archeology of tourism in Yellowstone National Park*. New York: Springer.

Sevier, R. P. (2010). Mississippi River 1866 Memphis to White River. The Mississippi Digital Map Library. Retrieved December 9, 2010, from http://usgwarchives.net/maps/mississippi/

Shrader, D. H. (1993). *Steamboat legacy: The life & times of a steamboat family*. Hermann, MO: Wein Press.

Washingtonmo.com. (2009). The history of Washington, Missouri. Retrieved June 18, 2010, from http://washingtonmo.com/history/

Chapter 14
The Mini-Fleet of Emerald Bay: Recreational Vernacular Watercraft

John W. Foster, Charles Beeker, Deborah Marx, and Sheli O. Smith

Introduction

Tourism played, and continues to play, a large role in the evolution and use of vernacular watercraft. Today, paddleboards, kayaks, skiffs, dinghies, and canoes are all present in recreational sports. Originally, the same small watercraft served very different roles in the survival and networking of historic peoples. The role of small vernacular watercraft, however, is not a recent phenomenon. They were also present in historic tourism related to water-oriented resorts. Historic antecedents of today's recreational crafts are important links to their non-recreational origins. The historic versions help explain a number of this boat genre's evolutions.

The nineteenth-century resort on Emerald Bay in alpine Lake Tahoe is a good example of the array of small vernacular watercraft incorporated into recreation (Fig. 14.1). Emerald Bay, situated on the western side of Lake Tahoe, in the Sierra Nevada Mountains on the border of California and Nevada, is and has been an American tourist destination for well over 150 years. The fjord-like configuration of

J.W. Foster (✉)
California State Parks, 8654 Amber Oaks Court, Fain Oaks, CA 95628, USA
e-mail: parkarky@yahoo.com

C. Beeker
Center for Underwater Science, Indiana University, Bloomington, IN 47405, USA
e-mail: cbeeker@indiana.edu

D. Marx
Stellwagen Bank National Marine Sanctuary,
175 Edward Foster Road, Scituate, MA 02066, USA
e-mail: deborah.marx@noaa.gov

S.O. Smith
PAST Innovation Lab, 1003 Kinnear Road, Columbus, OH 43212, USA
e-mail: sheli@pastfoundation.org

A.M. Evans (ed.), *The Archaeology of Vernacular Watercraft*,
When the Land Meets the Sea, DOI 10.1007/978-1-4939-3563-5_14

Fig. 14.1 Emerald Bay is located on the western shore of Lake Tahoe on the California Nevada border. The 1894 El Dorado County Map delineates the owners of the property around the scenic bay

the deep bay with its surrounding alpine forests and crystal blue water has beckoned visitors for centuries. Ample evidence of Native American habitation surrounds the narrow bay. Letters and journal entries reveal that as soon as the lake was discovered by westward moving immigrants, people began to visit Emerald Bay specifically for recreation (Nesbitt 1989). Documentary evidence of real estate transactions over the last century and a half detail the growth of the tourism industry in the area, specifically the Emerald Bay Resort/Camp, as well as the importance of the vernacular craft to the area.

Rise and Fall of the Emerald Bay Resort/Camp

In 1844, John Freemont charted Lake Tahoe and began the area's constant inhabitation by Europeans. The Lake became a haven of the wealthy for relaxation and recreation. Thus began the summer resort business, which in turn supported the working class who provided services for the vacationers and travelers. In 1865, the first recorded property transaction took place in the Emerald Bay Resort/Camp area. Records indicate a transfer of property from John L. Eckley to J.C. Tucker on 20 September 1865; the property reverted back to Eckley a month later for unknown reasons (Nesbitt 1989: 6).

Three years later, on 15 August 1868, Eckley sold his property to stagecoach magnate Ben Holladay Jr., nicknamed "Croesus of the Coast" and "Napoleon of the Plains." The property included Fanette Island and the land that wraps around Emerald Bay. Holladay built a Victorian Gothic Revival two-story, five-room summerhouse (known as the "Cottage"), and it was the first private house on Lake Tahoe. The continued importance of water transport on the lake is reflected by the pier, boathouse, small house on Fanette Island, and another small house near the shore for his caretaker, Richard Barter ("the Hermit"). Holladay suffered economic hardships and the government seized his property due to his debt (Nesbitt 1989: 6–7). In 1880, a sheriff's sale sold the Holladay property, totaling 500 acres, to Dr. Paul T. Kirby, of Virginia City. By 1884 Kirby and his wife Lucy built a summer resort (where Vikingsholm is today), which included a hotel, cottages, tents, and a steamer landing (Fig. 14.2). The resort was named the Emerald Bay Resort and prospered during the second Comstock load (Nesbitt et al. 1990: 8–10). Paul Kirby died in 1888 leaving the property and the newly opened resort to his wife, Lucy Kirby, who continued to expand the business. On 17 December 1888, Lucy Kirby established the Lake Tahoe Emerald Bay Post Office and became its first Postmaster (Mountain Democrat 2000).

Lucy Kirby's Emerald Bay Resort became a well-known vacationing spot that attracted the *Sacramento Daily Union*'s eye in 1888 and again in 1890. In both newspaper mentions, boats and water sports are highlighted as resort amenities (Sacramento Daily Record-Union 1889; Sacramento Daily Union 1890).

> … on the western extremity is located L. N. Kirby's Emerald Bay Resort. The waters, protected as they are from the wind, renders boating and fishing always a safe and pleasant pastime and makes the place a favorite with the ladies. Boats are free to guests, and a fine yacht the *Mollie Brown*, is at the service of pleasure parties.
>
> At the western extremity… Mrs. Lucy N. Kirby has located what is now popularly known as the Emerald Bay Resort. Here is a small hotel, with a large parlor and dining room, surrounded by numerous nicely furnished cottages. Boats and fishing tackle are free to guests, and these with swings, hammocks, croquet, and climbing, furnish an abundance of outdoor exercise.

Lucy Kirby and her second husband, Russell Cowles Graves, continued to operate the resort until Lucy started selling off parts of her property in 1892. Graves continued to sell off the property to numerous people and in 1896 Lucy Kirby sold her remaining property to her husband. At this time, it is unclear whether Lucy got

Fig. 14.2 Early Emerald Bay Resort with Steamer Landing (Photo by George Oliver published in *The Saga of Lake Tahoe*, 1957)

divorced or died because records indicate that a few years later Russell Cowles Graves deeded his property to his new wife, Margaret L. Graves (Nesbitt et al. 1990: 10, 14).

In 1907 Russell and Margaret Graves began construction of the Emerald Bay Resort/Camp on the northwest shore of Emerald Bay; the location of the Graves' resort was not the same as the Kirby's original Emerald Bay Resort, which was located where Vikingsholm is today. The Emerald Bay Resort/Camp prospered during this period. A highlight of any visitors stay at the resort was an excursion onboard the steamer *Tahoe*, which docked at the resort's own pier (Scott 1957: 129). In 1913 a highway around Emerald Bay, originally called the Meyers-McKinney Road (Highway 89 today) was completed that provided increased accessibility to the Emerald Bay Resort/Camp.

For over 70 years, the only access to the land comprising Emerald Bay Resort was by water. Following Nelson Salter's purchase of 30 acres from Graves in 1914, with the specific intent of expanding the newly opened Emerald Bay Resort/Camp, the various private property owners agreed to construct a road that connected their resort to the new highway that belted the entire lake. The road was completed in 1916. At that time, a survey map indicated that Salter's resort included cottages, tents, an express depot, post office, dance pavilion, and steamer landing. Improvements to the property included the addition of two and three room cottages, additional tents, and a butcher shop enhancing the existing dance pavilion and steamer landing (Scott 1957: 131). Salter's frontage along Emerald Bay eventually grew to 1000 ft. Under Salter's ownership the Emerald Bay Resort/Camp saw increased business, and the resort's advertising brochures brought people from all over the United States (Lake Tahoe Camp Brochure nd).

> The camp is electrically lighted throughout, and accommodations[*sic*] 150 and is equipped with cozy cottages, with or without bath and carpeted board floor tents varying in size for the accommodation of one or many. It consists of an office, dining room and kitchen, store, cottages, tents, bathroom and lavatories. Each tent is supplied with a double iron bedstead, bedsprings, top mattress, dresser, pillows, blankets, clean linen, wash stand, chairs, and crockery as completely as a hotel.

The Emerald Bay Resort/Camp supplied the latest advancements in technology such as long distance telephone service, telegraph, daily mail, and express service. Salter operated a 40-ft long launch equipped with a 20 hp engine and electric lighting for excursions around the lake day and night. He also provided guests with rowboats and outboards for their leisure activities around Emerald Bay. Other sources of amusement at the resort included bathing, fishing, mountain climbing, plane rides, movies, bingo, campfires, dancing, tennis, horseback riding, picnics, and croquet (Lake Tahoe Camp Brochure nd).

In 1947, Salter sold the Emerald Bay Resort/Camp to Joseph Watson who in turn sold the property to the State of California in 1953 and 1954. Watson leased back some of the land from the state and continued to operate the Emerald Bay Resort/Camp for a few more years, but with fewer amenities than during Salter's ownership. In 1957, the State of California, who owned the property, removed the furniture from the Emerald Resort/Camp and took it to Sacramento for auction. A Department of Commerce map from 1957 recorded the existence of the Emerald Bay Resort/Camp on the map, however 2 years later, in 1959, the resort officially closed. In keeping with the trends in parks at the time, the resort buildings were removed to return the area to a natural look and the remaining boats were sunk in place (Scott 1957: 131).

Remnants of the Emerald Bay Resort/Camp

In 1953, Thomas G. Mapel, C. Bernard White Jr., and Walter F. Wilmette completed an inventory of the Emerald Bay Resort/Camp's structures for a property appraisal. The results of this inventory give an excellent idea of the contents of the resort prior to its destruction by the state 6 years later. A list of buildings and structures included two jetties (Nesbitt et al. 1990: 47–48). No boats or watercraft were mentioned, yet few pictures of the resort exist that don't have recreational craft shown (Fig. 14.3).

In 1961, the State of California approved plans for a boat-in campground on the site of the former Emerald Bay Resort/Camp and construction began a year later (California Parks and Recreation: Emerald Bay Unit nd).

Twenty-eight years later, the California Parks and Recreation Dive Unit, made up of archaeologists and biologists, began to systematically study the collection of small watercraft located in the shallow shoreline waters near the historic Emerald Bay Resort. Although the resort and pier had long since been demolished and

Fig. 14.3 Emerald Bay Resort with lots of recreational watercraft in the photo (Courtesy of D. Marx)

replaced with a boat camp and new recreation pier, the cold, fresh water of Emerald Bay protected the sunken boats. Throughout the 1990s and into the first decade of the twenty-first century, California State Parks with the assistance of Indiana University located and mapped 11 small watercraft representing six styles. They range in style from small sailboat to kayak, and in construction materials from metal hull to wooden hulls. Some exhibit features denoting they were motorized or at minimum carried an outboard engine, while others were obviously only rowed. Other distinctive features, such as live-bait wells, denote the primary recreational use of the vessel and shed light on the common features of small watercraft used in everyday life. Although some of these features are no longer used today, such as bait wells, they explain how boatwrights creatively solved problems such as keeping bait alive in a boat for extended periods of time (Fig. 14.4).

The small boats reveal attributes of naval architecture in their simplest form. The Emerald Bay "Mini-Fleet" as the collection is called, displays curved hulls, hulls with hard chines, square transoms and hourglass transoms, interior stringer systems, and both curved and plumb bows. The Mini-Fleet also reveals the prevalent employment of secondary use. Among the vessels is a lifeboat, possibly off one of the steamers that regularly called at Emerald Bay Resort. Pictorial evidence captures the sailboat in use for recreation but it was also large enough to carry supplies necessary to run the resort or ferry garbage to the opposite side of the bay where there is substantial evidence of it being dumped near two abandoned barges.

In each instance of the small vessels, there is evidence that the boats were sunk at their mooring. It has been suggested by the historian Nesbitt that this happened in

Fig. 14.4 Live bait wells were a great solution on the relatively calm waters of the lake (Photo by John Foster)

or around the time that California Parks took over the property. This would explain why the majority of boats are aligned along a similar bathymetric depth, where they would have been away from shore yet easily retrieved for use by resort staff or visitors. None of the boats showed any sign of registry, but several had evidence of numbers on their stern. Sport divers in the area recalled taking brass numbers off the sterns of boats located in close proximity to the modern day boat camp (James Duff 1990, pers. comm.).

Over the years that the Emerald Bay Resort operated, many small craft of varying sizes and shapes plied the waters of the bay. Resort guests used the small boats either for pleasure cruising around the protected waters of the bay or for fishing. Numerous historic photographs show the boats moored in front of the resort and in use by the guests (Marx 2004). The collection of styles reflects the wide array of small boats on Lake Tahoe from the earliest times. The older boats of the Mini-Fleet are wooden. Many of the styles were later produced in metal, or a combination of metal and wood. Most likely the majority were built on Lake Tahoe. Since timber in the area was plentiful, the presence of metal boats represents modern trends away from traditional wooden boat building but not away from the traditional styles. Small boats are a tradition on Lake Tahoe that dates back into prehistoric times. Obexer's, one of the oldest chandleries and boat building establishments on the Lake, is still building boats, and today Lake Tahoe boasts one of the largest fleets of classic wooden speed boats in the world (Van Etten 1992: 105) (Fig. 14.5).

Fig. 14.5 Lloyd Garrison built fishing boats for Tahoe Tavern between 1917 and 1921. Note the large, single plank sides. Lumber was plentiful (Courtesy of William Garrison)

The following sections describe the types of architecture exhibited within the Mini-Fleet, revealing six styles of watercraft in use at Emerald Bay Resort. Pictorial evidence indicates that the hard chine skiffs were popular from the early days of the resort and remained popular until the end of the resort (Table 14.1; Fig. 14.6).

EBMF01 Fishing Boat with Bait Well

Fishing made up a great deal of the leisure boating on Lake Tahoe. Historic images show lots of small fishing boats either pulled up onto the shore or moored just off shore in the sandy shallows of the Lake's edge.

The EBMF01 is a small, carvel built fishing boat with an hourglass transom and rocker keel. Twenty-nine small ribs are regularly spaced along the inside of the hull. Two, centrally located, thwarts provided lateral strength to the small hull as well as a convenient seat for passengers. There is a clamp along the inside of the hull above the ribs. Three stringers ride close together over the ribs and just to both sides of the keel providing longitudinal strength for the small boat. A 17 in. breast board added strength to the bow.

The bait well is centrally located between 6 and 9 ft. and the ribs are truncated on both of its sides. The walls of the bait well rise vertically, to within 5 in. of the top of the hull, ending just underneath the thwart. Random spaced holes pierce the hull inside the bait well along both sides of the keel letting water into the bait well, where live bait could have been stored in the well. The 10 in. height of the bait well

Table 14.1 Scantlings of Emerald Bay Mini-Fleet vessels

	Scantlings										
Site No.	LOA[a]	Breadth[a]	Depth[a]	Chine	Stem	Transom	Hull material	No. of ribs	Propulsion	No. of thwarts	No. of stern seats
EBMF01	15.6	4.18	1.84	Hard	Straight cant	Wineglass	Wood	29	Outboard motor	2	2
EBMF02	13.36	1	1	Hard	Straight cant	Double-ended kayak	Metal	?	Paddled	1	0
EBMF03	13.108	4.24	1.96	Round	Curved	Y-back	Metal/wood	17	Rowed	2	0
EBMF04	15.12	4	1.108	Hard	Straight cant	Trapezoid	Wood	11	Rowed	1	1
EBMF05	15	4.396	1.96	Round	Straight cant	Round	Wood	37	Inboard/Outboard motor	0	1
EBMF06	17.96	4	1.96	Round	Curved	Round	Wood	14	Sail	0	0

[a]Measurements in tenths of feet

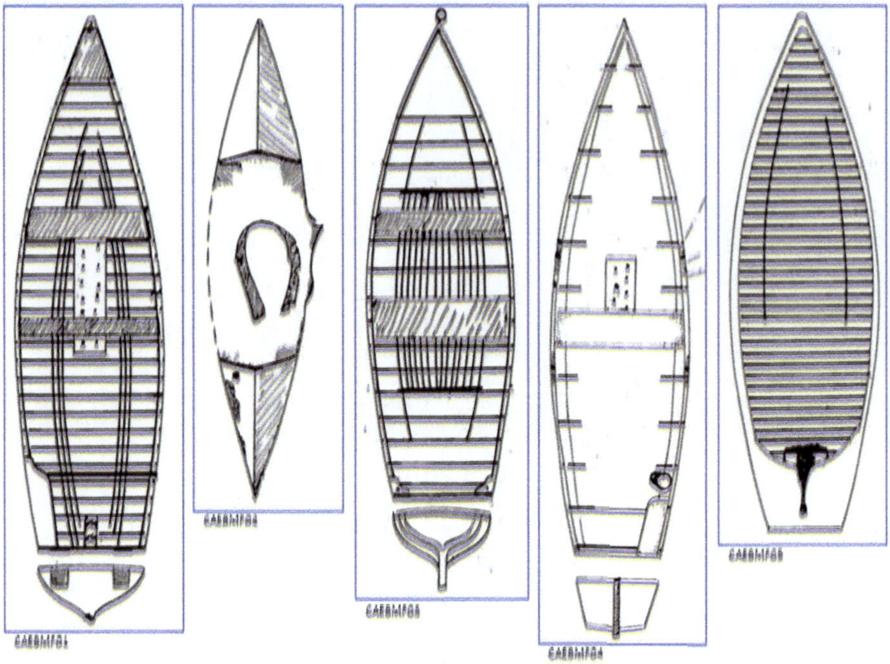

Fig. 14.6 Five of the styles were rowed, paddled, or motorized (Drawing by S. Smith)

reflects the draft of the boat. When loaded with people and gear, the boat drew less than 10 in. of water.

Two batter boards are affixed to the outside of the transom for mounting an outboard engine. A small bench mounted inside along the port aft quarter indicates the position of the driver and most likely the preferred side of mounting the outboard engine. Just forward of the transom a thin bulkhead is still present. Presumably, the bulkhead kept the outboard engine splash that sloshed over the transom from running into the amidships and swamping the boat. A sturdy knee runs out 11 in. along the keel and 12 in. up the transom for additional strength.

The boat shows little deterioration. The top starboard rail is ragged possibly reflecting where modern mooring lines have rubbed across it wearing away the softer waterlogged wood. However, the hull has not broken open and is still firm to the touch underwater.

EBMF02 Kayak

Kayaks were introduced during the early years of Mission California when Eskimos were brought to California with their skin boats to assist in sealing.

Historic images reflect how popular the small craft were on Lake Tahoe. The Tahoe kayak has a traditional double-ended shape, but it is covered with a thin metal instead of skins and had either a canvas or wooden center cover. The center covering on EMBF02 is gone except for the metal band that fit around the opening for the paddler. Historic images of kayaks at Emerald Bay show both types of kayaks.

The peaked centerline along the top of both ends allowed water to roll off the boat when small waves broke over the bow or stern. The sides and bottom of the boat are sharply angled reflecting the process of wrapping a thin covering over a rigid interior frame. The Tahoe kayak is silted-in so its interior structure is not readily discernible. General kayak construction, however, has a series of six stringers that run fore and aft from end to end of the craft giving the boat distinctive edges at the centerline, top, and bottom, and along the top and bottom sides. Full circle frames are set systematically throughout the length of the boat providing lateral strength to the double-ended shape.

EBMF03 Y-back Rowboat

The Y-back metal hull rowing boat represents a later era in leisure boating on Lake Tahoe. Early rowboats were made of wood, but metal boats were well suited to the freshwater lake and cheaper to construct by the middle decades of the twentieth century. Metal hulls are still popular for fishing boats on the Lake.

The Emerald Bay Y-back represents the transitional period when wood and metal were combined. Although the hull on this particular rowboat is metal, all other structural pieces are made of wood. The rowboat sports a wooden, hourglass transom with a distinctive, raised wood "Y" attached on the exterior, thus the name, Y-back. Whether the "Y" is a decoration or structural addition is unclear. Two wooden thwarts that doubled as seats are located centrally just under the caprail. Along the bottom of the hull from just forward of the first thwart running aft to just behind the second thwart are a series of closely spaced one-inch square stringers that ride on top of the frames. These footboards protected the thin metal skin in the bottom of the boat from puncture by gear or shoes while the open work construction allowed any water in the bottom of the boat to move freely.

In the bow and stern are small wooden knees that lend structural support to major junctures of the hull. The three knees are also drilled, but it is unclear what the holes could have been used for other than mooring lines. It is possible that the rowboat could be sailed and that the holes represent standing rigging placement. The sailboat (EBMF06) has similar knees in its stern and bow that are attributed to standing rigging attachment.

The Y-back is designated as a rowboat primarily because there is no evidence for an inboard engine or batter board for an outboard engine mount nor are the indications for sailing strong enough to place the boat in that category. It appears that oarlocks were once locked between the two thwarts, but they are no longer present.

Fig. 14.7 A 4-in. diameter drainpipe leans against the hull just forward of the stern seat. The weight of the pipe would have been ample to hold fishing poles out of the way of the boat occupants

The metal of the hull still shows traces of green and white paint along the upper edges, but the metal is beginning to deteriorate and in the bow of the boat there is a large hole on the lower portside.

EBMF04 Hard Chine Fishing Boat

The simplicity of style is undoubtedly what makes the hard chine fishing boat not only popular historically but still popular today. The historic visual record is peppered with examples of relatively easy to construct hard chine fishing boats (Fig. 14.7).

This style of hull does not require traditional rounded frames but only a minimal number of wide, thin planks to create the hull—usually two on the bottom and one or two on the sides. Eleven pairs of angled knees that tie the sides of the boat to the bottom planks make up the ribs of this boat style. Unlike in the rounded hull boat with bait, the ribs of the hard chine fishing boat do not extend across the bottom of the hull. Instead, it is the thwart/seat and a thin thwart near the stern that primarily keep the hull together. A single thwart provides the majority of the lateral strength in the middle of the hull's length. Two sets of oarlock cleats forward and aft of the thwart reflect the ability to row forward or backward.

Like rounded hull fishing boat EBMF01, the hard chine fishing boat also has a centrally located, live-bait well. The 2-ft long bait well begins forward of the thwart and extends under the thwart to its aft edge. This bait well is not as finished as the one in EBMF01; the top edge of the box is flat and the box is forward of the seat instead of straddling the two thwarts.

Along the stern starboard quarter of the boat there is a small seat that runs fore-and-aft and a thin board doubling as support for one end of the seat and lateral support for the hull in the stern. A ceramic pipe, four in. in diameter resembling the common drainpipe, is leaning against the interior side of the hull just forward of the seat. The positioning of the pipe suggests that it was used to hold fishing poles. Small thin boards are affixed to the floor of the hull but their purpose is not clear, and there is no alignment or other indicators that suggest an inboard engine.

The transom of the fishing boat is angled like the hull. A 2 by 2 in. sternpost is attached to the center axis of the transom. There are no batter boards on the transom which may indicate that outboards were not attached to the boat, but the starboard seat in the stern suggests that at some point an outboard was set up on the boat and the seat provided a convenient place for the helmsman to sit.

EBMF05 Lapstrake Motorboat

The lapstrake motorboat has the most complex hull of the miniature fleet. Lapstrake hulls were common around Lake Tahoe, but it was not until the first quarter of the twentieth century that inboard engines began to appear on Lake Tahoe. The lapstrake hull was stronger and thus could accommodate the weight and vibrations of small one cylinder engines placed either near the stern or centrally in the larger launches. These boats did not go fast but they could easily pull numerous smaller rowboats behind them.

The boat has a plumb stem (Fig. 14.8). The keel of the boat rocks up toward the stern providing a full body with narrowing stern. This hull shape makes it easier for the boat to push through the water and keeps the boat from throwing a large wake. A caprail runs the entire length of each side with a 2 ft. long, rub or batter board along the outer starboard side of the hull. Thirty-seven ribs evenly spaced provide lateral strength along the hull and each is notched appropriately to accept the lapping hull strakes. The boat has a hemisphere transom. The stem has a metal protective band running up along the wooden stem's outer edge. A metal ring for mooring is attached to the bow.

Symmetrically placed to either side of the central axis on the transom and above the water line are two small portholes. The portside hole is patched with a square piece of metal, but the starboard hole is still open. Since there were no other holes lower in the hull, it appears that originally one port provided access for the propeller shaft while the other provided an exhaust port. Possibly, a later refit or new engine supplanted the need for a lateral exhaust port.

The stern is partially covered but has cutouts that most likely provided access for the tiller and rudder. Just inside the stern sits the engine block. The base of the engine block follows the curvature of the hull while the upper face is cut out to accept the engine and keep it from shifting with the vibrations of the running motor. The curvature of the stern cover reflects the splashboard that once graced the edge of the cockpit. Launches like this had an oval cockpit with a splashguard defining the perimeter. The board is gone but the cut of the stern cover defines it.

Fig. 14.8 The plumb bow of the motorboat, sports a protective metal bumper band and a one and a half inch metal mooring ring

The boat still bears marks of numerous coats of paint on the outer hull. It seems that at one time the hull was painted green; however, the final coat of paint is white. White paint seems to have been the universal color of the Emerald Bay miniature fleet as the final color of the hull. The hull is in excellent condition considering it sits within the modern mooring field of the boat camp.

EBMF06 Sailboat

It is most likely that many boats plying the waters of Tahoe used collapsible masts and sails. Once outside Emerald Bay, sails would have been extremely useful in cutting down traveling time up and down the Lake. Numerous historic images show lapstrake-built boats with forward placed masts sporting a fore-and-aft sail.

The miniature fleet sailboat is located quite a distance from the other boats and has the most damaged hull. At some time in the recent past, a large tree trunk fell onto the submerged hull and broke off the first four feet of the bow as well as laid open the hull splitting it along the keel centerline. The transom is semi-intact. The hull is of lapstrake design with a plumb stem and hemisphere stern (Fig. 14.9).

At first glance, the sailboat is similar to the lapstrake motorboat (EBMF05) although slightly longer. The difference, best illustrated by the bow, is mainly in the way the hull has been reinforced to support wind power rather than engine propulsion. The bow was broken off and lay upslope of the main hull section, but it is intact with fore-and-aft, inch-thick planking that runs aft 4 ft. from the prow. A breast knee ties the two sides of the boat together and 2, 4 in. thick deck beams lend additional support. The second deck beam is cut to accept the 1.5 in. thick deck planks.

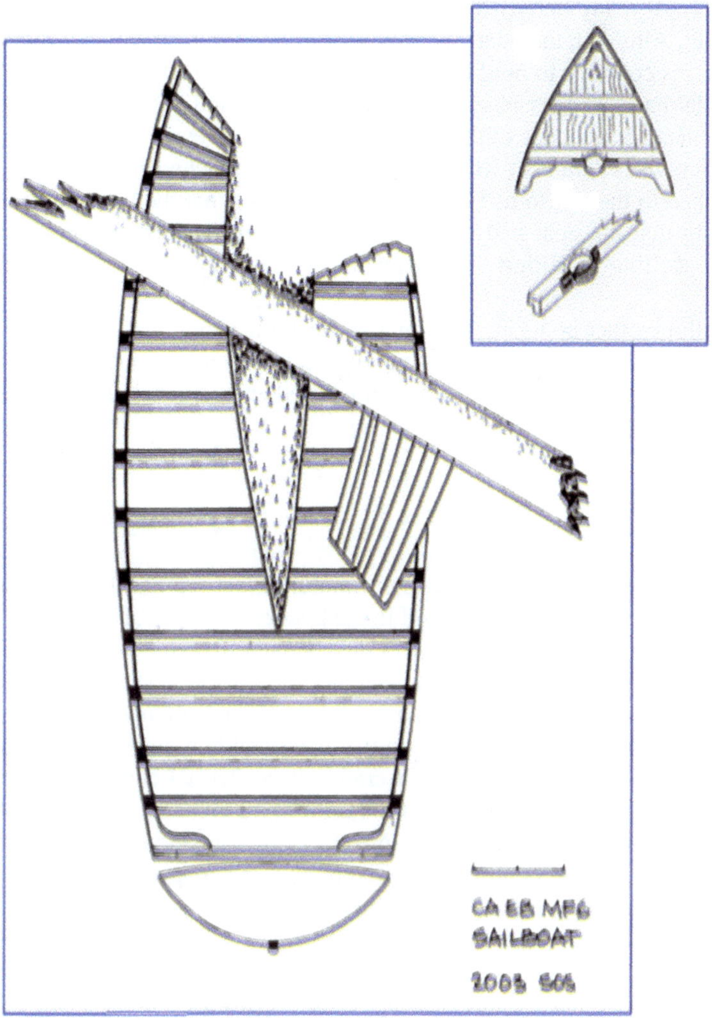

Fig. 14.9 Only one of the watercraft was a sailboat (Drawings by S. Smith)

Central in the span of the aft deck beam is a 5 in. hemisphere cutaway. A matching band of metal forms the second half of the circle. Together the deck beam and metal band form a mast partner. Nails tacked to the top of the deck beam reflect where a collar once sat around the mast and suggests that the mast was rarely lowered. Two strong lodge knees finish off the forward deck, adding additional support to the mast partner beam.

The forward edge of the lapstrake hull reveals a plumb bow but the stem is no longer attached or perhaps it is buried under the bow fragment upslope. The main hull has sustained the most damage and no longer holds the hull's curvature.

There are still many details in the main hull and the stern, although damaged, is semi-intact. Although the boat has been split in two for the majority of its 17 ft. and 8 in., the ribs continue to hold the lapstrakes of the hull together.

The ribs have wedges inserted between the hull and the frame so that they fit tightly to the uneven shape of the lapping hull planks. The caprail along the top edge of the hull reveals details of where cleats were once attached. The oarlock cleats are still attached to the transom. Their presence suggests that the boat could be sculled or possibly that the boat had two, small, tandem, detachable rudders. Having the choice of sculling or sailing would have made the boat versatile and useable in all weather.

Like all the other boats in the miniature fleet, the sailor is also painted white. Its significant distance from the other small boats, however may suggest that the sailor was in use longer than the rest of the miniature fleet and not sunk at the same time.

Conclusion

The boats of the Emerald Bay Resort Miniature Fleet represent a cross-section of small boats from an important period in the history of Lake Tahoe in particular and American leisure boating in general. Often passed over for larger more complex vessels, the small craft that plied the waters of Lake Tahoe shed light on daily life and the important resort nature of many of Tahoe's communities. The fleet reflects how people spent their leisure time on the water and how important fishing was to leisure pursuits. Moreover, the small craft reflect gradual changes in technology. The boats range in construction materials from wood to metal and vary in style from simple to complex. The collection also reflects the transition from human through wind to motor-powered boating. In all likelihood, the boats date back to the golden years of Emerald Bay's Resort in first quarter of the twentieth century.

The fleet is an assemblage of variously powered vessels: the paddled metal kayak (EBMF02); the wind powered day sailor (EBMF06); the originally rowed but later outboard powered hard chine fishing boat (EBMF04); the outboard powered wooden fishing boat with the live bait well (EBMF01); and the inboard powered lapstrake motorboat (EBMF05).

In addition to representing a spectrum of propulsion, the 6 craft illustrate a variety of styles—the ubiquitous, simple hard chine fishing boat, a more complex carvel hull fishing boat, two lapstrake hulls, and a kayak—ranging from the simplest to the very complex. Such range also sheds light on the various occupations of people living in the communities around Lake Tahoe. While a hard chine fishing boat would be easy to construct, a lapstrake boat would have required advanced boat building skills. The choice of the different hull types represents the styles best suited for the short chop waters of Lake Tahoe. It is also important to note that only the lapstrake boat had an inboard. In reviewing the historical images, the majority of launches with inboard engines are lapstrake construction (Van Etten 1992). Early, one cylinder engines vibrated substantially and thus had to have robust engine blocks to sit upon and the hulls of the boats carrying the engines had to be able to withstand the continued vibrations of the engines.

The six small boats of the Mini-Fleet are constructed from a variety of woods and metals, all well preserved in the fresh waters of Lake Tahoe. The different soft- and hardwoods, the metals, and fasteners testify to the availability of products among Tahoe communities and illustrate the trade network that linked Tahoe with Reno and Sacramento. There is also an example of a transitional construction type boat, EBMF03. The Y-back rowboat has a metal hull but all other structural components are wooden, including the elaborate Y-back transom. The change from building all wood boats is often attributed to the shortage of wood product and in the case of large vessels this was undoubtedly the primary reason, but in the case of small craft the shift from one wood to metal is more likely a reflection of technology trends. After World War I, boat building began to shift culminating in the introduction of fiberglass hulls. Since none of the miniature fleet is a fiberglass hull, it is reasonable to conclude that all the vessels were constructed prior to World War II.

Individually, the boats of the miniature fleet are interesting but their historic value increase significantly as a combined collection deserving study and preservation. The Emerald Bay Resort Miniature Fleet and the two barges across the bay represent 90 % of the boat styles used for leisure and work on Lake Tahoe in the later part of the nineteenth century and first half of the twentieth century. Thus, the sunken fleet of Emerald Bay is truly a Californian treasure.

References

Marx, D. (2004). *Emerald Bay State Park Boat Camp Shoreline survey 2003*. Bloomington: Underwater Science Program, Indiana University.

Mountain Democrat. (2000, May 5). *Mountain Democrat*, Vol. 149.

Nesbitt, P. E. (1989). *History of Emerald Bay State Park*. Report. California Department of Parks and Recreation.

Nesbitt, P. E., Evans, N. H., & Kelly, J. L. (1990). *The cultural resources of Emerald Bay State Park*. Report. State of California Department of Parks and Recreation, Cultural Resources Planning Division.

Sacramento Daily Record-Union. (1889, July 6). *Sacramento Daily Record-Union*, Vol. 61(114).

Sacramento Daily Union. (1890, June 15). *Sacramento Daily Union*, Vol. 2(5).

Scott, E. B. (1957). *The saga of Lake Tahoe* (Vol. 1). Lake Tahoe, NV: Sierra.

Van Etten, C. (1992). *Lakers & launches*. Pasadena, CA: Welsh Graphics.

Chapter 15
Getting a Handle on the Georgia Strait Handliner, a Vernacular Boat from British Columbia

Charles D. Moore

Introduction

Handlining, or hand-trolling, for salmon was a viable, if marginal, commercial fishery in the Georgia Strait for the first 50 years of the twentieth century. The method was simple. One or two hooks with bait or lures were trolled through the water in the hopes of hooking a salmon. Though perhaps weighing 50 lb (25 kg) or more, these hard fighting fish were then landed by hand. The one-person watercraft were sometimes dugout canoes, sometimes flat-bottomed skiffs, but the preferred craft, at least by the 1930s, were smooth-planked boats, sharp at both ends, and less than 15 ft (4.6 m) in length. Oars were used when fishing. Small sprit sails could be used when traveling to the grounds. The handliners' boats had to be nimble, seaworthy, and capable when working around tidal rips and in short seas off headlands, able to beach safely in light surf, and above all, be easy to row (Moore 1992: 225). The boats were also distinguished by their string seats, better known as "dry ass" seats. This feature was not a luxury in a climate known for its rainfall, when rowing up to 40 miles (64 km) a day in wet clothes could leave the fisherman crippled with septic blisters and boils.

Salmon handlining was a method employed from northern California to Southeast Alaska. The boats used in California, particularly off Monterey, were entirely distinctive, while the boats used from Washington to Alaska, including the Georgia Strait, were at least superficially similar (Moore 1992). The Strait of Georgia saw a concentration of up to 700 fishermen active in row boats in the 1930s (Evans 1975: 210). Some of these rowboat fishermen were known to make a 2200 km round trip to the northern coast of British Columbia in a season, but most stayed within the 200 km length of the Strait (Fig. 15.1). A few fished year-round, staying near home over the winter months, but most fished from May to Mid-September, traveling

C.D. Moore (✉)
Golder Associates Ltd., 2nd Floor, 2920 Virtual Way, Vancouver, BC, Canada, V5M 0C4
e-mail: chmoore@golder.com

A.M. Evans (ed.), *The Archaeology of Vernacular Watercraft*,
When the Land Meets the Sea, DOI 10.1007/978-1-4939-3563-5_15

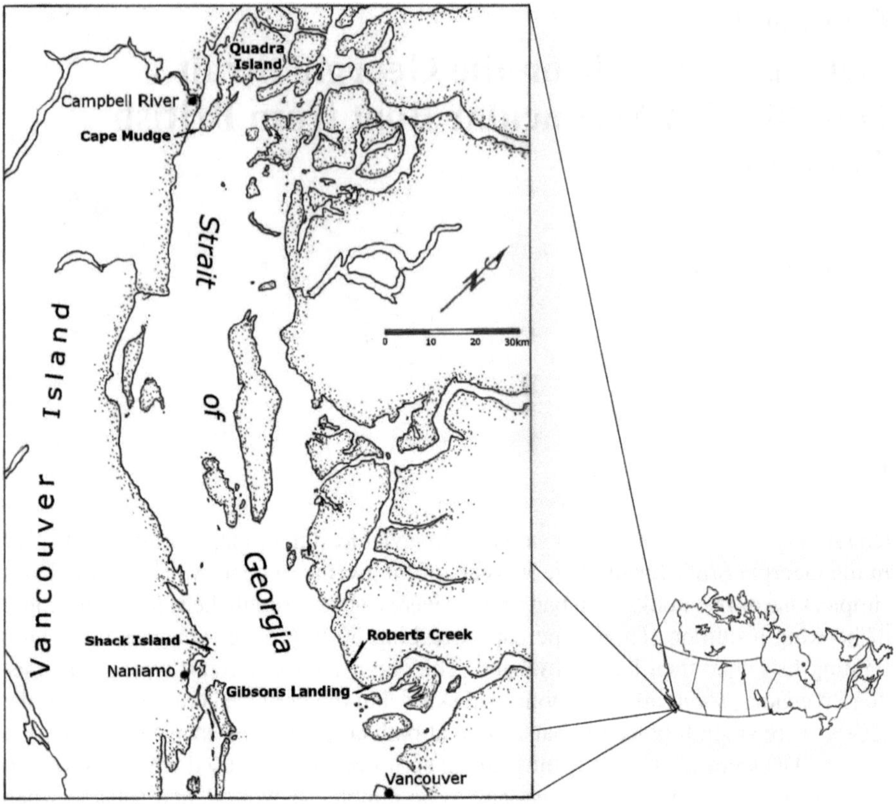

Fig. 15.1 The Strait of Georgia is a body of water over 200 km in length and up to 30 km wide in southern British Columbia that constitutes an inland sea between Vancouver Island and the continent, or the "Sunshine Coast" as the eastern shore is called for most of the area shown here (Drawing by author, 2006)

from ground to ground around the central and northern parts of the Strait. Cape Mudge, at the southern extremity of Quadra Island was the most popular destination. The shores near the seasonal fishing grounds became dotted with camps comprised of tents and driftwood shacks set up by the fishermen. Yet the handliners are virtually absent in the accounts of government and the salmon fishing industry which focus on gillnetting and seining production for the canneries, or trolling from bigger gas boats after they were introduced. Today, even within the fishing communities, few remember the men or the boats of the handline fishery.

Professional boat builders located in nearby cities were said to have built boats for the fishery in its early years but, perhaps as early as 1920, low fish prices put new professionally built boats out of the range of most fishermen (Templeton 1977). Many fishermen began building their own boats, perhaps using planks and timbers cut from a single cedar (*Thuja plicata*) log salvaged from the beach, and other materials acquired as inexpensively as possible (Templeton 1977). Some became semiprofessional by building boats for other fishermen.

Both the men and the boats may be called "handliners," although the fishermen apparently used no term to distinguish their boats from other rowboats or skiffs. Recent observers have suggested that the handliners (the boats) evolved from watercraft used on the lakes of the eastern interior of the continent, or the "peapods" of the Atlantic Coast. In the latter case it has been claimed that the handliners are shaped just like the peapods but more lightly built. The transmission may have been through immigrant boat builders, who, like Andy Linton, a pioneer boat builder in Vancouver, were trained on the East Coast (Simson 1983: 172). Conversely, it has also been suggested that both peapods and handliners may descend from aboriginal canoes in their respective regions (Moore 1992: 254–258).

Vernacular small craft are defined by a set of attributes unique to vessels found in a specific location. The results of a study of nine boats thought to have been built for and used in the handline fishery of the Georgia Strait between the late 1920s and the late 1930s are presented in an attempt to define a vernacular type and address other questions. It is believed that a systematic study of the boats' structural characteristics and form might reveal information about the type's origins, cultural context, the shared conceptions of its builders, regional variations, and the nature of the fishery it served.

The Handline Fishery in the Strait of Georgia

The term "handliner" is a relatively modern one (Morley 1958). In the 1930s, there were simply "rowboat fishermen" who fished from "rowboats" or "skiffs." The few popular published accounts of the fishery are almost all set in the 1930s, and the fishermen are white, fiercely independent men, who eked out a living in a marginal fishery, proud to work for no man, perhaps putting a few dollars aside towards buying a gas boat, or just avoiding the Depression-era work camps (Morley 1958; Evans 1975; Forester and Forester 1975: 74–75; Trower 1976).

Gas-powered trollers first appeared around 1915. Power trolling boats were capable of running several lines at once in deep water, and had great ease of mobility. Yet the rowboat fishermen still held their niche. Working the near-shore grounds among the kelp beds, the handliners coexisted with the technologically more advanced gas boats for another 30 years. From more remote locations, powered collector or buyer boats delivered the fish to railheads or other locations where the product would be kept iced or refrigerated.

Common to "creation" stories found in the official fishing histories, whether set in California, British Columbia, or Alaska, the introduction of handlining around 1900 is typically credited to a local white fisherman who first conceived the idea of using a brass lure for fishing salmon (Smith 1895: 223; Cobb 1916: 27; Hyman 1966: 10, 18, 20; Damron 1975: 43, 48). The fishery became economically attractive about this time with the introduction of refrigerator cars on the transcontinental railways. These provided year-round access to eastern markets for both fresh and mild-cure Pacific salmon (Damron 1975: 34).

The handlining fishing method was not introduced by Caucasian fishermen, however. Salmon have been caught by this method for thousands of years by the aboriginal inhabitants of coastal Northwest America. While fishing for salmon with nets provides more fish with less effort, it is possible only seasonally when salmon school in preparation for spawning. The advantages of the hook and line fishery were better quality fish delivered year-round. In the wake of mid-nineteenth-century settlement in the Pacific Northwest, many First Nations fishermen were able to earn cash by delivering troll-caught salmon for the fresh fish markets active in new town sites swelling with immigrant populations (Smith 1857: 153; Pemberton 1860: 28).

Handline fishing technology did not change significantly in 1900. Gradually, metal hooks replaced the old hooks of bone, while brass plate replaced bone and shell for lures, and cotton or linen fishing line replaced lines of human hair and cedar bark twine. Bait was still harvested with a "herring rake," and the salmon were still landed by hand, with one or two lines set out from each boat. Oral histories confirm that aboriginal fishermen continued to be active in the handline fishery until after the Second World War.

The traditional watercraft used by aboriginal fishermen in the Georgia Strait were dugout canoes, carved from cedar and expanded. Ethnographic studies state that Coast Salish peoples used canoes in the 15–20 ft (4.6–6.0 m) range for trolling (Waterman and Coffin 1920: 18). Oars quickly replaced paddles on fishing canoes, which were used by some First Nations fishermen through the 1940s, while others used and built plank boats (Knight 1978: 57–58; Frank Hackwood 2005, pers. comm.). Of the plank boats used in the Georgia Strait handline fishery between 1900 and the late 1920s no description is known, although double-enders used as handliners in Southeast Alaska were introduced by the Davis family at Metlakahtla in 1905 (Loken 1981: 11). Photographs show similar boats in Vancouver and Victoria harbors about the same time, although no association with handlining can be made from these images.

Surviving Handliners

Nine boats were selected for study, all of which were said to have been built for handlining. All nine employed smooth plank construction, were sharp at both ends, had straight keels, typically had provision for a fish bin aft, and used string seats. In contrast to similarly-sized, commercially-built boats of the time, the thwart knees were all of wood (not metal), and the breast and stern hooks were made from relatively substantial grown crooks where decks were not used (Table 15.1).

None of the nine boats is dated precisely, but are generally attributed to the early- to mid-1930s, with one said to have been built in the late 1920s. Four are in museum collections (two are in the Vancouver Maritime Museum; one is in the Sunshine Coast Museum and Archives, Gibson's Landing; and one is in the Museum at Campbell River), while the remainder are in private hands. For one of the latter,

Table 15.1 Summary of distinguishing characteristics for handliners in four locales on the Strait of Georgia

	Building locations on the Western Shore		Building locations on the Eastern Shore	
Characteristics	Shack Island	Quadra Island	Roberts' Creek	Gibson's Landing
Sample size	4	1 (+ photos)	2	2
Hull construction	Carvel	Carvel	Carvel	Strip
Hull symmetrical	Yes	No	Yes	No
Length (overall)	13′7 1/2″–14′0 1/2″	14′0″	14′6 3/4″–14′7 1/2″ (longest)	13′0 1/4″–13′3″ (shortest)
Beam (max)	48″–48 1/4″	44 3/4″	45 3/8″–46 1/2″ (greatest L/B)	42 1/2″–46 1/2″ (least L/B)
Frames	One piece—no floors	No floors	Floors	One piece—no floors
Keelson	No	Yes	Yes	No
End decks	Fore and aft—w. single batten	None	Fore deck—two battens	Fore and aft—no battens
Large hooks	None	Fore and aft	Aft only	None

access to survey was not possible, but data was gathered on its hull form, thanks to the examination of a replica that was built to the lines of the original in 1964.

Four distinct building areas around the Georgia Strait basin are represented in the sample (Fig. 15.1). From the western, or Vancouver Island shores, four boats were built on Shack Island near the city of Nanaimo, and one was built on Quadra Island near the city of Campbell River, which lies about 140 km north of the southern range of the studied handliners. The boats built on Shack Island were built by one family, the Luomas. They might be classified as semiprofessional builders. Their boats were reputed to be the best on the coast and they built many; but they were not formally trained, their shop was without electricity, and they continued to fish as well. Likewise, the Treadcrofts, a family of three brothers who built the boat from Quadra Island along with numerous other boats, might be classed as semiprofessional. From the eastern shore, an area known generally as the Sunshine Coast, two boats originated from Robert's Creek and two from Gibson's Landing. At least three of the Sunshine Coast boats were built by the fishermen who used them. Specific characteristics of construction shared by the boats originating in each locality set them apart from the others. Four subtypes may be defined (Table 15.1).

Both boats originating in Gibson's Landing have hulls built of cedar-strip construction. They are the only boats in the sample built with this method. At just over 13 ft (4 m) they are also the shortest of the sample. The frames are one piece without floors, crossing a keel without a keelson. The hulls are not symmetrical fore and aft. Small decks are located both fore and aft and are smooth, without batten finish (Fig. 15.2).

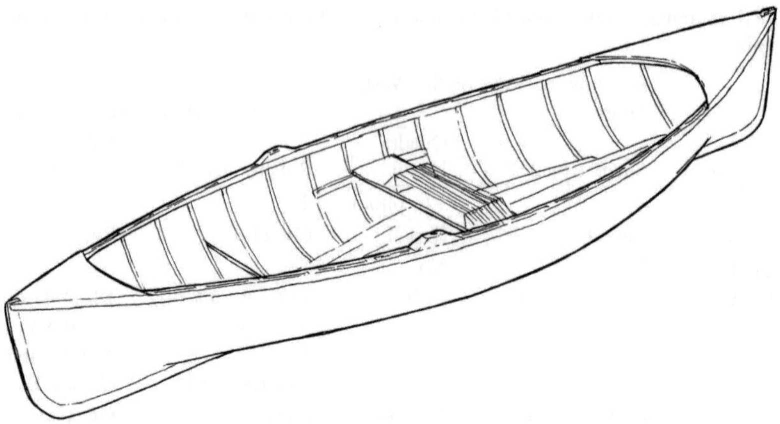

Fig. 15.2 A strip-planked handliner built in Gibson's Landing, now in the collection of the Vancouver Maritime Museum (Accession Number 1975.0023.0001; Drawing by author, 2006)

Robert's Creek is only 12 km away from Gibson's Landing by water, but both Robert's Creek boats are among the longest of the sample at about 14.5 ft (4.4 m). They employ conventional carvel construction on frames that have floors set over a keel with keelson. They are essentially symmetrical fore and aft. There is a foredeck, detailed with two battens; a substantial hook is placed at the aft end (Fig. 15.3).

All of the boats from the western side of the Strait are of middling length. The boats built on Shack Island are between 13.5 and 14 ft (4.1–4.27 m) long, and are attributed to the Luoma family of boat builders (Fig. 15.4). The Luoma boats seem to be symmetrical fore and aft. They are all carvel-built with one-piece frames lacking floors crossing a keel without keelson. They have decks fore and aft, detailed with a single batten each. It might be expected that these Luoma boats would be shaped the same, perhaps built over the same mould. Yet with respect to midsection shape and stem profiles, each of the four Luoma boats is unique.

The single 14 ft (4.27 m) boat from Quadra Island is carvel-planked on frames without floors crossing a keel with a keelson (Fig. 15.5). While sharp at both ends, its end profiles are quite different, with a plumb stem and swept stern-post. There are no decks fore or aft; their place is taken by exceptionally substantial two-part hooks. Two other boats matching the general description of the Quadra Island boat were observed on nearby Cortez Island in the 1970s (Richard Blaghorn 2005, pers. comm.).

Comparative Analysis

In addition to construction details, data was collected to reflect size, overall proportions, midship shape, and scantlings. Important in determining what specific data would be collected was the speed in which documentation was acquired. So that field data could be collected in about 1 hour, a gauge was built from extruded

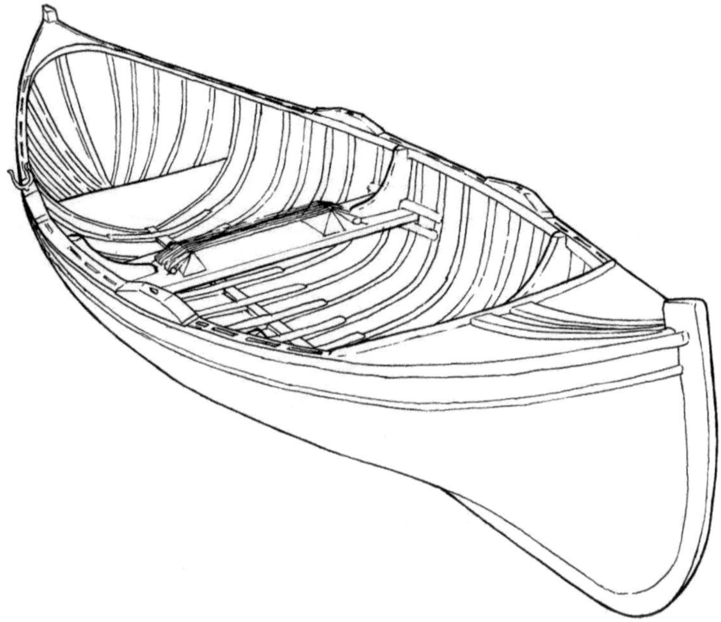

Fig. 15.3 A handliner built in Robert's Creek, now in the collection of the Sunshine Coast Museum and Archives (Drawing by author, 2006)

Fig. 15.4 A handliner built on Shack Island by the Luoma family in the 1930s and still being used today (Drawing by author, 2006)

aluminum to facilitate the taking of midsection shape. It was also important that comparative data could be readily extracted from published plans of other types. Twenty-three data fields were established (Table 15.2). Five relate to size, six relate to overall proportions, three to midship shape, five to scantlings, and four to construction details. Eric McKee's study of British working boats (1983) was the principal source determining this selection.

Comparative data was also collected from 31 other small craft falling within the 12–16 ft (3.65–4.87 m) range, mostly sharp fore and aft, including: two lap-strake, professionally built, recreational boats from the Georgia Strait area; a Davis boat (handliner from Southeast Alaska); an oyster skiff and a dugout from Washington State; three guideboats; three other eastern lake boats; three St. Lawrence skiffs;

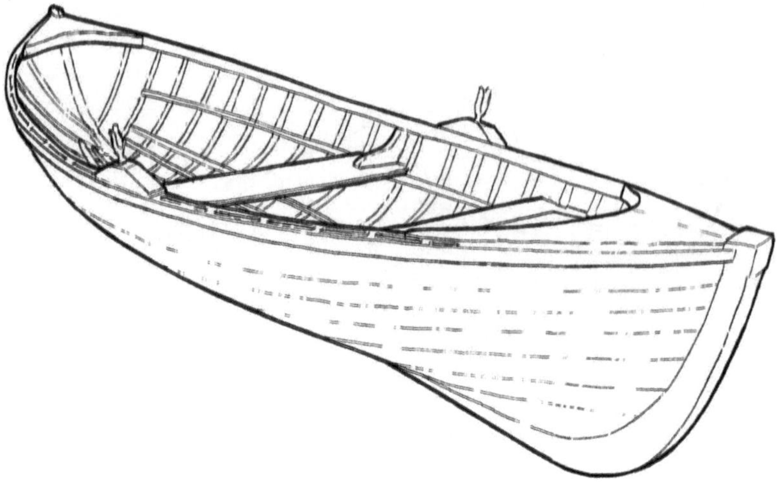

Fig. 15.5 A handliner built on Quadra Island, perhaps in the 1920s (Drawing by author, 2006)

eight peapods; and nine whitehalls. Some of these comparative data were collected in the field by the author, but most were gleaned from various purchased plans or from published sources (Chapelle 1951; Loken 1981; Fuller 2002). Due to space consideration, only a small selection of these data is presented here.

The range of variation in construction detail for handliners is comparable to variations within the peapods, or "gunning skiffs" as boats of the type are identified in Nova Scotia. For example, exactly half the peapods sampled were not symmetrical fore and aft, contrary to the commonly stated opinion that most were (Fuller 2002: 58). Peapod and handliner scantlings are actually quite similar, although the use of oak for the backbone and frames would certainly make the peapods somewhat heavier and more rigid boats. On the other hand, differences in shape between the two types are considerable.

Table 15.3 presents data summarizing hull shape in 11 selected fields that include comparative data sets from three St. Lawrence skiffs and six Maine peapods. The data in each category are based on the characteristics of nine handliners, and eight peapods, the two extreme dimensions for each type being discarded in each data field.

There is great cohesion within the handliner group with respect to midship section shape and overall proportions. The ranges are exclusive in several areas such as beam and depth where the handliner falls neatly between the St. Lawrence skiffs and peapods without overlap. In other areas there is minimal overlap including length-to-beam and beam-to-depth ratios, the three types falling into three clear groupings.

Perhaps the general area where the handliner is least consistent is in profile. This reflects the great difference in sheer line shown, for example, between the handliners of Gibson's Landing and those of Robert's Creek. The full ends are interesting, perhaps echoing a distinctive feature of the sailing gillnetters once used to catch salmon near the estuaries of major rivers of western North America (Moore 1992: 290–291).

In midsection shape, with respect to beam-to-depth ratio, bilge radius (expressed as a percentage of beam), and deadrise, the handliners are the most cohesive, or

Table 15.2 Definitions of 23 data fields used to describe and compare vernacular boat shape

	Abbreviation	Definition
Size	LOA	Length overall, to outside of stem and stern-post
	Beam	Maximum breadth, to outside of sheer wale
	Depth	Depth at lowest point of sheer to bottom of keel
	HT End	Height from top of stem to bottom of keel (line extended forward)
	Sheer	Difference between depth and HT End
Proportions	L/B	LOA divided by beam
	B/D	Beam divided by depth
	≡Ends	"Y" or "N," yes or no, is the forward half of the boat the same shape as the aft
	Ent Shr	Half angle of entry at the sheer (in plan view)
	Ent WL	Half angle of entry at the estimated light loaded waterline
	Twist	Difference between Ent WL and Ent Shr
Midship shape	Rise	Angle relative to horizontal of the bottom between keel and turn of bilge (e.g.: less than 7° = "flat"; more than 18° = "Vee")
	Bilge	Fullness of bilge where the bilge radius amidships is expressed as a percentage of beam (e.g.: less than 10 % = "hard"; more than 33 % = "slack")
	MSect	Percentage of available area (moulded depth × beam) taken by midsection shape of the hull (e.g.: over 85 % = "full"; under 70 % = "easy")
Scantlings	Keel	Dimensions of keel, sided/moulded in decimal feet
	K'lson	Dimensions of keelson (or "covering board") sided/moulded in decimal feet
	Frame	Dimensions of average frame, sided/moulded in decimal feet
	Fr. Spcg.	Average spacing (room and space) of frames
	Plank	Average thickness of plank
Construction details	Sprung	"Y" or "N," yes or no, is the keel sprung or straight
	Keel Frm	"B," for beam keel, if the keel is deeper than it is wide, or "P" for plank keel if it is wider than it is deep
	Flrs	"Y" or "N," yes or no, indicates presence or absence of floor elements
	P Style	"C," "L," or "S" indicating clenched, carvel, or strip planking style

Note: This selection has been tailored to handliners and similar boats, but is substantially based on Eric McKee (1983: 78–81)

show the least variation within each of the three types considered. Of particular note are their markedly flat floors.

Due to the variations of construction detail and styling among the handliners, and the lack of parallels in shape, no immediate kinship with other watercraft types is evident. It may be further noted that these variations, for instance sheer shape and deck details, are among the least functional of the characteristics considered. While

Table 15.3 Selected comparative data for handliners, St. Lawrence skiffs and peapods

		Handliners (7)	St. Lawrence Skiffs (3)	Peapods (6)
Overall dimensions	LOA	13′3″–14′6 3/4″	14′0″–16′0″	14′3″–16′0″
	Beam	44 3/4″–48 1/4″	38″–42″	52″–57 1/2″
	L/B	3.38–3.78	4.42–4.57	3.18–3.61
	Depth	16 3/4″–20 1/2″	13 1/4″–15 1/2″	21″–25 1/2″
Profile	Depth	16 3/4″–20 1/2″	13 1/4″–15 1/2″	21″–25 1/2″
	HT End	22″–27 5/8″	23 1/4″–26 3/4″	26″–34 3/4″
	Sheer	5 1/4″–11″	10″–11 1/4″	5″–9″
	Ent Shr	26–34°	18–23°	25–29°
Midship shape	B/D	2.67–2.79	2.71–2.99	2.24–2.47
	MSect	79–89 %	79–86 %	76–84 %
	Bilge	19–28 %	11–29 %	18–31 %
	Rise	0–3°	7–18°	7–10°

Note: This table presents data from 11 selected fields comparing the data sets from nine handliners, three St. Lawrence River skiffs, and eight peapods, with the maximum and minimum extremes discarded in each field from the handliner and peapod sets

they may not demonstrate a material cultural link with a geographic source or cultural group, they may be seen as expressions of highly localized tastes or individuality, distinguishing subtypes.

In contrast, there was clearly a generally shared mental template reflected in some of the most functional characteristics, such as beam and depth proportions. These aspects help define the Georgia Strait handliner. Beam and depth, as well as the ratio of the two, are also readily understood to relate to the environment. The more exposed coastal environment typically demands a boat that is proportionally beamier and deeper. If we rank the three environments: the coast of Maine is the most exposed, followed by the Georgia Strait, and finally the Thousands Islands area of the St. Lawrence River. The ranking of the three characteristics follow this scheme, with the peapods having the greatest beam, and depth and lowest beam-to-depth ratio. The handliners fall predictably between the ranges of the other two types.

Consistency of proportions and midship section shape suggests that the handliner traditions were established before the 1920s. Did this have its root in the professional boat builders who were said to have built for the fishery in its early years? Examination of two professionally built boats that were built on the West Coast at the same time as the later handliners suggests not. They show a beam and length-to-beam proportions that are outside the range of the handliners. Furthermore, professionally built boats of the period were almost universally of lap-strake construction, typically employed exotic woods, used relatively light breast hooks and metal thwart knees, and there is no example known using a string seat.

The professionally built boats were also more like the eastern boats with respect to deadrise angle, as were the Davis handliners in Alaska. Other West Coast types like the Shoalwater Bay Oyster skiff of Washington exhibit even more deeply V'd hull forms. So where did the distinctively flat floors of the handliners originate? A functional argument, that handliners were often beached under difficult circumstances may be dismissed because the recreational boats were beached as well, as were the Davis boats and most peapods. A clue may be found in the traditional West Coast dugout canoes. These were generally flat-floored, but particularly the Coast Salish canoes built in the Georgia Strait area (Jennings 2002: 104–107). It is interesting, although not conclusive, to observe that the Micmac, Maliseet, and Passamaquoddy bark canoes said by some to have inspired the peapods had as steep a deadrise as it is possible to put into a bark canoe, including the use of a unique keel strake in one of these craft to facilitate the creation of deadrise (Adney and Chapelle 1983: 63, 72, 73, 74, 75, 76, 82).

Discussion

The almost complete disappearance of the boats used for handlining from the official histories of commercial salmon fishery following the introduction of powered trollers is as much an economic phenomenon as a technological one. The handliner fishermen and boat builders existed below the economic level necessary for full recognition among the professional classes of their trades. The introduction of powered trollers was approximately coincidental with a sharp downturn in the market value of salmon in 1920–1921. This was the economic environment faced by large numbers of veterans recently returned from World War I. Within a decade, the economic crash of 1929 and global depression that followed further limited the earning potential for fishermen whose ranks were swelling with the otherwise unemployed. Reduction in resource value may contribute to greater inequity of income among fishermen (Smith 1976). In their increasingly marginalized state the handline fishermen adopted certain habits of cooperation, many of these appearing as adaptations of First Nations precedents.

In addition to the fishing method, the seasonally mobile nature of handlining between the months of May and October echoes First Nations traditions of moving to seasonal camp locations. The handliner shacks would often be placed on ancient summer camp sites, and the old canoe runs, some 100–150 m in length, would be cooperatively maintained as skidways for the handliner's boats.

Despite the popular characterization of handliners as fiercely independent and self-sufficient, the competitive aspects of this mode of life seemed to be limited to the fishing grounds. Besides maintaining skidways, camp-life as it is described in oral histories is markedly cooperative, boats were landed and carried up with assistance from other fishermen, the boats usually traveled between grounds in groups, an early fishermen's union was organized on the beaches, and cooperative fish buyer operations were attempted.

The boats from the eastern side of the Strait in particular seem to epitomize the cooperative spirit in construction with individual accomplishment expressed in nonfunctional flourishes. Particularly for boat-building by nonspecialists, skills need to be learned from more experienced members of the community. Oral histories speak of an "old Norwegian" who mentored the younger builders in Robert's Creek (Richard Blaghorn 2005, pers. comm.). One of those builders, Hubert Evans, encountered trouble with the law by passing that boat-building art to young men fleeing the government work camps. Oral histories from Gibson's Landing speak of a number of boats that were built over winter in a shared workshop space set up in a disused glue factory on the waterfront. All of these boats were built with strip construction, not a common building method at the time, but an excellent one for amateur builders. Each builder thought he was building according "to his own design" (Will Thompson 2005, pers. comm.), yet, based on the sample of surviving boats, there was a community conception of how a Gibson's boat should be built and appear, in addition to the broader conception of what form a handliner should take. This conception of form was evidently shared as well by the semiprofessional builders represented in the study sample, the Luomas and Treadcrofts.

Besides the already noted flat floors, the handliners of the 1920s and 1930s seem to have little in common with traditional Coast Salish trolling canoes. With a profile corresponding to Waterman's "Type C" including long overhangs fore and aft, these canoes were narrow, universally described by immigrants as a challenge to operate due to their narrowness, and exceeded 15 ft in length in order to adequately support the weight of a grown man (Waterman and Coffin 1920: 18, 21). They were not well suited for use with oars that were said to be commonly used on trolling canoes around the turn of the century.

The critical clue as to the handliner's origin may lie in a small and poorly documented type of dugout canoe. Three examples of this type of dugout canoe exist in the museum collections of the Vancouver Maritime Museum and the Pacific County Museum, Washington. All three were salvaged and reused by Caucasian coastal residents in the first half of the century, with the general understanding that they had been abandoned by Coast Salish peoples who lived in the vicinity. Other examples of these dugouts may be noted in photographs (Fig. 15.6). These canoes present a simple profile without any stylistic characteristics relating them to traditional First Nations canoe types, which, presumably, has contributed to them being generally ignored. They are between 10 and 13 ft (3.0–3.9 m) in length and are proportionally much wider than traditional canoes. They are all sharp fore and aft, and mount oarlocks. In short, they have a distinct resemblance to the double-ended, flat-floored, handliners built with planks. At least one report suggests that canoes like these, rather than canoes of the longer, more traditional form, were already being used for trolling by 1890: US Fish Commissioner Joseph Collins noted that within the Coast Salish territories of Puget Sound, trolling canoes were typically 10 ft long, with a fairly generous beam of 2 1/2 to 3 ft (0.75–0.9 m) and had a value of $10 (Collins 1892: 21). Within the Georgia Strait, Collin's colleague, Tanner, did not comment on canoe length but observed that First Nations craftsmen created trolling canoes with "great attention to symmetry of outline, and much care and ingenuity in workmanship. They cost, when new, from $5 to $20 each (Tanner 1890: 52)".

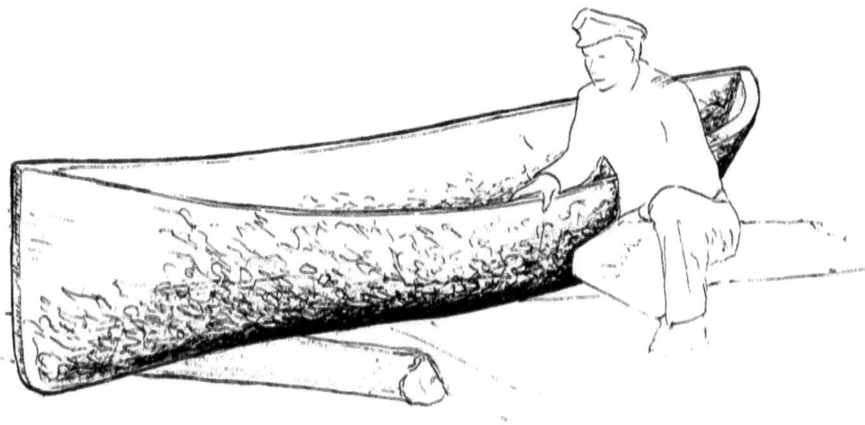

Fig. 15.6 An abandoned dugout canoe of First Nations manufacture that shows the nontraditional proportions and size of a canoe built for trolling under oars (Drawing by author, 2006, after a photograph by Derek Todd, n.d., Forester and Forester 1975: 47)

Innovation within the dugout tradition was likely a response to the introduction of oars, given the advantage of breadth for rowing purposes. Canoes built with greater beam were possible due to the availability of wide diameter trees complemented by log-expanding capabilities. With increased buoyancy and the need for the canoe to be handy with one man, a canoe that was simultaneously shorter and proportionally much wider would follow. It may be useful to observe that, far from degenerating, dugout canoe manufacture in the Pacific Northwest flourished in the 100 years following contact. The trade of canoes between First Nations increased through this period due in part to metal tool use reducing the time required for manufacture (Durham 1960: 14, 78). In particular, small trolling canoes along with medium-sized sealing canoes came into demand in the 1890s (Knight 1978: 57). Only in the 1920s with the increased availability of inexpensive milled wood and fastenings did dugout manufacture fall sharply off with most surviving First Nations craftsmen converting into plank boat construction.

There was considerable First Nations involvement in plank boatbuilding in the Georgia Strait, including on the southern Gulf Islands and Cape Mudge (Knight 1978: 57–58). Caucasian fishermen were also known to occasionally use dugout small craft of aboriginal manufacture, particularly in the first two decades of the twentieth century, but even into the 1930s (Morris 1996). At least in the Nanaimo area in the 1930s, they also purchased planked handliners from First Nations builders located there (Hackwood 2005, pers. comm.).

The surveyed handliners, in adhering to a form that included flat floors and a relatively narrow range of depth-to-beam and beam-to-length proportions, may embody aspects of a truly indigenous vernacular already tuned to the demands of the local environment. This suggests that the resulting form was an innovation, spurred by the introduction of oars and perhaps influenced by European-style plank boats, but established by First Nations craftsmen within a dugout canoe-building tradition.

Living Heritage

In the spring of 2005, a replica of Hubert Evans' boat from Robert's Creek was built by local boat builder Larry Westlake with the support of the Sunshine Coast Museum and Archives. Along with two other boats it then completed a 150 km rowing trip up coast from Gibson's Landing. The trip demonstrated to modern scholars the necessity for cooperation when landing these craft in a region where a few hours of ebbing tide can transform a narrow sandy beach into a miserable rocky landing and 250-m overland carry. Outreach along the route encouraged some of the last people with first-hand knowledge of these boats to come forward and tell their stories, and led to new leads on where additional boats might be found. Remarkable in these stories is the pride with which the story tellers speak of the boats as creations of their fathers or grandfathers (typically), and how, to some individuals, the boats remain significant symbols of family and the local community.

There is a renewed interest in building and sailing these boats. The construction and sailing of replicas provides a unique opportunity for a new generation to experience the rowing characteristics of these historic boats, and feel first-hand the ease of sailing a well-balanced craft without centerboard or rudder. Experience also demonstrated not only the expected advantages of the string seat during inclement conditions, but also the surprising energy-saving stability it offered the user rowing in rough conditions.

More people will have the chance to experience a string seat. While the builders and fishermen themselves may not have possessed the terminology to distinguish the handliner, much less define the four local subtypes identified here, an advantage of systematically collecting data from the surviving boats is that the mental template that surely existed can be presented in the form of a conceptual envelop. Within these parameters various replicas may be built that properly conform to the historical conception of a handliner. Following their identification through a preliminary survey, it is now possible to select the best representatives of each subtype for complete documentation and the production of complete lines and construction drawings. A set of plans for a Roberts Creek boat is available for sale now through the Sunshine Coast Museum and Archives. Two more sets are in production, and it is hoped that additional boats will be documented in the coming years.

Conclusion

The nine boats surveyed provide a glimpse of a preindustrial tradition where concepts of small craft form and boat-building techniques were passed down manually, giving physical expression to both regional and more localized traditions. It is not clear from the written record or from discussions with descendents of builders to what degree the builders and users were conscious of the distinct traditions. The nine boats studied provide a sample size sufficient to demonstrate the preeminence of form over construction detail in defining this regional type. Through comparative analysis there is no suggestion that the observed attributes, of either detail or form, could be traced to a specific immigrant group or exotic source. On the contrary, a hypothesis for a significant First Nations contribution to the type, based on analysis of form, warrants testing through further identification and documentation of small dugout canoes that may have been used in the fishery.

By examining the material culture of these watercraft, aspects of maritime culture of coastal British Columbia in the first half of the last century become accessible. It may offer a correction to assumptions that boats built with planks belong to entirely exotic and introduced traditions. Furthermore, the history of the fishing industry that most people, including those working in the industry today, accept is one of technological determinism: the introduction of gasoline-powered trollers around 1915 meant the end of rowboat fishing. A handful of popular histories reveal the survival of the handline fishery, but from an individualistic perspective where white fishermen used or built any sort of "rowboat" they thought might work. Recorded in the oral histories of the 1930s is evidence of some First Nations involvement, insight into the profoundly cooperative aspects of the culture within the region, and a perspective of the deep personal attachments of some community members to the boats. This is the world view embodied by the watercraft. The typological examination, by revealing a fairly broad range of plank boat construction detail and finish according to subtype overlaying a firm and cohesive foundation of type based on a particular form, suggests an establishment of type dating to the late nineteenth century and centered within the aboriginal dugout tradition. Despite the apparent lack of substantial social links between the communities of white builder/fishermen and their First Nations counterparts, throughout the "peri-historic" period of the handline fishery, marked and defined as it was by the economic downturns of the 1920s and 1930s, the two coastal communities had in common an economically marginal existence with many shared cultural aspects being drawn to a significant degree from First Nations traditions.

Acknowledgements I would like to acknowledge the assistance of funds granted by the British Columbia Heritage Trust in support of the Museum Small Craft of the West Coast Fisheries Research and Documentation Project, a project supported by the Britannia Heritage Shipyard (Richmond, BC) and the Vancouver Maritime Museum (Vancouver BC). Following the Trust's dissolution, additional funding was gratefully received through a John Gardner Grant provided by the Traditional Small Craft Association of Mystic, CT. Thanks are due as well to the Sunshine Coast Museum and Archives for sponsoring the building of one handliner replica and the first "Migration" of handliners from Gibson's to Lund in 2005, and to Larry Westlake for his sustained passion and dedication to documenting and building handliners (the boats).

References

Adney, E. T., & Chapelle, H. I. (1983). *The bark canoes and skin boats of North America*. Washington, DC: Smithsonian Institution Press.

Chapelle, H. I. (1951). *American small sailing craft: Their design, development and construction*. New York: W.W. Norton.

Cobb, J. N. (1916). Pacific coast fishing methods. In *Pacific Fisherman yearbook*, Seattle, WA (pp. 19–33).

Collins, J. W. (1892). The fishing vessels and boats of the Pacific Coast of the United States. In *Bulletin of the United States Fish Commission, 1890*, Washington, DC (pp. 13–48).

Damron, J. E. (1975). *The emergence of salmon trolling on the American north-west coast: A maritime historical geography*. Doctoral dissertation, Department of Geography, University of Oregon, Eugene, OR.

Durham, B. (1960). *Canoes and kayaks of Western America*. Seattle, WA: Shorey Books.

Evans, H. (1975). Day of the hand troller. In H. White (Ed.), *Raincoast chronicles: First five* (pp. 210–214). Madiera Park, British Columbia, Canada: Harbour.

Forester, J. E. & Forester, A. D. (1975). *Fishing: British Columbia's Commercial Fishing History*. Saanichton, B.C.

Fuller, B. A. G. (2002). *A catalogue of small boat plans from mystic seaport*. Mystic, CT: Mystic Seaport.

Hyman, F. J. (1966). *Early fishing on the Mendocino Coast*. San Francisco, CA: J. Porter Shaw Library, National Maritime Museum.

Jennings, J. (Ed.). (2002). *The canoe: A living tradition*. Toronto, Ontario, Canada: Firefly.

Knight, R. (1978). *Indians at work: An informal history of native Indian labour in British Columbia 1858–1930*. Vancouver, British Columbia, Canada: New Star Books.

Loken, M. (1981). *The Davis boat* (Traditional small craft of the Northwest, Vol. 2). Seattle, WA: Seattle Center for Wooden Boats.

McKee, E. (1983). *Working boats of Britain: Their shape and purpose*. London: Conway Maritime Press.

Moore, C. D. (1992). *Salmon fishing boats of the North American Pacific Coast in the era of oar and sail*. Master's thesis, Department of Anthropology, Texas A&M University, College Station, TX. Retrieved from http://nautarch.tamu.edu/anth/abstracts/Moore.htm

Morley, A. (1958). The lonely hand-liner. In R. Watters (Ed.), *British Columbia: A centennial anthology* (pp. 139–144). Toronto, Ontario, Canada: McClelland & Stewart.

Morris, R. (1996). Sandy Jones and the Kawasemi: the last of their kind. In P. A. Robson & M. Skog (eds.), *Working the tides: A portrait of Canada's West Coast fishery* (pp. 34–39). Madeira Park, British Columbia, Canada: Harbour.

Pemberton, J. D. (1860). *Facts and figures relating to Vancouver Island and British Columbia*. London: Longman, Green, Longman and Robert.

Simson, J. (1983). Gastown: Everybody knew everybody. In H. White (Ed.), *Raincoast chronicles six/ten* (pp. 168–175). Madiera Park, BC.

Smith, P. (1857). The fish and fisheries of the Pacific Coast: Salmon fishing in California and Oregon. *California Farmer, 8*(20), 153.

Smith, H. M. (1895). Statistical report on the fisheries of the Pacific Coast of the United States in 1894. In *Bulletin of the United States Fish Commission, 1894*, Washington, DC (Vol. 14, pp. 233–288).

Smith, C. L. (1976). Intracultural variation: Decline and diversity in North Pacific fisheries. *Human Organization, 35*(1), 55–64.

Tanner, Z. L. (1890). Explorations of the fishing grounds of Alaska, Washington Territory, and Oregon, during 1888, by the U.S. Fish Commission Steamer Albatross. In *Bulletin of the United States Fish Commission, 1888*, Washington, DC (Vol. 9, pp. 1–92).

Templeton, J. (1977). Interview on three tapes. Part of the Genesis III Project, by the Campbell River & District Historical Society. Sound and Moving Image Division, Provincial Archives of British Columbia, Victoria, BC.

Trower, P. (1976). Skiffs, gillnets and poverty-sticks. In H. White (Ed.), *Raincoast chronicles: First five collector's edition* (pp. 230–234). Madeira Park, British Columbia, Canada: Harbour.

Waterman, T. T., & Coffin, G. (1920). *Types of canoes on Puget Sound*. Indian Notes and Monographs, 3. Seattle, WA.

Index

A.M. Evans (ed.), *The Archaeology of Vernacular Watercraft*, When the Land Meets the Sea, DOI 10.1007/978-1-4939-3563-5

MIX
Papier aus verantwortungsvollen Quellen
Paper from responsible sources
FSC® C105338

If you have any concerns about our products,
you can contact us on
ProductSafety@springernature.com

In case Publisher is established outside the EU,
the EU authorized representative is:
Springer Nature Customer Service Center GmbH
Europaplatz 3, 69115 Heidelberg, Germany

Printed by Libri Plureos GmbH
in Hamburg, Germany